Varieties of Scientific Realism

Evandro Agazzi
Editor

Varieties of Scientific Realism

Objectivity and Truth in Science

Springer

Editor
Evandro Agazzi
Universidad Panamericana
Mexico City
Mexico

ISBN 978-3-319-51607-3 ISBN 978-3-319-51608-0 (eBook)
DOI 10.1007/978-3-319-51608-0

Library of Congress Control Number: 2017930276

Printed on acid-free paper

This Springer imprint is published by Springer Nature
The registered company is Springer International Publishing AG
The registered company address is: Gewerbestrasse 11, 6330 Cham, Switzerland

Contents

Introduction: The Conceptual Knots of the Realism Debate

Evandro Agazzi

1 The Concept of Cognition

A couple of elementary 'facts of life' stimulate our reflection. The first is that "all humans want by their nature to know", as Aristotle says at the beginning of his *Metaphysics*. The second is that, in order to understand what we 'see', we normally introduce something that we do not 'see'. These two facts are testified by a basic *evidence* which we do not understand here in its reductive empiricist sense as a structure of sensible perceptions, but rather in the more pregnant sense of 'phenomenological evidence', by which we refer to the capability of immediate conscious apprehension of facts with which human beings are endowed. In other words, these facts are not 'supported' by the said phenomenological evidence, but are the *content* of that evidence. This is why they are certainly accepted by commonsense but cannot be qualified as 'naive' beliefs of common sense (with the implicit understanding that they are 'uncritical'). This cannot avoid, however, that the meaning of the statements expressing the phenomenological evidence depends on the meaning of the terms occurring in these statements (i.e. on the sense of the concepts associated with the corresponding linguistic terms). In the case of the first 'Aristotelian' statement the concept whose sense has to be clarified is "to know". This sense must be counted among the most primitive in any linguistic context since it denotes the specific action thanks to which a certain entity establishes a peculiar relation with the 'world', an action in which the entity which *knows* 'identifies' itself in a certain sense with the entity which is *known*, though remaining ontologically distinct from it. Or, to put it differently, an action through which an entity is capable of 'interiorizing' or 'assimilating' other entities without destroying them. (Therefore, this action is different, e.g., from feeding, in which a plant

E. Agazzi (✉)
Interdisciplinary Center for Bioethics, Panamerican University,
Mexico City, Mexico
e-mail: evandro.agazzi@gmail.com

E. Agazzi (ed.), *Varieties of Scientific Realism*,
DOI 10.1007/978-3-319-51608-0_1

interiorizes and assimilates the substances of the soil by destroying their original compounds, or animals feed on plants or other animals by eating them and destroying them). The medieval philosophy had elaborated the concept of *intentional identity* to characterize that peculiar way of identification and interiorization that preserves ontological distinction, which is typical of knowing and is present only in certain animals and, at its highest level, in humans.

We shall use the term "cognition" to denote this sense of knowledge understood as an *action*, because in English "knowledge" is rather used to denote the result of such an action. In this most basic concept of cognition, which is present in common sense and ordinary language, we can easily recognize the idea of the ontological independence (i.e. the independence of existence) of the known entity from the action of the knowing entity. If we agree to call "reality" (for our present purpose) the target of the knowing action (cognition), we can recognize in this ontological independence of reality from cognition the situation defended by *ontological realism*. The negation of this thesis (i.e. ontological anti-realism) can succeed only by showing that the concept of knowing as used in ordinary language is inconsistent (a little like the concept of a square circle) because the alleged independence of reality from cognition does not obtain, is simply an arbitrary unconscious imagination and illusion: what is actually the case is that reality is posited by the knowing action. This is the thesis advocated by *idealism*, which is the direct negation of ontological realism (also called sometimes "metaphysical realism"). Leaving aside the conspicuous difficulties one finds in the cumbersome speculations of the idealist philosophers when they try to prove that reality is posited by thinking and is ultimately ontologically identical with thinking itself, a more modest but radical criticism can be leveled against the idealistic proposal: in order to show that the *real* relation between reality and thinking is that of an ontological dependence, one must consider this relation as something that thinking 'considers' as distinct from its own action, that is, that such consideration is not an action of self-consciousness. This amounts to saying that the idealist position, if taken seriously, is self-defeating since it requires for its defense that the relation thinking-reality be considered as a content in itself 'about which' we develop our 'considerations' (that are nothing but acts of thinking).

2 Representations

The result of cognition are the *representations* which, due to the 'dual' nature of cognition itself, are intended to be representations *of* reality in the following 'neutral' sense: representations are the ways in which reality is *present* within the different forms of cognition, but we must also recognize that representations have their own ontological status: they are different from nothing, hence they are also part of reality, though at a different level (or of a different *kind*) with respect to the reality of which they are representations and which we can conventionally call "the world". The transition from general reality to representations marks the step from ontology to epistemology, the last being considered in a broad sense as the domain of cognition.

At this junction, however, a subtle and unperceived historical change occurred in the way of conceiving knowledge. The classical and common-sense answer to the question "What do we know"? was "We know the world"; but suddenly, at the beginning of the 17th century, the answer became "We know our representations", a statement for which no phenomenological evidence nor argument were offered, but was accepted as obvious by the most influential philosophers of that time, after having been expressed with special force by Descartes. The strange situation was that, on the one hand, the specific aim of cognition was still considered to be that of knowing the world but, on the other hand, it was admitted that we only know our representations and must try to prove (investigating only the domain of representations) that they actually present how the world is. To put it differently: our knowledge consists of representations, and reality 'in itself' is supposed independent of our representations. Therefore, the proposal to *know* how reality is independently of our *knowledge* sounds almost contradictory, and this explains why the efforts for solving this ill posed problem were sterile. Less radical forms of this proposal, however, can be and have been advanced.

An early solution of the difficulty was offered by *idealism*. Representations were usually called "ideas" by the philosophers of the 17th century, and some of them maintained that reality is not something that lies beyond representations, but is the very content of representations. In such a way the ontological independence of the world from cognition was denied, and for this reason since that time idealism was presented as the opposite of realism. The traditional 'intentional identity' of cognition and reality was conflated in an ontological identity. The most famous expression of this original idealism is George Berkeley's statement *esse est percipi* ("to be is to be perceived"), but no less significant is the monistic philosophy of Spinoza, whose VII Proposition of the II Book of his *Ethics* states *ordo et connexio idearum idem est ac ordo et connexio rerum* ("the order and connection of the ideas is the same as the order and connection of things"). His doctrine was a clear (and recognized) prefiguration of the classical German transcendental idealism of the 19th century. Paradoxically, the fact of having 're-qualified' reality as the content of thinking permitted to consider idealism also as a form of realism with a clear advantage in principle: if we are able to explore correctly the world of the ideas we will ipso facto determine the structure of reality, and this is *epistemological realism.*

3 The Error

Leaving aside, for the moment, the idealistic position that tries to by-pass the bipolar nature of cognition as an action whose aim is to represent the features of an ontologically independent world, we note that the elementary constituents of such representations are expressed in statements which are qualified as *true* when the cognitive action is successful. Unfortunately, cognition often fails to attain its goal and this is made evident by the common experience of mistakes and errors in our cognition. An error is expressed in a statement that does not represent the structure of the world to which it makes reference and, paradoxically, the existence and recognition of errors is one of the most convincing arguments for ontological realism. Indeed, how can I be sure that the world has a structure (i.e., that it is endowed with certain definite properties and relations)? The answer is that if it had no structure, whatever statement could be affirmed about it. Hence, if certain statements cannot be affirmed, because they are false, because they express an error, this means that they did not represent the structure of the world, hence the discovery of errors (and the possibility of correcting them) testifies of the ontological independence of the world from cognition.

Truth is not a property of reality, but of representations, and consists precisely in the fact that a representation 'corresponds' to reality. But how can we have representations of what is not real? Moreover, how can we discriminate true from false representations? Parmenides had attributed to reason the exclusive capability of attaining truth by a clear understanding of the ultimate and simplest characteristic of reality, which is *being*, so that *non-being*, having no existence, cannot even be thought of and spoken of. All statements which common sense spontaneously accepts because they are supported by sense perceptions are—according to Parmenides—confined to the status of pure *opinions* that are almost certainly false since they would entail the existence of non-being (like the belief in the existence of change and of the multiplicity of beings). Protagoras, however, taking seriously Parmenides' thesis that non-being cannot be thought of and spoken of, maintained that all opinions are true, since they can be thought and discussed. A way out of this difficulty was offered by Plato by first clarifying that a negative statement is not the affirmation of a non-existent, but the affirmation of something "other" than what is being stated. In this way the existence of true opinions was duly recognized, and Parmenides' distinction between opinion and truth was replaced by the difference between opinion and knowledge: knowledge was defined as true opinion supported by arguments providing its reasons (in modern terms we can say by a "justification"). This fundamental step amounted to linking knowledge with *reality* (due to the requirement of truth), and in addition with *certainty* (due to the requirement of rational justification).

Already in antiquity, however, the availability of criteria or methods for securing both requirements was strongly debated. Sense perceptions were easily shown to be an insufficient ground, owing to their subjective privacy and also to well-known examples of sense illusions, while the possibility of finding intellectually evident first principles from which the justification of particular domains of knowledge would logically follow appeared as no less problematic. This is why skepticism surfaced from time to time during the history of Western thought.

4 The Role of Reason

The search for certainty was not the only—and probably even not the principal—motivation for that elaboration of rational justifications that was required for knowledge in the classical Western tradition. The principal motivation was rather the aim to *understand and explain* that is typical of human cognition and which produces the second of the two evident 'facts of life' that were mentioned at the beginning of this Introduction: in order to understand and explain what we see, we introduce something we do not see. We intend this 'seeing' in a very general sense, as meaning 'ascertaining' in the different ways in which we could use this notion and the result of which are *descriptive statements.* For many practical purposes such an elementary cognition is sufficient, but in many other circumstances we are led to put forth certain *how-questions* or *why-questions*. For example, a detective who is trying to 'reconstruct' the dynamics of a murder occurred in a room tries to imagine (on the ground of the factual evidence at his disposal) *how* the killer can have come into the room, with what kind of weapon and from which position he could have assailed his victim, how he could have left the room without being seen by anyone, etc. All these (and other) are reasonable conjectures about *unobserved* facts that must be checked by means of additional factual evidence regarding not the murder in the room but some of the imagined circumstances. If the controls are positive and with significant consilience, the result of the investigation might lead to the identification of the killer and, to complete the investigation, it will be also necessary to answer some why-questions, in order to explain *why* that individual arrived at killing his victim. Dozens of similar examples can be found in our everyday relations with other people or in the use of machines. It was this fundamental cognitive attitude that produced the birth of Western philosophy, when the desire to understand and explain was addressed to the whole of reality, to the regularities appearing in the sky and on the surface of the Earth, to the course of human events and to the sense of human existence. No wonder, therefore, that this cooperation of ascertaining, understanding and explaining was explicitly recognized as constituting the proper structure of *knowledge*. A cooperation that it would be restrictive to qualify as a correlation between sense perception and reasoning: it would be more appropriate to speak of a synergy of 'empiricity' and 'logos', where by empiricity we mean not only the content of sense perception, but also, for instance, the content of a historical document, of a dream, of a feeling (i.e. whatever we could consider as a datum, as a factual information), and by logos we mean the various activities of the mind through which empiricity is interpreted, elaborated, understood, made 'intelligible' and explained. We can express this fact by saying that human knowledge walks on two legs, empiricity and logos, and that, thanks to this synergy, it becomes broader and deeper. But what do we mean by 'broader' and 'deeper'? If understood seriously this means that previously unknown *entities* have been brought within the domain of our *knowledge,* and also that previously unknown *properties* of already known or newly discovered entities were brought into the domain of knowledge. No restriction is implicit in this process, in particular it is not supposed that the newly discovered entities or properties have to belong to the same kind of empirical evidence as the one of the explained facts. Sometimes this can be the case: for example, if the killer is discovered, he will be an entity of the same kind as the killed person; or, when the new planet Neptune was discovered thanks to the prediction based on a theoretical inference within Newtonian celestial mechanics, it was an entity of the same kind as the already known directly observed planets, and was also empirically observable. In other cases this may not happen: for instance, if in order to discover the killer it was necessary to suppose his intention of killing that particular person, and we had independent evidence and arguments to support this hypothesis, the said intention is a mental entity whose nature is very different from that of the other facts ascertained through the empirical evidence; or the particular structure of the genetic code of an animal which explains the particular color of its eyes has a chemical nature very different from the perceptual nature of the color. For the same reason, if eventually it is concluded that a certain unobservable entity introduced to explain a certain empirically ascertained fact does not actually exist, this does not depend on the nature of the entity, but on the inadequacy of the explanation. For example, in antiquity lightning was explained as a weapon thrown by angry gods and the existence of such gods was later denied not because they were supposed to be supernatural invisible entities, but because their existence was considered an inadequate explanation of atmospheric phenomena. Also the existence of phlogiston was denied after a certain time, despite that it was supposed to be of the same nature of the usual physical entities, because the explanation it provided for chemical phenomena was inadequate.

Despite the above considerations, a certain diffidence remains regarding the cognitive purport of reason because its intervention in the process of cognition is often considered as a kind of intromission, of manipulation, of shaping that possibly distorts the genuine representation of the world that, on the contrary, is supposed to be faithfully mirrored in the perceptual moment of cognition, due to the 'passivity' or 'receptivity' of the senses. This is probably the implicit feeling that supports the preference for empiricism.

5 Receptivity

Even Kant was influenced by this perspective and in his *Critique of Pure Reason* explicitly stressed the receptivity of the senses as opposed to the constructive action of the understanding, and credited the senses with the capability of providing the "intuitions" which (though still remaining 'internal' to the knowing subject as pure "appearances" or "phenomena") offer the indispensable ground for knowledge, because the a priori transcendental conditions for "thinking"—imposed on the phenomena by the structure of the intellectual categories—have only the function of securing to knowledge the indispensable characteristic of universality and necessity. This is why Kant could declare himself at the same time "empirical realist" and "transcendental idealist": due to their receptivity the senses—though remaining unable to attain the "things in themselves"—could preserve at least a 'limited' independence of the content of knowledge from the knowing activity of the subject.

This ingenuous solution, however, opens at least two questions: (i) does receptivity eliminate the risk of subjective bias in knowledge? And (ii) is the specific constitution of the knowing subject an insurmountable obstacle to our cognition of reality? The answer to the first question is negative: the idea that sense perceptions 'come from' the external world and are 'received' by the subject in its own particular way is rather spontaneous for common sense and was defended by various philosophers since antiquity; in medieval Scholastics was even expressed as a general principle: *quidquid recipitur ad modum recipientis recipitur* ("whatever is received is received in the way of the recipient"). Many philosophers took this fact as the cornerstone of subjectivism: the same food tastes pleasant to a person and unpleasant to another one, or even to the same person when she is healthy and when she is sick, so that flavor is not an intrinsic property of the food, but depends on the receptive subject. The same can be repeated for numberless properties we attribute to reality and for several other kinds of judgments. Starting with Protagoras and going on during the whole history of Western philosophy this way of thinking supported what we can call an "anti-realist" position in epistemology, which was taken even by thinkers who accepted the idea of the receptivity of the subject not only at the level of sensations, but in cognition in general.

On the other hand, one can admit the obvious fact that the constitution of the different sense organs is a necessary condition for knowing certain features of reality: one needs eyes for seeing colors, and ears for hearing sounds. However, one cannot see sounds or hear colors, and this means that every particular sense organ can reveal only certain specific properties of reality. Or, symmetrically, this means that certain properties of reality can be detected only by specific sense organs. This is a "realist" position in epistemology fully compatible with the receptivity of cognition (one could say that the different cognitive capabilities 'detect' or 'give access' to different aspects of reality). Moreover, it is also possible to maintain that not the receptivity, but the active and constructive parts of cognition are those that offer the ground for saving the universality and necessity of knowledge, thus being the best defense against subjectivism and skepticism. This is notoriously the core of Kant's "Copernican revolution" in which, instead of assuming that "all our knowledge must conform to objects" it is claimed that "objects must conform to our knowledge". The reason why this move does not amount to subjectivism is that—according to Kant—the contribution of understanding to knowledge does not consist in *representations* but in *functions* that organize sense representations and, moreover, are not capabilities of the individual minds, but universal features of reason, being conditions for thinking as such.

6 Transcendentalism

The Kantian approach shifted the focus of epistemology to the study of the *possibility conditions* of our knowledge which, in particular, determine what are the modalities and the limits of our cognition and, consequently, of knowledge itself. These possibility conditions that are presupposed by any cognition and independent from its possible contents are, in this sense, a priori or *transcendental* and determine the cognitive value of our representations. Therefore, this study of the cognitive capabilities of reason (often called also *criticism*) does not coincide with the arbitrary assumption that what we *know* are our representations, but simply accepts that we *have* representations and we investigate what cognitive value they may have. In this sense the "methodical doubt" of Descartes was a starting point of this 'critical' orientation in the search for *certainty* (a certainty, however that did not concern the fact that we 'have' a given representation, but that it represents reality). No wonder, therefore, that the subject matter of this kind of investigation is the broad world of representations, but always with the explicit or implicit aim of evaluating their cognitive value, and for this reason the problem of realism tacitly boils down to this research. This is why, for example, at the core of the phenomenological thinking we find the effort of overcoming the doubt by attaining a primordial evidence that is at the same time a (conscious) revival of the Cartesian approach and a recovery of the classical notion of the intentional identity of cognition and reality, along with the transcendental view of framing the possibility conditions of our knowledge.

The study of the possibility conditions of knowledge (that is, the transcendental point of view) has broadened considerably in the last hundred years, especially through the contributions of the hermeneutic movement and of the linguistic, sociologic, pragmatic approaches that, in a first stage, have favored anti-realist views, but more recently are showing interesting developments in the direction of various forms of realism.

7 The Realm of Science

Regarding the knowledge of the physical world (or "Nature" for brevity) pre-modern philosophy shared a generalized ontological realism: Nature consists of a display of "substances" existing in themselves and independently from human knowledge. These substances have properties and relations that have no independent existence because they exist only *in* the substances, but are independent from our knowledge as well. We can know a substance only by knowing its accidents, which *appear* to us, whereas the substance has an *essence* which we try to know in our endeavor to understand and explain the appearances (often called *phenomena*). Therefore, the appearances were not understood in the negative sense often implicit in ordinary language, as something misleading: appearances were what is *manifest* of reality, and for this reason a commonly accepted methodological imperative was that of "saving the appearances" (or "saving the phenomena"): therefore, that was also a position of *epistemological realism* regarding the phenomena, that aimed at being extended also to the substances, by inferring the knowledge of the essence from the knowledge of the accidents.

The scholastic philosophers were aware that there are different possibilities of "saving the appearances" by proposing what today we would call models, or images, or representations of the underlying substance, so that the selection of the correct representation had to be secured by means of much more fundamental and broader *philosophical* considerations.[1] For this reason they distinguished—e.g. in astronomy—the models that could make easier the calculation of the celestial movements (and were used by the 'mathematical astronomers') from those which were considered to be *true* representations of the *real* structure of the universe (and whose elaboration was the task of the 'philosopher-astronomers'). When Galileo, thanks to his astronomical discoveries, was convinced that the Ptolemaic system was wrong, he admitted that the rival Copernican theory was true and attributed to Copernicus himself this persuasion, against those who maintained that he had only wanted to offer a more efficient mathematical model.[2] This was also the position of Newton, who explicitly used the term "phenomena" in the sense of what is "manifest", while refraining from admitting in the natural science the "hypothesis" of "hidden qualities" allegedly belonging to the essence of the physical bodies. The impressive harvest of discoveries rapidly acquired by the new natural science of mechanics convinced scholars that this was at last a model of what science in general should be and this is the well-known declaration made by Kant in the Preface to the second edition of the *Critique of Pure Reason.*

How could that cognitive success be explained? Half of the answer was given by Galileo himself when he explicitly proposed to restrict the investigation of the physical bodies to the study of *a few accidents* (called by him "affections") giving up the pretension to capture "by speculation" the *intimate essence* of bodies.[3] This position was shared by Newton as well. The second half of the answer consisted, in the case of Galileo, in a selection of those 'accidents' that could be the target of investigation, and he discarded those that are ascertained by sense perceptions (and as such depend on the peculiarity of the sense organs), limiting scientific research to those "real accidents" that are mathematically expressible. From this followed that the "great book" of Nature is written in mathematical characters and can be red only by using mathematics. Coming to the concrete application of these proposals, the study of a physical phenomenon consists in formulating a "supposition" about its structure by proposing what we would call in modern terms a 'mathematical model' of it, and then artificially preparing an experimental set up in which this model is tested. If the test is negative, the model (though mathematically sound) is rejected; if the test is positive, this model is accepted as a representation of the *true* structure

[1]A particularly clear declaration in this sense regarding one of the most advanced sciences of that time (astronomy) is expressed by Thomas Aquinas: "Though, if these suppositions are made, the appearances were saved, one should not say nevertheless that such suppositions are true, because possibly the appearances regarding the stars can be saved according to some other way not yet understood by humans" (Aquinas, *De coelo et mundo*, 2, 12, 17).

[2]See the following statement contained in the first letter to Marcus Welser on the solar spots: "The philosopher-astronomers, besides trying to save at any rate the appearances, try to investigate—as the greatest and most marvelous problem—the true constitution of the universe, since such a constitution exists, and it exists in a way which is unique, true, real, and impossible to be otherwise, and worth being put before any other knowable question by the speculative minds, owing to its greatness and nobility". (Galilei, *Opere* V, p. 102).

[3]See the third letter to Marcus Welser in Galilei, *Opere* V, pp. 182–188).

of that phenomenon.[4] In this balanced synergy of empiricity ("sensible experiences") and logos ("mathematical demonstrations") one easily sees the relevant cognitive role attributed to reason: the mathematical conjectures are not the result of empirical generalizations, but are introduced by reason and are the prerequisite of the experiment. Moreover, a great achievement such as the principle of inertia introduced by Galileo was attained by a chain of logical arguments, despite the fact that empirical generalization would have led to its rejection, since there is no one single case in which we can observe that a body on which no force is acting remains in a state of rest or indefinite rectilinear motion. Owing to the above sketched reasons Galileo inaugurated modern natural science according to the ontological and epistemological realism of the tradition.

This 'second half' of Galileo's answer is only partially shared by Newton. The concrete construction of his physics remained substantially in keeping with the Galilean approach, but with some not negligible differences that can be qualified as the expressions of his rather strict *empiricism*. For instance, mathematics remained in his work a powerful methodological 'instrument' for the rigorous presentation of the physical knowledge but without any ontological commitment, and in his famous *Scholium Generale* of the *Principia* he discarded hypotheses that pretended to be more than propositions "deduced from the phenomena" and "generalized by induction".[5] Despite this, he made various efforts in his other investigations in order to find out some ontological ground for the explanation of the property of gravitation but unsuccessfully, so that we can say that he was an ontological realist (since admitted that science investigates a reality independent of our cognition) but an epistemological agnostic (since he remained doubtful about our possibility of going beyond the ascertainment of phenomena). Nevertheless, he cannot be qualified as an epistemological 'anti-realist', since he never maintained that the aim of science is just that of finding theories that are simply 'empirically adequate', to put it in the terms of the contemporary debate.

The Galilean and Newtonian views of science remained paradigmatic for the development of natural science until the end of the 19th century, with an increasing favor accorded to realism, both ontological and epistemological. This could happen because mechanics quickly became the provider of a minimal metaphysical ontology for natural science, by supporting either the 'atomistic' view that the ultimate ground of things are material particles moving in empty space, or the 'continuistic' view of a material impalpable substratum filling the whole of space in which physical actions propagate and causally produce ascertainable phenomena.

[4]See for example Galilei, *Opere* VIII, pp. 202–203.

[5]Here is the celebrated text: "I have not as yet been able to discover the reason for these properties of gravity from phenomena, and I do not feign hypotheses. For whatever is not deduced from the phenomena must be called a hypothesis; and hypotheses, whether metaphysical or physical, or based on occult qualities, or mechanical, have no place in experimental philosophy. In this philosophy particular propositions are inferred from the phenomena, and afterwards rendered general by induction" (Newton 1726, Engl. transl., p. 943).

The advantage of these views was that both of them gave an intuitive 'visualizable' picture of physical reality and could be expressed in adequate mathematical formulations, so that each one could be considered as offering the *true* portrayal of reality: the disadvantage was that they were incompatible and only one could be true. For a certain while they seemed to be 'empirically equivalent'; later on they could be considered as separately suitable for different domains of physics; finally both appeared inadequate to fully account even for the empirical phenomena investigated in the particular branch of physics where each of them had been considered suitable (viz., thermodynamics for the atomistic view, and electrodynamics for the continuistic view). This situation produced a crisis of epistemological realism because it indicated that even the most sophisticated physical theories did not succeed in being *true* of their intended domains. Still, both views shared the common background framework of 'classical' mechanics; but with the advent of quantum mechanics and relativity theory at the beginning of the 20th century even that background was challenged and anti-realism became the most favorite attitude in the philosophy of science. Ingenuous efforts of saving at least a restricted version of realism regarding not physical reality 'in itself' but the 'objects' of science were undertaken through a recovery of the Kantian doctrine of transcendentally founded objective knowledge of phenomena. Nonetheless, a distrust in the genuine cognitive powers of reason in favor of an almost exclusive confidence in observation became the dominant philosophical understanding of science.

The story that we have taken the liberty of roughly outlining in its most salient lines (because is well-known) seems to deprive of any plausibility the spontaneous conviction that we are in a situation of *progress* with respect to our ancestors because *we know more and better* than them. For, if we try to explain why we are in such a more advanced cognitive situation, it is inevitable that we refer to the advancements of knowledge realized in the different *sciences.* The same conviction is expressed, by the way, in the often repeated statement that the knowledge attained by the different sciences in their respective fields during the last hundred years is greater than the knowledge accumulated during the entire history of human kind. To see that this is not just a naïve impression of common sense it is sufficient to remember that such a respected philosopher as Wilfrid Sellars has thematically contrasted the "manifest image" of the world accepted by common sense with the "scientific image", the first being wrong and the second *true*.[6] Without overlooking the criticisms that can be addressed to certain parts of Sellars' doctrine, it is undeniable that its core is correct, and this because we must be able to recover the requirement of *truth* for scientific theories as something that is not reducible to the generic requirement of *objectivity* but is rather strictly related with a better understanding of the concept of objectivity itself. This in particular requires overcoming the 'strict empiricist' position that has inspired the main-stream philosophy of science of a good portion of the 20th century, in order to give back to reason its

[6]See Sellars (1963).

legitimate credentials in the progress of scientific knowledge, and it is not by chance that the renewed favor for realism that has been occurring in the last few decades often rehabilitates the notion of truth for scientific theories that seemed out of place regarding science especially within the analytic philosophy of science of the first half of the 20th century.

8 The Content of This Volume

There are many different kinds of new scientific realism and it would not be reasonable to pretend to 'make the point' on them. It seems more useful to present a limited sample of some of the most significant and best characterized of them, while a comprehensive view can be offered by outlining certain debates occurred in the history of Western epistemology which have constructed the conceptual background and framework of the present discussions on scientific realism. In this tradition we find, for example, the notion of the empirical underdetermination of theories (which is amply present in contemporary discussions on scientific realism); or the concept of the 'internal' constitution of an entity as a source for the explanation of its empirically ascertainable behaviors (mirrored in concepts such as those of propensity, disposition, or capacity, largely used today in the philosophy of natural and human sciences); or the difference and correlation between how-questions (*quomodo*) and why-questions (*propter quid*) that is implicit in the concept of 'mechanistic explanation' that is proposed today as a complement to the nomological-deductive model of explanation defended by the analytic philosophy of science; or the widespread acceptance of the explanatory force of the principle of causality, coming after a very large period of time in which the concept of cause was considered spurious in science and such as to be replaced by the neutral concept of correlation; or the taboo concepts of teleology and holism (that are rescued today within the system theoretic approach to complex realities); or, finally, the frank admission that certain basic metaphysical concepts and principles are legitimately used in the meta-scientific reflections. The foregoing sections of this Introduction aimed precisely at sketching such a historical and theoretical framework in which the continuity and the novelties of the present debates on scientific realism could emerge. This reason is also supported by the fact that references to the history of philosophy and science occur significantly in several of the contributions of this work.

After the *Introduction,* the book contains a Prologue and three Parts.

The *Prologue* opens with a chapter by **Mario Alai**, *The Debates on Scientific Realism Today: Knowledge and Objectivity in Science*, which considers the discussions on scientific realism as a well-defined and specialized domain of the philosophy of science, that has been characterized in a very large amount of books and papers, and offers a detailed survey of this production according to a clear classification that helps very much the understanding of this composite mosaic and the often subtle reasons of its inner articulation. Beside its great informative value, this chapter has the important function of compensating, thanks to its practical completeness, the unavoidable limitation due to the restricted number of chapters in which only certain of the most salient positions on scientific realism are presented in this work. Moreover, it has the merit of being an updated portrayal of the state of the art in this domain, which was badly needed considering the rare occurrence of such synthetic overviews.

The second chapter of this Prologue is the contribution by **Evandro Agazzi**, *The Truth of Theories and Scientific Realism*, in which the sciences are considered in their theoretical-conceptual and linguistic aspect as well as in their practical-operational aspect. In such a way the legitimate referential sense of the concept of truth is rescued, which can be applied also to scientific theories, and this justifies attributing a genuine 'reality' also to the referents of the theoretical concepts, within their respective 'regional ontology'. This chapter aims at outlining the thematic horizon in which the further contributions of this work are inscribed.

The chapters of the following three parts are organized according to a minimal logical order in which contributions are distributed according to the more general issues of the scientific realism debate, the different typologies of such realism, and its consideration in a few specialized sciences. For this reason the contributions concern in part the specialized domain of the realism debate, in part more general issues of philosophy of science that are significantly but indirectly related with the realism debate, in part the study of concrete questions regarding realism occurring in certain natural and human sciences and in mathematics.

The articles in **Part One** deal with general problems and arguments in the discussion between scientific realism and antirealism. **Alan Musgrave** *(Strict Empiricism versus Explanation in Science)* presents the explanatory task of science and philosophy as the decisive reason for realism, especially in his own original version of the *inference to the best explanation* and the *no miracle argument* as inferences to the *reasonableness of believing that theories are true*. He then contends that van Fraassen's and Stanford's "surrealist" accounts cannot explain scientific success, and counters Laudan's historical objections by arguing that the novel predictions of false theories either were not novel or can be explained by partial truth.

In his contribution **Bas van Fraassen** *(Misdirection and Misconception in the Scientific Realism Debates*) points out that current debates on scientific realism are often affected by misrepresentations of what the respective positions are. Once the issues are clarified, it turns misrepresentaions out that the question is not an ontological one (what there is), but about what is science, what are its adequacy criteria, and what epistemic and doxastic attitudes toward theories we should take. Realists and empiricists can cooperate in solving these questions, taking inspiration from Weyl, Glymour, and Suppe.

Against the background of the model-theoretic conception of theories, **Michel Ghins** (*Scientific Realism: Representation, Objectivity and Truth*) discusses the criteria by which scientific theories can be taken to represent unobservable entities objectively, and the difference between the faithfulness of representations and the truth of propositions.

Marco Buzzoni (*Robustness and Scientific Realism*) argues for a technical-experimental interpretation of *robustness*, which bridges between robustness conceived as the support coming to hypotheses from independent but convergent sources (Whewell, Wimsatt, Hacking, Glymour, Kosso, etc.), and robustness conceived as the stability of an engineered or biological system. This intermediate notion can also provide an improved *no miracles argument*, escaping Hacking's charge of begging the question.

In earlier works **Thomas Nickles** (*The Temptations of Scientific Realism: Cognitive Illusions, Objections, and Replies*) studied some cognitive illusions which may underlie the confidence in realism: that the course of scientific research has now come to its final goal, that our theories are fully mature and non-problematic, that no revolutionary or long-term evolutionary change is for-seeable in the future. As a result, he now suggests a more guarded agnostic and pragmatist position, defending it from various possible objections. Moreover, he points out that all the moves and attitudes which seemingly make realism more robust are also available to pragmatists.

Gerhard Vollmer (*Why do Theories Fail? An Argument for Realism*) is also critical of the most common argument for realism, that from the success of scientific theories: confirmation by success is an instance of the 'affirming the consequent' fallacy, so it may well happen that theories are successful yet false. The strongest argument for realism, instead, is its capacity to explain unsuccess: theories fail, when they so do, because the world is different from what they say. Since antirealists lack an equally plausible explanation, realism is to be preferred.

In the last chapter of this Part **Fabio Minazzi** (*The Epistemological Problem of the Objectivity of Knowledge*) discusses the general question of the objectivity of human knowledge from the point of view of logical neo-realism, but significantly drawing also on Husserl's concept of "regional ontologies", Bachelard's notion of *ontogenèse*, the tradition of criticism, and Agazzi's analysis of objectivity in its epistemic contexts. This allows him to distinguish the kinds of knowledge provided by different disciplines, their values and limits, while criticizing scientism and bringing out *rigor* and public *intersubjectivity* as requisites for all subject areas.

Some recent conceptions of scientific realism are discussed and defended in **Part Two**.

Stathis Psillos (*Scientific Realism and the Mind-Independence of the World*) defends the view that the realist claim of mind-independence is captured by what he calls 'the possibility of divergence', viz., the possibility of a gap between what there is in the world and what is issued as existing by a suitable set of epistemic practices. The realist commitment to mind-independence is split into two components: irreducible existence and objective existence, and it is shown that various versions of anti-realism compromise one or both of these conditions.

The role of metaphysics is also scrutinized by **Steven French** (*Structural Realism and the Toolbox of Metaphysics*). In his view philosophers of science should not dismiss it, but draw from it whatever "tools" and conceptual devices may serve to articulate their conceptions. In particular, the paper provides examples of how certain metaphysical "maneuvers" may help in framing a structural realist account of science.

Alberto Cordero (*Retention, Truth-Content and Selective Realism*) discusses the best way to develop selective realism, currently the most plausible realist position. He argues that Saatsi's, Vickers', and Votsis' strategy, committed only to the minimal components indispensable to derive predictions, is unnecessarily weak, and neglects the realist import of explanatory success. Hence, he proposes a naturalistic generalization of that strategy, whereby the theory-parts worth of realist commitment are selected through the confirmation criteria actually employed by scientists, including both predictive and explanatory power, and absence of reasonable doubts.

In the last paper of Part Two **Hans Lenk** (*A Scheme-Interpretationist Actionistic Realism*) puts forward a *scheme-interpretationist* and *actionistic* form of scientific realism: we can grasp the world and act on it only within and relative to interpretive perspectives and methodological schematizations. This is equally the case in action, action-orientation and formation, cognition and recognition, representation, depiction, cognitional modeling or abstract modeling, and active interventions such as experiments and everyday agency. His proposal is then spelled out through various distinctions: primary interpreting schemata (biologically or even genetically fixed) versus secondary schemata (variable); what is ontologically basic versus what is only methodological-epistemological; what is real "in itself" versus what is only socio-culturally or virtually real. As a result, he explains in which sense reality "in itself" can be recognized, but only indirectly.

Finally, the contributions in **Part Three** concern realism in some particular sciences or disciplines. From the point of view of logical semantics, scientific theories are ordered sets of propositions, whose models are abstract algebraic structures. But through and beyond these abstract models theories must be connected to the real concrete world. **Jan Wolensky** (*The Semantic Definition of Truth, Empirical Theories and Scientific Realism)* shows how this can be achieved through the concept of empirical valuation, thus allowing to phrase scientific realism in terms of the semantic theory of truth.

Dennis Dieks and **Roland Omnès** debate the realism question in quantum mechanics. Dieks (*Realism and Objectivity in Quantum Mechanics*) considers the general antirealist argument that in principle any body of data may be accounted for by numberless empirically equivalent theories. Realists have replied that in actual scientific practice cases of empirically equivalent theories are very rare or inexistent; when two theories are equally compatible with the evidence available at some particular time, new evidence can soon break the tight; and theories may enjoy different confirmation in spite of having the same empirical consequences. However **Dieks** points out that these replies encounter serious difficulties in quantum mechanics, where many incompatible interpretations are nonetheless strictly empirically equivalent. These instances of empirical underdetermination, therefore, are much more telling for the realism debate than many standard philosophical examples.

Another great obstacle to realism in quantum mechanics is considered to be the incompatibility of Schrödinger's equation with the collapse of wave functions.

However **Omnés** (*Is Uniqueness of Reality Predicted by the Quantum Laws?*) draws attention on the "local entanglement", a phenomenon directly deriving from the same equation but scarcely discussed in the literature. It has recently been found to have some interesting properties, which in the presence of fluctuations in the environment could generate the collapse. If conclusively proved, this would reconcile Schrödinger's equation with the wave-function collapse, so solving a major problem in the philosophy of quantum mechanics.

The next two contributions concern cognitive sciences. **Jean Guy Meunier** (*Theories and Models: Realism and Objectivity in Cognitive Science*) explains that in disciplines like psychology, philosophy and computer sciences the methodological and epistemic ideal of objectivity is less easily pursued than in the "hard" sciences. In fact, cognitive sciences are often related to their objects by a hierarchy of models: a *conceptual* model couched in the natural language, a *computational* model, and a *simulation* model, which instantiates the computational model in a physical computer process. One must therefore look at the relations between these models and reality in order to understand what exactly is the import of scientific realism when dealing with these disciplines.

Instead **Vladislav Lektorski** (*Realism as the Methodological Strategy in Cognitive Science*) argues for realism as a methodological strategy and the adequate interpretation of *situated* and *embodied cognitive science*, where actions and operations play a key role in linking cognition to the real world. He criticizes Fodor's "methodological solipsism" and Varela's attempt to overcome the realism-idealism dichotomy. Against this background he also discusses Gibson's notion of affordance and some prominent philosophical positions: *entity realism*, *constructive realism*, *externalism*, in particular *active externalism*, and the *activity approach* in Russian psychology and epistemology.

Amparo Gómez (*Mechanisms, Capacities, and Entity Realism in Social Sciences*) believes that scientific realism needs to rely on a metaphysics of causation. Today causes are mainly conceived in two alternative ways: as mechanisms, and as dispositions or powers. The former conception must explain how are mechanisms able to cause, and according to Gómez they do so because they realize properties, hence powers or capacities. Thus, in a sense, the two rival views of causality become mutually complementary. This however involves a particular treatment of powers, similar to Mumford's, Chakravartty's and Bird's, and alternative to Ellis' *new essentialism*.

The last two papers in the volume concern mathematics. While in the physical sciences realism conceives objects as representation-independent, **Gerhard Heinzmann** (*Objectivity in Mathematics: The Structuralist Roots of a Pragmatic Realism*) points out that mathematical entities are more naturally considered as existing and being so-and-so only to the extent that they are represented and proved by us. Thus, while in the physical sciences there can be a strong objectivity but no assurance of truth, here "objectivity" is proof, hence it guarantees truth. Moreover, there seems to be no role for explanation or explanatory evidence, here. According to Heinzmann, however, a pragmatic interpretation of mathematical practice might show that there are "explanatory proofs" after all: viz., proofs which derive the content of the conclusion from that of the premises by intuitive "seemings" and topic-specific mathematical representations.

Also **Reinhard Kahle** (*Mathematical Truth Revisited: Mathematics as a Toolbox*) discusses the specificities of truth and existence in mathematics, often conceived as conditional or relative to a framework, like in Bernays or Carnap. Thus, even "non-standard" structures may be admitted as useful, rather than as true in some absolute sense.

Acknowledgements More than half of the chapters of this book are revised versions of invited papers presented at the Congress of the International Academy of Philosophy of Science (IAPS) held on September 23–25, 2015 at the University of A Coruña (Spain) on the theme *Scientific Realism: The Problem of Objectivity and Truth in Science*. The program of that conference was jointly prepared by Evandro Agazzi (President) and Wenceslao Gonzalez (Member) of the IAPS. A special thanks goes to the latter also for the impeccable organization of the meeting, and to the University of A Coruña and the Deputación of A Coruña for their financial support. Mario Alai (Member of the IAPS) has greatly helped in the editing of this volume.

References

Aquinas Th., *In libros Aristotelis De caelo et mundo expositio*, transl. by F. R. Larcher, Fabian R., and P. H. Conway, *Exposition of Aristotle's Treatise On the Heavens.* 2 vols. Columbus, Ohio: College of St. Mary of the Springs, 1964. E-text, http://dhspriory.org/thomas/DeCoelo.htm.

Galilei G., *Opere: Le opere di Galileo Galilei*, Edizione Nazionale a cura di A. Favaro, Firenze, Barbera, 1929–1939; repr.1964–66, 20 vols.

Newton I., (1726), *Philosophiae Naturalis Principia Mathematica*. Engl. Transl. by B. Cohen and A. Withman, University of California Press, 1999.

Sellars W., (1963) *Science, Perception and Reality*, Routledge and Kegan Paul, London.

The Debates on Scientific Realism Today: Knowledge and Objectivity in Science

Mario Alai

Abstract Debates on realism in science concern two main questions: whether theoretical knowledge is possible, and whether it is objective. Today, as in the past, the possibility of theoretical knowledge is often denied because of the empirical underdetermination of theories. Realists rely on explanatory power, theoretical virtues, and instrumental but theory-free observation to solve this problem. Besides, they use the *no miracles argument* for the truth of successful theories. Antirealists, however, deny that explanation is either necessary or possible, and that is a cue to truth. Moreover, they reject realism and the cogency of the no miracles argument by the pessimistic induction from the falsity of past successful theories. Some realists reply that there is a radical discontinuity between past science (largely off-track) and current science (basically sound). But this reply is at best insufficient, and most realists prefer to restrict their commitment to selected parts or features of theories, both past and present. Forms of "selective realism" are *entity realism*, *structural realism*, *deployment realism* and *semirealism*, but also the *verisimilitude* research program and the *restricted-domain* approach. Realists need criteria to identify the true components of theories, and a noteworthy candidate is essential involvement in functionally novel and surprising predictions. The second main question is a special instance of the old debate between realists and relativists or idealists: according to antirealists science cannot be objective, because of its inherently "perspectival" nature, characterized by a priori and subjective factors. On the contrary, *perspectival realists* argue that the specific "viewpoints" within which scientists must work do not prevent them to discover objective features of reality.

M. Alai (✉)
Università degli Studi Carlo Bo, Urbino, Italy
e-mail: mario.alai@uniurb.it

E. Agazzi (ed.), *Varieties of Scientific Realism*,
DOI 10.1007/978-3-319-51608-0_2

1 Knowledge, Objectivity, and the Earlier Debates on Scientific Realism

Debates on realism in science have taken place since the Antiquity (Alai 2008), and they concern two different questions: (I) whether the unobservable entities posited by theories can be known (that is, whether we can have beliefs about them, and these beliefs can be both true and justified); and (II) whether any knowledge we have of them is objective or not. Question (I) arises from the doubt that since observation is the basis of all our factual knowledge, unobservable entities cannot be known; questions (II) arises from the doubt that since scientific representations are inextricably laden with, or distorted by, the subjective, idiosyncratic and a priori features of human cognition and scientific practice, they cannot convey any reliable information on how their objects are "in themselves". The same two questions arise even concerning different subjects (ordinary material objects, the mind, etc.), and antirealism on knowledge is generally termed 'skepticism', while antirealism on objectivity is called 'idealism', 'subjectivism' or 'constructivism'.

These two questions are logically independent, so they have been mostly discussed separately, and one can be realist on one and antirealist on the other. For instance Putnam held in (1978b) and (1981) that we can know unobservable entities quite as we know ordinary observable objects, but this is not knowing an absolutely mind-independent reality; a similar stance has been taken by some "perspectival realists" (see below). Nevertheless, certain powerful arguments for realism work in both debates, so that one may find it hard to be realist on knowledge but not on objectivity, or vice versa (Alai 2005, 2006: 214–216, 2009, forthcoming). Moreover, as we shall see, the two debates somehow converge, in the sense that some of the most plausible forms of realism on each question (selective realisms on the one hand, and perspectival realisms on the other) basically agree on the possibility of objective but partial knowledge. Here I shall examine the state of contemporary debates mainly on the knowledge question, and more briefly on the objectivity question.

Since the Antiquity and up to the beginning of the last century, the typical antirealist stand about scientific knowledge was that we can at most aim at "saving the phenomena", i.e. form true and justified beliefs about observable entities. A frequent corollary was instrumentalism: to the extent that theories apparently describe unobservable entities, they should not be interpreted literally as descriptions, because they actually are (or should be) just practically useful instruments, computing devices to predict future phenomena or directions for technological applications, and as such neither true not false.

However, instrumentalists should explain why theories are so useful and successful, if they don't offer an at least approximately true description of the underlying reality; moreover, what justifies the claim that the literal interpretation of scientific statements as descriptions is wrong, and that scientists which aim at finding the truth are wrong and should pursue a different goal? (Alai 2006: 220). The only possible justification would be the claim that saving the phenomena is the best we can do, but even so instrumentalism seems to be a dubious and idle corollary.

In fact, it isn't a live option anymore, except perhaps as concerns quantum mechanics (e.g., Wigner 1967).

Logical positivists tried to bypass the epistemological debate on truth and justification by a "linguistic turn": their verificationist theory of meaning entailed that, appearances notwithstanding, scientific theories did not actually speak of unobservable entities; the meaning of theoretical terms was entirely reducible to possible observations, so in principle they could be replaced by a purely observational vocabulary (Carnap 1923).

However, eliminativism was abandoned when it became clear that non-observational terms cannot be completely defined by observational terms, as it happens already with simple "dispositional" terms (such as 'soluble' and 'fragile') and the irrational values of physical measurements (Carnap 1936: § 7; Hempel 1952: II). Moreover, non-observational terms play a necessary role in systematizing experience, predicting future observations, and leaving the way open to the discovery of new properties (see Hempel 1958); moreover, their elimination would leave numberless unexplained coincidences (Smart 1963: 39; Psillos 1999: 72–73). Finally, verificationism itself was abandoned in the second half of the XX Century, especially as a consequence of Quine's (1951) criticisms.

Therefore van Fraassen (1980: II, 1) explained that linguistic questions are no longer live issues between scientific realists and antirealists: antirealists may grant that theories must be read literally, as purportedly true descriptions of unobservable entities, but deny that we have compelling reasons to believe that they are true. According to his "constructive empiricism" all we need to believe is that a theory is empirically adequate, i.e., it "saves the phenomena". Thus he brought back the debate to the epistemic question on which it had focused for about 20 centuries, and in so doing he set the agenda for most subsequent discussions on the knowledge question up to this day (Alai 2006: 217–220).

2 The Empirical Underdetermination of Theories

Antirealists believe that since observation is the only basis for factual knowledge, unobservable entities cannot be known. This is made evident by the empirical underdetermination of theories (see Stanford 2013): they are introduced as explanations of the observable phenomena, but for any set of given phenomena many incompatible explanations can be found; therefore, it is claimed, we cannot know which one is true. This has been one of the main arguments against realism since the Ancients up to our days (Celsus 1935: §§ 28–29; Alai 2008: §§ 3–4).

The history of science, however, would seem to refute this argument: cases of actual competition among empirically equivalent theories have been very few, and eventually they have been decided to everybody's satisfaction (e.g., undulatory vs. corpuscular theory of light; Ptolemaic vs. Copernican vs. Tychonian cosmology). There are mathematically intertranslatable theories, like Newtonian mechanics (based on force) versus Lagrangian or Hamiltonian mechanics (based on a principle of minimal action); or theories introducing fields and theories using action at distance with retarded potentials (Putnam 1978b: 133), but they look more like different formulations of the same theory. There seems to be a genuine conflict between Heisenberg's matrix mechanics and Schrödinger's wave mechanics (Putnam 1978c: 555; Friedman 1983: 165 ff.; Fano 2005: 166), and between standard quantum mechanics and Bohmian mechanics; but one day we might find new empirical consequences which allow to decide between them. In general, "it has never been shown that for any theory there exist non trivial and minimally plausible alternatives" (Psillos 1999: 168); underdetermination does not look like a concrete problem in scientific practice, but a merely in-principle risk not to be taken too seriously (see Laudan and Leplin 1991, 1993).

Yet, isn't it a logical fact that there are infinitely many possible theories compatible with any body of data? So, it has been suggested that our failure to consider many of those theories (including perhaps the true one) has to do with our conservatism and lack of imagination, and that many unconceived alternatives "exceed our grasp" (Stanford 2006).

3 Explanation and Theoretical Virtues

This, however, is just part of the story: scientists do not look simply for theories that are compatible with the phenomena, but that are *true, explanatory* and *fecund.* So, theories must be *empirically testable, plausible* (i.e., consistent with the largest possible background of accepted beliefs), *simple, non-ad hoc*, and provide models of the underlying causes, which at the same time *explain* the phenomena, *unify* them, and *predict* new phenomena as effects of the same causes. So many possible theories which could be *imagined*, are not even considered because they lack these *theoretical virtues.*

In other words, the evidence for a theory does not coincide with its empirical consequences: for instance, trivial and ad hoc consequences don't confirm (Laudan 1996, ch. 3; Laudan and Leplin 1991, 1993; Psillos 1999, 169–176), and certain phenomena can confirm a hypothesis even if not entailed by it (Psillos 1999: 170). In Bayesian terms, different hypotheses have different prior probabilities, so even if supported by the same data, they get different posterior probabilities (Fano 2005: 166; Psillos 1999: 163). Theoretical virtues have confirmatory power (Glymour 1980; Kosso 1992; Psillos 1999: 171–176) and supply the necessary guidance to all *ampliative* inferences: even the mere inductive projection presupposes the uniformity of nature, and follows the rule of simplicity. The same holds for analogical reasoning, while abductive reasoning exploits the explanatory power of hypotheses.

Antirealists deny the confirmatory power of theoretical virtues: simplicity is evidence for truth only if nature is simple, or *uniform*; but this can be shown only by circularly assuming that our theories, which depict it as simple, are true. Assuming that consistency with other accepted theories is evidence for truth presupposes the *petitio principii* that accepted theories are themselves true: the entire body of our scientific beliefs might be coherent, but false.

This reminds us of the Humean criticism of ampliative reasoning: its validity cannot be proved a priori, since it is a factual question, nor *a posteriori*, since an inductive proof (from the past success of these methods) would be circular. Yet, this is not enough, if scientific antirealism must differ from general Humean skepticism (Alai 2006: 228–230). Moreover, we use basic patterns of deductive inference, like *modus ponens* and *modus tollens*, without requiring a non-circular proof of their validity; so, why shouldn't we do the same for equally basic forms of inductive inference? Besides, many antirealists use them in inferences from observed facts to yet unobserved ones; in particular, only ampliative methods allow us to claim that a theory is *empirically adequate*—as opposed to merely compatible with known data (Lipton 1991, 154 ff.). Some authors reply that ampliative methods are reliable when inferring from observed to *unobserved-but-observable* entities, but not when inferring to *unobservable* entities. However, this discrimination is unjustified, since the validity of inferential patterns does not depend on the subject-matter (Alai 2010: § 3).

Some antirealists deny that we need explanations, or that explaining is a task for science (Duhem 1906). However the search for explanations is natural for human beings and a spur to inquiry (Aristotle *Metaphysics,* 982b); it is typical of ordinary knowledge, and science is just a development of ordinary knowledge (Rescher 1987: 36–41). The desire to explain is not mere curiosity, but a search for consistency and self-correction: not for any event we ask why or how it happened, but only for those which in the light of our preconceptions should not or could not have happened. Hence, finding an explanation involves correcting some mistake or learning some new crucial information. In fact, the search for explanations produces new discoveries and is a key to the self-corrective character of science.

Van Fraassen (1980: 25) claimed that, when explaining observable regularities through unobservable ones, we leave the unobservable regularities unexplained; therefore we should limit our search for explanation to the observable level. However unobservable regularities may in turn be explained by deeper unobservable regularities, and these by even deeper ones, etc. At each level explanation gains generality, simplicity and information: fewer laws, entities and properties are employed. This does not necessarily launch an infinite regress, since wherever we stop we have a correct, informative and interesting explanation even if the *explanans* is left itself unexplained. The longest the explanatory chain gets, the more it becomes informative and interesting. Furthermore, often observable regularities are not explained by further regularities, but by postulating *entities*, such as genes, viruses, etc. (Aronson 1984).

4 The *No Miracles Argument*

If we accept the ampliative methods, in particular the Inference to the Best Explanation, then not only we can hope that our selection of theories among the infinitely many logically possible ones is reliable, but we have a meta-argument to the effect that best theories are in fact true or partly true, and scientific method is truth-conducive, hence scientific realism is right: as noted by Putnam (1975a, 73), this is the only explanation of the stunning predictive and applicative success of science which does not make it a miracle (or if you will, a miraculously lucky coincidence). This *No Miracles Argument* (NMA) has been used in various versions also by Smart (1968: 150), Grover Maxwell (1970: 12), Musgrave (1985), Lipton (1991), Niiniluoto (1999: 197), etc. (See Psillos 1999: 70–81).

It has been objected that the NMA is circular or question-begging, for it presupposes the reliability of abductive reasoning, which is rejected by antirealists (Laudan 1981: 45–46; Fine 1984: 85–86: see Ladyman 2002: 218–219). In this sense, the choice between realism and antirealism is a *stance* (van Fraassen 2002; Psillos 2011a, b). But as argued above, realism seems to be better grounded and more consistent with our epistemic practices at large.

4.1 Novel Predictions

The NMA presupposes that truth is the only (non-miraculous) explanation of success. But certain forms of success can be explained otherwise: success in accounting for known phenomena might be simply due to the theorist's puzzle-solving ability (ingenuity, imagination and patience). This explanation is not available for the prediction of novel phenomena. However, if the predicted phenomena are similar to already known phenomena, the theorist might have inductively extrapolated them from the previously known ones, and then built the theory with an eye to accommodate them. For instance, it would be hardly surprising that a theory T based, among other data, on certain chemical and physical properties of carbon-12 and carbon-13, predicted analogous properties for carbon-14. Again, if T predicts new and heterogeneous phenomena, but they are a priori probable, this success might be simply credited to good luck. In all these cases there is no need to assume that T is true. For example, many false astronomical theories could predict the existence of a new unknown planet somewhere in the universe. But the prediction of a new planet (Neptune) with precise mass and orbit in a particular region, as licensed by Newton's gravitation theory, showed that there was some truth in it. Equally, quantum electrodynamics predicted that the magnetic moment of the electron was $1159652359 \times 10^{-12}$, while that obtained by experiment is $1159652410 \times 10^{-12}$ (Wright 2002), and such concordance cannot be due to luck (Alai 2014a §§ 3.2–3.3, b, §§ 4–5).

Therefore the NMA properly applies only to the prediction of novel, surprising (heterogeneous from all that was known before) and bold (a priori improbable) phenomena (Musgrave 1988; Psillos 1999). Further examples of predictions fulfilling these constraints are Fresnel's bright spot and dark spot prediction; the new chemical elements predicted by Mendeleev; Einstein's predictions of the retard of clocks in motion and the bending of light; the background radiation predicted by the Big Bang theory; etc.

This means that novel predictions confirm more than mere a posteriori accommodation; but many objected that confirmation is a logical relation between a hypothesis and its empirical consequences; therefore prediction should have no advantage over accommodation. The ensuing debates (see Alai 2014a: §§ 1–3) have shown that in order to count for the NMA a new phenomenon NP needs not be *historically* novel (i.e., not known by the theorists, or not used by them in building the theory): even if NP was used, it is enough that it was not used *essentially*. That is, it is enough that the theory T was plausible independently of NP, i.e., that it was already the best possible account of a body of *old* phenomena OP not including NP. In fact, in that case it would be a miraculous coincidence if T, build to accommodate OP, were false, yet at the same time predicted the highly improbable NP, radically heterogeneous from OP. Since this notion of novelty does not concern a relation between NP and the theorist, but the relations of NP to T and OP, it is compatible with the logical nature of confirmation. Novel predictions so understood confirm for the same reason for which the consilience of non-*ad hoc* predictions by independent theories confirm those theories: the unlikeliness that they are just a lucky coincidence (ibi: § 4).

4.2 Objections to the NMA

Since success is essentially the truth of the (novel) prediction 'NP', it has been objected that it has a trivial explanation, which entails nothing about T: 'NP' is true simply because things are as it states (White 2003; Rees 2012: 302). Yet, we have a real puzzle: how could the theorist find, among the numberless possible theories compatible with the old phenomena OP, one predicting the heterogeneous and improbable new phenomenon NP, without essentially using it? This is the real *explanandum* in the NMA (White 2003; Laudan 1984a: 92; Alai 2014a: 299, c: 50). It has also been noted that the *explanans* cannot be that the theorist found a true theory: for the *true* theories entailing NP are just a tiny subset of those entailing NP, so we would explain a puzzling fact by an even more puzzling one. On a careful analysis, however, the actual *explanans* is that scientific method is truth-conducive (because the uniformity and causal structure of nature makes ampliative inferences reliable), *hence* the theorist found a true theory (Alai 2014a: § 1, c: § 5).

Problems with the NMA were raised also by Lipton (1991), Howson (2000) and Magnus and Callender (2004), for whom the impossibility to know the relevant base rate prevents to claim that successful theories are probably true. Paul Hoyningen-Huene (2011) argued that the version of the NMA based on novelty is falsified by what he called "transient underdetermination".

Some antirealists claim that the success of T can be explained differently: for instance, because T is empirically adequate, i.e., compatible with all the phenomena (van Fraassen 1980, 12); or because all phenomena are *as if* T were true (Fine 1984; Leplin 1987). But this does not explain why T predicts NP, for T may be *compatible* with all phenomena without predicting any of them, in particular without predicting NP. For instance, a theory with no empirical content would be empirically adequate. For Stanford (2000, 272 ff.) T is successful because it makes predictions very similar to those of the true theory; but this would be like saying that T predicts NP because it predicts a number of things including NP. So, it would not be an explanation, but just a repetition of the *explanandum*. Other variants of the empirical adequacy explanation, like Lyons' *modest surrealism* (2002, 78), and Fine's *instrumental reliability* explanation (1986, 1991), have the same flaw. In general, no "as if" account really explains: "the hypothesis that it is raining explains why the streets are wet—but 'The phenomena are as if it were raining' does not" (Musgrave 2006–2007).

Besides, none of these *expalanantia* explains how the theorist found a theory that predicts NP without essentially using it. Realists explain that T was found thanks to the scientific method, which is fecund and truth-conducive; but antirealists cannot answer in a similar way. In fact, there are just two *methods* to find theories predicting new phenomena: one is by finding true and fecund theories (the realist method); the other is by inductive extrapolation from known phenomena (Laudan 1984a, § 5; Rees 2012, 302; Wright 2013, chs. 3, 4; Alai 2014d, 125–126), but this is impossible if the new phenomena are radically heterogeneous from the old ones (Alai 2014c: § 6). According to Fine (1984) and Hacking (1983: 64) the success of T is just the obtaining of the phenomena predicted by T, hence it is already explained by T itself. However, advancing a hypothesis as an explanation is accepting it as true, hence their "deflationistic" explanation cannot dispense with the assumption that T is true. Moreover, it doesn't explain how T was found (Alai 2014c § 4).

Stanford suggests that even a radically false theory which saves the phenomena is likely to make novel predictions, thanks to "the systematic relationship among phenomena within the same domain of inquiry" (2000: 281). However, if "systematic relationship" means that the phenomena are homogeneous, there is no novelty in that field; if instead it means that they are connected by unobservable underlying mechanisms, then it would be a miracle if T predicted NP without getting those mechanisms right.

Van Fraassen's (1980, 40) "natural selection" explanation was that our theories are successful, simply because the unsuccessful ones were dropped. However, this explains just (1) why we have *only* successful theories, not (2) what makes these theories successful nor (3) how they were found. When asked "why birds can fly?" one cannot just answer (1) "because those which did not fly were wiped out by natural selection". It should also be explained that (2) birds fly thanks to wings, feathers, hollow bones, etc. (Kitcher 1993; Alai 2014c, § 7).

5 Indirect but Theory-Free Observation

Theoretical hypotheses are a creation of human mind, introduced as the best explanation of certain phenomena, and tested by their observable consequences. Antirealists object that (a) any explanation is underdetermined and (b) confirmation by consequences is the invalid scheme of *adfirmatio consequentis*. If one points out that some theoretical entities are *instrumentally* observed (e.g., by the electronic microscope), they reply that since the reliability of those instruments can be established only on the basis of theories, theories cannot be confirmed by them.

However, some theoretical entities can be observed by instruments which do not presuppose theories (Alai 2010: § 4). Van Leeuwenhoek, a draper, invented the optical microscope by observing that the lenses he used to ascertain the quality of his fabrics magnified small but observable objects by 200 or 300 times. Thus, when through his lenses he saw things which escaped the naked eye (like bacteria, spermatozoa, and muscular fibers), he could safely assume that those were real entities, and approximately measure their actual size. In so doing he relied merely on his eyes, the uniformity of nature (e.g., the assumption that the ratio of the lens images to their objects did not vary at different scales), and elementary mathematics.

Perrin was able to measure Avogadro's number, hence the volume and size of molecules, by procedures which also presupposed nothing but the uniformity of nature and easy computations. For instance, he dropped a droplet of oil on a water surface covered by talc powder: the drop expanded, pushing the talc aside. Eventually it reduced to the thickness of one molecule and became invisible, but its surface coincided with the talc-free area, and could be measured. Thus, by dividing the volume of the droplet times the surface, he found the value of about 1 μμ for the diameter of molecules (Perrin 1913: § 32). With similar methods Millikan was able to measure the charge of the electron, and those results finally dispelled all the doubts on the atomic structure of matter.

Another example is the chemical composition of the stars: it cannot be observed directly, but it can be recognized by observing the same spectra which are emitted by samples of different elements in the laboratory. Probably cases like these are more common than it might seem, and in front of them stubborn antirealists must pay a high price: doubt the uniformity of nature, and give up induction in science and everyday life (Alai 2010: § 4).

Perrin's measurement of Avogadro's number also exemplifies the confirming power of Whewell's "consilience of inductions", because he reached the same results by different methods (Kosso 1992: ch. 9). Those procedures presupposed various theories, but, since they were mutually independent, the agreement of their results would be a miraculous coincidence unless they had some real grasp of the world. So, this reasoning is just another variant of the NMA.

6 The Historical Objections to Realism

The most severe objection to realism, beside empirical underdetermination, is that in the history of science most theories sooner or later have been found to be false, and none that is still accepted is older than 100 years or so. This is the premise of the so called "pessimistic meta-induction" (PMI) (Putnam 1978a: 25):

(PMI_1) all past theories were completely false, even the successful ones;
(PMI_2) there is no radical epistemic or methodological difference between past and present theories; therefore
(PMI_3) most likely also current and future theories are and will be completely false; therefore
(PMI_4) unobservable entities cannot be known.

(PMI_2) is often left implicit, although it is crucial to the argument. (PMI_1) also works as a premise to Laudan's (1981) refutation of the NMA, called by Lyons (2002) the "meta-modus tollens" (MMT):

(MMT_1) past successful theories were completely false; therefore
(MMT_2) truth is not the only explanation of success; therefore
(MMT_3) the NMA *is flawed.*

Even restricting the notion of success to novel predictions will not save the NMA from the MMT: as pointed out by Lyons (2002), many ancient false theories made important novel predictions.

7 Resisting the Historical Objections: The Discontinuity Strategy

Some scholars denied the cogency of the PMI: for Lewis (2001), Lange (2002) and Magnus and Callender (2004) it is based on inductive fallacies, like the NMA. Doppelt claimed that it is incoherent (2011: 310, 2013: 48–49, 2014: 282–283), but he seems to be wrong (Alai 2016: § 7). More often the PMI is countered by denying either of its premises: the "discontinuity strategy" rejects (PMI_2), while the "selective strategy" rejects (PMI_1).

The discontinuity strategy adds that since there are radical differences between past and present science, no inductive inference from the former to the latter is possible; therefore also (MMT_2) applies only to past science, and the NMA remains cogent for current science.

Hardin and Rosenberg (1982) claimed that many false theories cited by Laudan were not products of *mature science*, so from their falsity we cannot infer to the falsity of theories in mature science. Yet, there were many false theories even in mature science (Newton's gravitation theory, Fresnel's wave theory of light,

Rutherford's model of the atom, etc.). Devitt (1984: 143–149) argued that progress in scientific methodology, observation instruments and experimental technology continuously improve the reliability of theoretical science, hence past failures should not be projected onto the present or the future. Still, even from the Antiquity to the XVIII Century scientific methodology, instruments, etc., had greatly improved, but all theories accepted in the XVIII Century have been subsequently rejected.

Gerald Doppelt (2007, 2011, 2013, 2014) argues that old superseded theories were radically false and their success cannot be explained by assuming that they were partly true, while current best theories are completely true and will never be superseded or corrected by better theories. However, this "is an illusion owing to our particular historical perspective" (Nickles, this volume), which prevents us to see that progress began already in the past, and it will go on in the future. *Pace* Hegel, Marx or Fukuyama, we are not at the end of history (Nickles forthcoming[b]), and science is always fallible. Moreover, Doppelt can't account for the success of past theories, nor for the failures of current ones (Alai 2016).

Nonetheless, the discontinuity strategy might be right that contemporary science is *much more* mature than Newton's theory (or any of the false theories cited by Laudan), exhibiting a *greater degree* of sophistication, unification and coherence, and that from the XVIII Century to now there have been much greater methodological and technological improvements than from the antiquity to the XVIII Century. For instance, ether and phlogiston were never measured, while we can measure many properties of unobservable entities (Dorato 2007: 181).

Fahrbach (2011: 1283) argued that "current best theories enjoy far higher degrees of success than any of the successful but refuted theories of the past". In fact "three-quarters of all scientific work ever done was done in the last 30–40 years", and the exponential increase in the amount, diversity, and precision of scientific data and computing power marks a sharp difference between science today and only a few decades ago. Therefore present theories are incomparable to earlier ones as to the number and diversity of the tests successfully passed.

Still, it is questionable whether this difference between past and present science is radically qualitative or merely quantitative, and in the latter case it couldn't block the PMI: at most it might require some caution in the inductive extrapolation from past failures, because "successful science" is not a unitary kind, but rather a set of practices bearing family resemblances (Bird 2015, § 3); or it might suggest that current theories will be superseded in a much longer time than past ones. At any rate, it cannot block the MMT, because if past theories were false but successful, truth cannot be the only explanation of success, even for current theories. Therefore the discontinuity strategy needs in any case to be supplemented by the *selective strategy*.

8 Resisting the Historical Objections: The Selective Strategy

The selective strategy rejects PMI_1 by holding that even in radically false and discarded theories there were *some* truths, which are still preserved today; hence even in current theories there are some (and presumably more) truths. Thus, selective realists are committed only to the truth of some parts of certain theories, and they stress the continuity between past, present and future science. If it can be shown that all the novel predictions of past false theories were due to their true parts, then also the MMT is blocked, and the NMA allows to argue that also the parts of current theories which produced novel predictions are true. There are different versions of selective realism, among which at least: *entity realism*, *structural realism*, *deployment realism*, *semirealism*, the *verisimilitude* strategy, and the *restricted-domain* approach.

8.1 Referential Continuity and Entity Realism

Realists employed Kripke's and Putnam's (1975c, d) causal theories of reference to argue that discarded theories were not completely off-track, because they posited the same entities we now believe in, although with different descriptions and sometimes with different names. For instance, the term 'ether' referred to whatever caused its introduction—say, the electromagnetic field (Hardin and Rosenberg 1982).

But this does not seem possible when the core descriptions associated to a term are completely wrong, or there exists nothing even slightly similar: there is nothing to which 'phlogiston' or 'caloric' might refer (Putnam 1978a: 25; Laudan 1984b: 160–161; Psillos 1999, 290–293); understanding 'ether' as 'electromagnetic field' may be overstretched (Worrall 1995). Holding that a term refers to whatever is the cause of the phenomena it is supposed to explain would trivialize reference: for example, Aristotelian natural places, Newton's gravitation force and Einstein's curvature of spacetime, all were supposed to be the cause of gravitational phenomena. A shared explanatory agenda cannot be confused with a shared explanatory ontology (Laudan 1984b: 161).

Psillos raised further problems (1999: 286–287), and proposed instead a causal-descriptivistic account, by which a term refers to the unique natural kind having the core causal properties assigned to it by the description, provided that the actual kind-constitutive properties are the causal origin of such description (ibi: 295). More recently, Schurz (2011) proved a correspondence theorem showing that a theoretical term originally intended to refer to an inexistent entity may *indirectly* refer to a real counterpart of that entity. For instance, in phlogiston theory 'dephlogistication' indirectly refers to the process of oxidation.

So, noting that different theories of reference have been proposed, motivated by conflicting intuitions, Votsis (2011a) suggests that perhaps the very concept of reference is not a monolithic one. At any rate, what 'reference' means is conventional, and scientists may intend to pick their referents in different ways: no doubt Mendel understood 'gene' very liberally, as *whatever* played a certain causal role, while Bohr may have understood 'atom' as something very close to the description he gave. Besides, probably the crucial question is not whether T's terms refer, but how much truth is found in T. It doesn't help much that T's terms refer, if everything T says on their referents is wrong. If a relaxed definition of reference is used, T can be radically false although its terms refer; if instead a strict definition is chosen, there can be a lot of truths in T even if its terms fail to refer (more on this below). So, reference may help, but it is neither necessary nor sufficient to resist the PMI (Alai 2006: 239).

According to a different approach, however false some theories may be, the entities they postulate must exist, because we currently *manipulate* them. For instance, in certain experiments electrons are sprayed to reveal the existence of quarks with fractional charges. We may have widely different beliefs concerning those electrons, but their existence is out of question: they are "here in front of us" (Hacking 1983: ch. 16). Equally, protons are produced and used to bomb the nuclei and study the trajectories of neutrons so expelled (Giere 1988: ch. 5).

Musgrave (2006–2007) objected that "entity realism is a hopeless form of realism", because the existence claims are empty without some description of properties and behaviour. Moreover, the manipulation argument is question begging: who says you are actually manipulating an unobservable entity, rather than merely performing certain macroscopic operations with certain macroscopic effects? (see van Fraassen 1985: 298; Nola 2002: 9).

To avoid these problems the argument can be reinterpreted as an inference to the existence of such entities as the best explanation of their observed effects (Nola 2002: 9–14), but then it supports also some theoretical assumptions about them. Otherwise, it can help to solve the Duhem-Quine indeterminacy of empirical control: machines incorporate the less problematic of our beliefs, which can be used to test the more problematic ones (Giere 1988). Even so, however, entities and beliefs about them go hand in hand. Summing up, the partial truth of theories cannot be confined to the existence of the entities they postulate (see also Dorato 2007: 183–184; Nanay 2016).

8.2 Structural Realism

In opposition to *entity realism*, *structural realism* (StrR) holds that only structures can be known, while entities cannot (Frigg and Votsis 2011). According to different versions, we cannot know the individual entities, but we can know their properties and relations; or not their intrinsic properties, but their first-order relational properties; or none of these, but the second-order structure of their relational properties.

This last was Russell's (1927) and Carnap's (1928) view (Ladyman 2014; French and Ladyman 2011).

Poincaré adopted StrR in reaction to the PMI: as theory T_1 is replaced by theory T_2, and T_2 by T_3, etc., the entities postulated by T_1 are substituted by different ones postulated by later theories (for instance, ether was substituted by the electromagnetic field); but the basic equations, tracking the underlying structure of things (e.g. Maxwell's equations) are preserved and are approximately true (Poincaré 1902: 160–162). Also the logical positivists maintained that we know only forms, not content (Schlick 1938), or structures, not objects (Carnap 1928, §§ 1, 6, 11, 16 etc.). Similar views were also held by Arthur Eddington, Grover Maxwell, Hermann Weyl (see Psillos 1999, 621–663; Ladyman 2014), and Ernan McMullin (1984).

Lately this position has been advocated by Worrall: "Fresnel completely misidentified the *nature* of light, but nonetheless it is no miracle that his theory enjoyed the empirical success that it did … because Fresnel's theory had… more or less the right *structure*". Thus, "there was continuity or accumulation in the shift [from Fresnel to Maxwell], but… one of *form* or *structure*, not of *content*" (Worrall 1989. See also Worrall 1994, 1995: 92–94). Therefore, showing that the success of discarded theories was due to their structural claims, StrR also vindicates the NMA against Laudan's MMT refutation. Besides, since incompatible but empirically equivalent theories may share the same structure, it has been suggested that StrR can also solve the underdetermination problem (Worrall 2011; Lyre 2011; French 2011).

More recently, the merely *epistemic* thesis that we can know only structures (EStrR), has been reinforced by the *ontic* thesis that there exist no objects but only structures (OStrR). The latter may be further detailed as claiming that (1) there are no individuals but only relational structures; or (2) relations do not supervene on the intrinsic spatio-temporal properties of their relata, or (3) individual objects have no intrinsic natures or properties; or (4) the identity of objects is ontologically dependent on their relations; or (5) individual objects are just constructs used to build approximate representations of the world (Ladyman 1998; French, Ladyman 2003a, b, 2014; Ladyman and Ross 2007).

OStrR is strongly suggested by contemporary physics: in the entangled states of quantum mechanics relations do not supervene on the properties of particles, and particles seems to lose their individuality, since there are no properties, even spatiotemporal, which allow to distinguish them from one another. Although they may be weakly discernible, Muller (2011) noted that this can be reasonably understood via a relationist conception of objects that supports OStrR.

Moreover, the traditional ontology of individuals, intrinsic properties and relations seems to be at odds with the nature of space, time and matter. The proper objects of contemporary physics are rather symmetries and invariants, and "elementary particles are hypostatisations of sets of quantities that are invariant under the symmetry groups of particle physics" (Ladyman 2014), or excitations of fields.

Many objections have been raised against StrR: do the mathematical equations preserved in theory change tell something about the underlying structure of the world, or merely about empirical regularities? (Laudan 1981: 237). Isn't the retention of equations just a convenient and labor saving pragmatic feature of scientific practice, due to the conservativeness of the scientific community? (Fano 2005). That equations tell us something about the structure of the world can only be shown by the NMA, which however supports at the same time the theoretical claims about entities (Psillos 1999: 152). In fact, for StrR structures are the only responsible for success. But mathematical equations by themselves can license predictions only when theoretically interpreted and supplemented with auxiliary hypotheses; so, the success of predictions confirms also those theoretical hypotheses (ibi: 153–155).

Besides, we cannot distinguish "between the *nature* of an entity and its *structure* such that we can … know its structure but not its nature", because "to say what an entity *is* is to show *how this entity is structured*" (ibi: 155–156). For instance, there was no structure of light on which Fresnel was right while being wrong on its nature: there simply were properties of light on which he was right and others on which he was wrong (ibi: 159).

Granted, the properties on which he was right were behavioral and relational properties (the ways of propagation) while he was wrong that the physical nature of light was molecules of ether. So, we can know a lot about the relations and behavior of the unobservable entities without knowing much about their physical nature. For example, Mendel built his theory of genes without having any idea about their physical instantiation. Yet, as research proceeds, we can often discover the very nature of our objects, as Einstein did with light, or Watson and Crick with genes (Psillos 160–161). That the contents of perception are not actual properties of thing, as stressed by Russell and Carnap, is clear; but the same may not hold for theoretical properties.

Moreover, structural realists must specify what exactly is the "structure" preserved across theories, and how is it represented. For Grover Maxwell, Worrall, and Cruse and Papineau (2002), it is the relevant equations, represented by the theory's "Ramsey sentence" (a formulation where all the theoretical terms are substituted by existentially quantified variables). However a number of problems arise concerning the adequacy of such representation (Demopoulos and Friedman 1985: 635; Psillos 1999: 63–69; Ladyman 1998, 2014). This is why, instead, Ladyman and Ross (2007) and French (2014) hold that what is preserved are symmetries and the associated group-theoretic structures, represented by the semantic formulation of theories.

Opponents of StrR also charged that it cannot account for the difference between physical reality and mere mathematical structures; that often also structure is lost in theory change; and that StrR only applies to physics (Ladyman 2014). Specific objections have been raised against the ontic versions of StrR, especially concerning the plausibility of the existence of relations without relata, and the extent to which physics univocally supports this view (ibid.). There is no room here to discuss them; besides, even if OStrR were wrong, one could still use EStrR as a form of selective realism able to resist the PMI.

Summing up, on the one hand the basic role of structures (in some sense of 'structure') in modern physics seems undeniable, and there are striking examples of structure preservation in theory change; on the other hand it has not been shown that no theory has ever correctly described the nature of entities, or that entities were never essential for novel predictions; moreover, although there have been attempts to apply StrR to biology (French 2011) and to the social sciences (Ladyman and Ross 2007; Kincaid 2008), it is far from clear that it can apply to all scientific disciplines. Therefore, while these questions are still hotly debated (Ladyman 2014), at present it may be reasonable to assume that StrR can account only in part for our realist commitments, and we should take a liberal position on which kinds of features (entities, or laws, or structures, or particular properties, etc.) theories can get right even if otherwise false.

8.3 Deployment Realism

Kitcher's and Psillos' "deployment" realism is not concerned with the *kind* of theoretical components which survive scientific change and deserve realist commitment (whether they are ontological or structural), as with their role: they are those essentially deployed in novel predictions."No sensible realist should ever want to assert that the idle parts of an individual practice, past or present, are justified by the success of the whole" (Kitcher 1993: 142). In fact if a component C were not deployed essentially in deriving prediction NP, there would be a different component C' not implying C, from which NP could have been derived; hence success could equally be credited to C', and the truth of C would no longer be the only plausible conclusion of the NMA.

Lyons (2002) criticized deployment realism by listing a number of novel predictions derived from components we now recognize as false. For instance, the idea of absolute acceleration was used in derivations from Newton's theory; the claims that phlogiston is the principle of heat and that "sulfuric acid was dephlogisticated sulfur" were involved in Stahl's prediction that the synthesis of phlogiston and sulfuric acid would result in sulfur; the postulate that charcoal is "high in phlogiston" and that inflammable air is pure phlogiston, were used in deriving Priestley's prediction that inflammable air would, like charcoal, turn calx into metal; the prediction that the rate of expansion is the same for all gases was derived from a number of false claims about caloric; etc. (Lyons 2002: 80–81).

But it has been replied that some of those predictions were true only under a charitable interpretation (e.g., by understanding 'phlogiston' as deprivation of oxygen), under which, however, also the claim used in deriving them turns out to be true. Other predictions were actually derived from false claims by chance, but this was possible because they were a priori probable (and we saw that only improbable predictions are evidence for truth). All the other false claims were not actually *essential* in the prediction, because they entailed some weaker and true claims which were sufficient to derive the same prediction (Alai 2014b). Lyons (2006) argued that the essentiality requirement should be dropped, but this would obviously deprive the NMA of its cogency.

8.4 Semirealism

Another version of selective realism is Anjan Chakravartty's (1998) *semirealism*. Like entity realism, he holds that we can know the existence of the unobservable entities with which we establish causal interactions; but unlike entity realism, he also grants that we can correctly describe their "detection" properties: "Detection properties are causal properties one has managed to detect; they are causally linked to the regular behavior of our detectors. Auxiliary properties are any other putative properties attributed to particulars by theories" (Chakravartty 2007: 47). So, we should be realist about detection properties, but agnostic about auxiliary properties (see Ivanova (ed.)). Bence Nanay (2013) proposed instead *singularist semirealism*, according to which "science is mostly right, not only about which unobservables exist, but also about their property tokens, but not their property types".

8.5 The Verisimilitude *Research Program*

Two traditional and influential approaches can also be considered as forms of selective realism. One is the "verisimilitude" research program initiated by Popper (Popper 1963; Oddie 1986; Niiniluoto 1987, 1998). Its key idea is that even false theories may be more or less "verisimilar", or "close" to the truth, since they include some true statements, or false statements with some true content. More verisimilar theories have more truth content and/or less false content. For instance, (1) 'All swans are white' and (2) 'All swans are black' are both false, but part of the content of (1) is the entailed statement (3) 'All swans except Australian swans are white', which is true and explains the predictive success of (1) (Musgrave 2006–2007). Also (2) entails a true claim, viz. (4) 'Australian swans are black', but that is weaker than (3), so (1) is more verisimilar than (2). "Approximate truth is a species of partial truth, since the approximations in question are logical parts of what we began with. 'It is 4 o'clock' logically implies 'It is approximately 4 o'clock' as well as 'It is 4 o'clock give or take 5 min'" (ibid.). Rescher (1987: Ch. 5) seems to follow a similar line when he distinguishes between forefront science, which is precise and never true, and "schoolbook science", which though vague and imprecise includes the true core of the forefront science.

8.6 *The* Restricted-Domain *Approach*

The other traditional and influential idea which may be interpreted as a version of partial realism is that theories must not be taken as true for all the phenomena and with absolute precision, but only for certain intended ranges of phenomena, or levels of approximation, or domains of entities, which are defined by the theory itself.[1] Therefore, even when rejected, they remain true within those limits. For instance, "The fact that we can use classical mechanics in creating many machines or for sending rockets into space certainly means that this mechanics is *true of its objects* and therefore 'tells a true story' about *certain aspects of reality*" (Agazzi 2014: 310–311).

If taken to an extreme this position might imply that all theories are analytic (Kuhn 1962, ch. 9), or that they are reducible to their empirical claims, or that they don't describe the actual world, but different worlds of our making, as claimed by Goodman (1978, chs. I, VII); for instance, it might mean that phlogiston theory was true of phlogiston, ether theory was true of ether, etc. On the contrary, theories and laws are intended to be true—period, true of everything. For instance, 'electrons have negative charge' is not intended to be true only of electrons, but of everything, for it has the universal form '$\forall x(Ex \rightarrow NCx)$'. If the class of intended applications is delimited a priori, it precludes the extension of the theory to new phenomena, depriving it of fecundity and heuristic power; if instead it is delimited a posteriori, following empirical failures, then it is ad hoc, and the theory risks to lose its empirical content. In order to restrict the domain of a law we need good reasons, beside an experimental failure; in particular, we need to find a different (specialized) law, which however has a universal scope, and above all must be embedded in a different theory. Moreover, the theories which Einstein called "principles theories" don't delimit any field of intended applications.

This approach is fine, however, if interpreted as the selective realist thesis that theories which are radically false overall, nonetheless include descriptions—not merely empirical, but theoretical as well—which are approximately true for certain domains, or aspects of reality, or scales, or levels approximation. There is no delimitation of the intended applications, either a priori or a posteriori, and when a claim is empirically refuted it is declared false. However, it may be found that it was partly true, since it entailed a weaker but true claim. Still, the latter must be supported and explained by a new theory. For example, the assumption that mass is inalterable was proven false—false of everything. However, it entails the true claim that mass is approximately inalterable at low speeds; yet, this limitation cannot be explained by Newton's theory, but only by Relativity. This is why Newton's theory is considered false—period. The theory of luminifer ether was intended for all the

[1]See Heisenberg 1955: 20; Agazzi 1969: 361, 368, 372 ff.; Agazzi 2014: 310–311, 403–407; Toraldo di Francia 1976; Dalla Chiara, Toraldo di Francia 1999: 93–96; Dorato 2007: 172, 203–204; Fano, Macchia 2015: 74. According to Kuhn (1962: ch. 9) this idea was "prevailing" in his time.

phenomena of light, but it is true of none. Rather than conceiving theories as completely true of a restricted domain, they may be seen as partially true of a universal domain. Partial truth also explains why different theories can be all true in the same field: they do not concern different realities, but different aspects of reality (Agazzi 2014, 405).

8.7 Local *Realisms?*

In view of so many different ways and versions of realism, Magnus and Callender (2004) argued that we shouldn't look for "wholesale" arguments for realism, but for a case-by-case defense. Saatsi (2016) suggests that realists should give up the idea of a general "recipe … capable of distilling the trustworthy aspects of a theory, applicable to *any* good, predictively successful mature theory". Instead, they should settle for an "exemplar realism" which focuses on specific, "local" reasons for realism. All these different recipes shouldn't be seen as competitors, but "as capturing the different possible ways that a theory can 'get the world right'" (French, this volume). Yet, why are all these versions forms of realism, what do they have in common? (this was Plato's problem of the universal). For Saatsi it is the general idea that science is successful because it gets something right about the world, but this needs not be exactly the same in all theories, contexts or disciplines.

We noted that *deployment realism* is already enough flexible as to which *kind* of components can be right, focussing more on how the particular true claims can be identified. But also different identification criteria have been proposed, and perhaps they are compatible, or suited to different contexts: being essential to derive novel predictions (for deployment realism); being confirmed by indirect but theory-free observation (as explained above); being the "minimally interpreted mathematical parts" of successful theories (Votsis 2011b); being the minimal sub-theories which are presupposed by the successful predictions and not empirically refuted (Peters (2014); being involved in predictive success, resistant to hostile probing and with outside support (Cordero, this volume); attributing properties that are in principle observable, measurable by distinct independent methods and causally produce the observed data (Ghins, this volume).

Votsis (2011b) and Peters (2014) argue that in order to save the NMA from the MMT we should be able to identify the particular true components of discarded theories *prospectively*, from the authors' point of view. In fact, if we could identify them only retrospectively, as those which are preserved in today's theories, we would beg the question of the cogency of the NMA by assuming that the currently accepted theories are true (see Stanford 2006: Ch. 6).

However, if we could do this for past theories, we should also be able to distinguish in current theories precisely which claims will be preserved forever and which ones will be discarded, which is impossible (Alai 2016: § 3; Nickles, this volume). Yet, realists can still make their point if they can (a) argue *in general* that a successful prediction must be derived from some true assumption, even without being able to pinpoint exactly which one; (b) for any particular successful prediction NP, show that indeed the theory includes certain assumptions $A_1 \ldots A_n$ from which NP can be derived, and that for all we know $A_1 \ldots A_n$ may be true, because they fulfill the just mentioned criteria; and (c) explain away each putative counterexample of true predictions apparently derived from false assumptions as sketched above.

9 The Objectivity Question

Thus far we have dealt with the former question dividing realism and antirealism, whether unobservables entities can be known. This question arises from the limits of our senses and cognitive apparatuses. There is however a second question: whether unobservables can be known *objectively*. It arises because knowledge is a function of two arguments, objective reality and the subjective factors which precede and "shape" our experience and conceptions; among these factors are on the one hand the peculiar scope and mode of operation of our senses, and on the other hand the different conceptual schemes, frames of reference, background beliefs, cultural biases, methodological allegiances, technological or environmental conditions of belief formations, etc. Antirealists maintain that these factors distort or completely filter out any objective information about reality and knowledge is purely subjective.

The objectivity question arises for knowledge in general, not just for science. Illustrious antirealist doctrines are Protagoras' and the skeptics' relativism; Kant's critical idealism; Nietzsche's doctrine that there are no facts, only interpretations (1886: § 14, 1901: § 540); the Sapir-Whorf thesis that language shapes the world (Whorf 1956: 213); post-modern gnoseology (Vattimo, Rovatti 1983; Ferraris 2012). Special arguments have been used to deny the objectivity of scientific knowledge in particular. Foucault (1966) claimed that man is built by human sciences, and Latour (1998) denied that Rameses II was killed by the bacilli of tuberculosis, because they were discovered only in 1882; in the late 1950s and 1960s Hanson, Kuhn and Feyerabend, the "new philosophers of science", held that meanings and experience are thoroughly theory-laden; Putnam (1978b, 1981) rejected "metaphysical realism"; Goodman (1978) claimed that we make the world, or actually, many worlds.

9.1 Perspectival Realism

Today "perspectivist" philosophers stress that science is always carried out within a "perspective" characterized by a priori factors like the above mentioned ones. Nickles considers "an illusion" the idea that we can gradually eliminate every human perspectival element and finally reach a completely objective science

(Nickles forthcoming[a]). However, perspectivists need not be subjectivist, relativist, or antirealist: one can acknowledge the perspectival character of knowledge while recognizing that objective reality plays a role at least as important as subjectivity. In fact, the two are not incompatible nor they limit each other, precisely as each argument of a function determines exactly the values without preventing the other to do the same. Of course, *just* looking at the values we cannot distinguish the two arguments; but in knowledge we are given both the value and the subjective factor, and this gives us the objective factor (Alai 1994: §§ 2, 3.6, 3.7). No wonder that subsequent research in the history and philosophy of science has shown that the radical relativism of the "new philosophers of science" was at best exaggerated, and since 1994 Putnam has retracted his antirealism (Putnam 1994: 489–494, 502–506, 516–517; Putnam 2012: chs. 1–4).

Thus it is possible to embrace forms of *perspectival realism* (roughly what Putnam called "sophisticated metaphysical realism"), as also urged by Votsis (2012). Basically, a perspective may determine either (a) the particular aspects of reality that are selected for representation, or (b) how those aspects are represented, or (c) both (Giere 2006: 14, Votsis 2012: 90). (a) is perfectly compatible with realism, and if (b) can be shown to boil down to (a) then we have a realist answer to the objectivity question. Moreover, although the knowledge question and the objectivity question are in principle independent, many take a realist stand on both (e.g., Devitt 1984: 22; Sankey 2008: 12–18).

Perspectival realism, however should be distinguished from doctrines which are misleadingly called 'realism', like Kant's "empirical realism" or Putnam's "internal realism" (Alai 1989, 1990), for they keep objectivity only by weakening it until it becomes compatible with subjectivism and antirealism (Agazzi 2014: 51–57).

9.2 *Agazzi and Dilworth*

A truly realist perspectivism, instead, was put forth by Agazzi in (1969). He called it "Gestalt view", a term also used for the "new philosophies of science" (Suppe 1977), but he showed that the a priori features of science highlighted by them could be taken into account without jeopardizing realism. This view has been presented and discussed in depth again in (Agazzi 2014). In a nutshell, each scientific discipline 'clippes out' its objects by particular empirical operations, which characterize its specific point of view, or "Gestalt" (Agazzi 1979: 42–44; 2014: ch. 2, pp. 83, 97, passim). Hence, "one and the same 'thing' can become the object of a new and different science every time a new specific point of view … is taken on it" (ibi: 84).

Thus we always work from within some particular perspective and on already structured materials; scientific objects are abstract and constructed, but the interactive operations by which they are "clipped out" ensure their reference to independent reality and an objective criterion of truth for claims about them. The intervention of the human subject simply brings to light different aspects of reality. "Under different conditions [and through different operations] reality would manifest itself under different aspects … but these too would be real" (ibi: 229). Therefore "(a) science attempts to represent a reality independent of science itself …; (b) what science states is an adequate representation of this reality 'as it is'" (ibi: 263).

A similar conception was proposed by Craig Dilworth in (1981). It was called 'perspectivism', it borrowed some of Agazzi's key insights, and in reaction to the new philosophers of science it provided criteria for evaluating scientific progress (1981: 84–88). On Dilworth's perspectivism scientific laws can be true and provide knowledge, while theories are not true or false, but serve the primary aim of science, i.e. *understanding* the laws (2015: 23).

9.3 Giere

New discussions on perspectivism have been spurred by Sosa (1991) and Giere, R. (2006), and the issue 84 (2012) of *Philosophica* collects various essays on them. Giere's "perspectival realism" is meant to provide "a genuine alternative to both objectivist realism and social constructivism …" (2006: 14–15). On this view perspectives play a key role both in scientific observation and in theorization. In each of them perspectives affect scientific outputs in two ways: first, both the human visual system and instruments are sensitive only to some kinds of input. Second, "the output is a function of both the input and the internal constitution of the instrument" (ibi: 14).

As noticed by Votsis (2012), the first way has no bearings on the realism/antirealism discussion, it simply entails that different theories have access to different aspects of reality (like for Agazzi). From the second way, however, Giere infers that science does not describe *objective* features of the world, both as concerns everyday macroscopic objects and theoretical unobservable entities: "truth claims are always relative to a perspective" (2006: 81). "So even the claim that the sky is blue is not an absolutely objective truth. Rather, the sky appears blue to normal human trichromats" (ibi: 123). We can only say: "According to this highly confirmed theory (or reliable instrument), the world seems to be roughly such and such, [but not] 'This theory (or instrument) provides us with a complete and literally correct picture of the world itself'" (ibi: 6). Yet, Giere rejects ontological constructivism, granting that the facts under investigation are objective (ibi: 81–82). Thus, his position resembles Kant's, holding that we cannot observe things in and of themselves, but things-*from-the-perspective of human-visual-apparatus, or things-as-represented-by such-and-such-instrument* (ibi: 43, 56).

However, he faces a dilemma: either the output of each perspective is just an alternative projection of the same content (like different maps of the same territory), but then they are mutually compatible, without any threat to objectivity; or one may be correct and other incorrect, but then this can be decided by empirical evidence, and if it cannot we get just another instance of the general underdetermination problem (Votsis 2012: 101-ff).

9.4 Some Latest Proposals

A more distinctly realist position is taken by Sosa (1991), followed by Massimi (2012). According to them we can know non-perspectival facts, but we know them perspectively, because the *justification* of our beliefs is perspectival: it "always takes place within an epistemic perspective, including not only first-order beliefs about body *x*, but also beliefs … about our perceptual system, cognitive faculties, measurement devices, and their reliability as sources of beliefs (Massimi 2012: 40–41; see Sosa: 145, 210, passim). The result is a sort of "perspectival coherentism", which makes justification and knowledge context-relative: "J. J. Thomson was justified to believe [that cathode rays were not electrons] from *his own* epistemic perspective as much as we are justified in not believing [that] from *our own* perspective" (Massimi 2012: 47). But this conception "does not open the door to epistemic relativism of Rortian type or to Kuhnian incommensurability, because it is objective facts that make those beliefs true or false" (ibi: 48). So, it is realist about truth, even if antirealist about knowledge. In (2016) Massimi further tries to show how truth itself can be 'perspectival' while remaining correspondence to objective states of affairs.

Teller's "panperspectival realism" maintains that "the world is too complicated for us to succeed in attaching specific referents to our terms" and "to get things exactly right". However, "our representations tell us about an independent world without securing reference by showing that the world is very like the way it is represented in a range of different, often complementary modeling schemes". Each of these schemes gives us understanding "from one or another 'angle'", so, "though never exact, these representations are of something extra-representational because they present the world modally as going beyond what is represented explicitly" (Teller: 1). Therefore "This counts as knowledge of how the world is (really)", and from this point of view "the theoretical and perceptual are on the same (inexact) footing" (ibi: 10).

In fact, Teller's perspectivism provides a further argument for realism on the knowledge question: since perception is as perspectival as theorization, and the former provides knowledge, so does the latter (ibi: 7). Thus Teller agrees with selective realisms that best theories are true, although not exactly (i.e., not completely) true, and he agrees with structural realism that only structures are objective, while entities are features of our representations.

Interestingly, the discussions on knowledge and those on objectivity converge to the conclusion that both can be achieved, but only partially. In fact, this kind of modest realism is compatible even with another form of perspectivism, Nickles' "nonrealism". According to Nickles (this volume, forthcoming[a]), we don't have sufficient evidence to believe that our best theories are true or nearly true, in the sense of providing us with a complete and final understanding of "what is really going on". A modest realist may agree, granting that probably no accepted theory is finally and completely true, so we are still far from "the whole truth" (if there is one), and perhaps we'll never get there. Yet, such a realist may hold that some partial truths about unobservables have been found by science in the past and they still keep accumulating, probably at an increasing rate.

Acknowledgements I thank EvandroAgazzi and Vincenzo Fano for valuable suggestions.

References

Agazzi, Evandro. 1969. *Temi e problemi di filosofia della fisica.* Milano, Manfredi (2nd ed. Roma: Abete 1974).

Agazzi, Evandro. 1979 Proposta per una nuova caratterizzazione dell'oggettività scientifica. In E.Agazzi, *Problemi di epistemologia contemporanea*, Lanciano: Itinerari, pp. 31–61.

Agazzi, Evandro. 2014. *Scientific Objectivity and its Contexts.* Cham: Springer.

Alai, Mario. 1989. *A Critique of Putnam's Anti-realism.* Ann Arbor, Mi.: U.M.I. Online: http://www.proquest.com/products-services/dissertations/.

Alai, Mario. 1990. L'equivoco del "realismo interno" di Hilary Putnam. *Rivista di Filosofia* LXXXI, 2: 263–290.

Alai, Mario. 1994. *Modi di Conoscere il Mondo.* Milano: Franco Angeli.

Alai, Mario. 2005. Dal realismo scientifico al realismo metafisico. *Hermeneutica*: 167–189.

Alai, Mario. 2006. Science and Non-Observable Reality. In *The Controversial Relationships between Science and Philosophy: A Critical Assessment*, ed. G. Auletta: 211–250. Vatican City: Libreria Editrice Vaticana.

Alai, Mario. 2008. Σώζειν τά φαινόμενα. Realistic and Antirealistic Attitudes in Natural Science. *Isonomia*, http://isonomia.uniurb.it/vecchiaserie/2008alai.pdf.

Alai, Mario. 2009. Realismo scientifico e realismo metafisico. In *Fisica e metafisica*, ed. Mario Alai, *Giornale di Fisica* 50, Suppl. 1: 19–27.

Alai, Mario. 2010. Van Fraassen, Observability and Belief. In *New Essays in Logic and Philosophy of Science*, ed. Marcello D'Agostino, Giulio Giorello, Federico Laudisa, Telmo Pievani and Corrado Sinigaglia, 663–675. London: College Publications.

Alai, Mario. 2014a. Novel Predictions and the No Miracle Argument. *Erkenntnis* 79, 2: 297–326.

Alai, Mario. 2014b. Defending Deployment Realism against Alleged Counterexamples. In *Defending Realism. Ontological and Epistemological Investigations*, ed. G. Bonino, G. Jesson, J. Cumpa: 265–290. Boston-Berlin-Munich: De Gruyter.

Alai, Mario. 2014c. Why Antirealists Can't Explain Success. In *Metaphysics and Ontology Without Myths*, ed. F. Bacchini, S. Caputo and M. Dell'Utri: 48–66. Newcastle upon Tyne: Cambridge Scholars Publishing.

Alai, Mario. 2014d. Explaining the Novel Success of Science. *Metascience* 23,1: 125–130.

Alai, Mario. 2016. Resisting the Historical Objections to Realism: Is Doppelt's a Viable Solution? *Synthese*: 1–24,
DOI 10.1007/s11229-016-1087-z.

Alai, Mario. Forthcoming. The Metaphysical Scope of the New Realism: Confronting Eco's and Parrini's Strictures". In *New Perspectives on Realism*, ed. L. Taddio, K. W Molìn: 9–31. Milano: Mimesis International.

Alai, Mario, Buzzoni, Marco and Tarozzi, Gino (eds.). 2015. *Science between Truth and Ethical Responsibility. Evandro Agazzi in the Contemporary Scientific and Philosophical Debate.* Cham: Springer.

Aronson, Jerrold L., 1984. *A Realist Philosophy of Science*. London: The Macmillan Press, New York: St. Martin's Press.
Bird, Alexander. 2015. Is there meta-scientific knowledge? Against both the no-miracles argument and the pessimistic meta-induction. Draft.
Carnap, Rudolf. 1923. Uber die Aufgabe der Physik und die Andwendung des Grundsatze der Einfachstheit. *Kant-Studien* 28: 90–107.
Carnap, Rudolf. 1928. *Der logische Aufbau der Welt*. Berlin-Schlachtensee: Weltkreis.
Carnap, Rudolf. 1936. Testability and Meaning. *Philosophy of Science* 3.
Celsus, A. C., 1935, *De Medicina*, Prooemium, Loeb Classical Library edition, vol. II., http://penelope.uchicago.edu/Thayer/L/Roman/Texts/Celsus/Prooemium*.html.
Chakravartty, Anjan. 1998. Semirealism. *Studies in History and Philosophy of Science Part A*29 (3): 391–408.
Chakravartty, Anjan. 2007. *A Metaphysics for Scientific Realism: Knowing the Unobservable*. Cambridge: University Press.
Churchland, Paul M. and Hooker, Clifford A. (eds.) 1985. *Images of Science: Essays on Realism and Empiricism*. Chicago: University Press.
Clarke, Steve and Lyons, Timothy D. eds. 2002. *Recent Themes in the Philosophy of Science. Scientific Realism and Commonsense*. Dordrecht: Kluwer.
Cruse, Pierre, and Papineau, David. 2002. Scientific Realism Without Reference. In *The Problem of Realism*, ed. Michele Marsonet: 174–189. Aldershot-Burlington: Ashgate.
Dalla Chiara, M. Luisa and Toraldo di Francia, Giuliano. 1999. *Introduzione alla filosofia della scienza*. Roma-Bari: Laterza.
Demopoulos, William, and Friedman, Michael. 1985. Critical notice: Bertrand Russell's *The Analysis of Matter*: its Historical Context and Contemporary Interest. *Philosophy of Science* 52: 621–639.
Devitt, Michael. 1984. *Realism and Truth*. Oxford: Basil Blackwell.
Dilworth, Craig. 1981. *Scientific Progress*. Dordrecht: Kluwer (Springer 2008[4ed.]).
Dilworth, Craig. 2015. The Perspectivist Conception of Science. In Alai, Buzzoni, and Tarozzi, (eds.) 2015: 21–25. Cham: Springer.
Doppelt, Gerald. 2007. Reconstructing Scientific Realism to Rebut the Pessimistic Meta-induction. *Philosophy of Science*, 74: 96–118.
Doppelt, Gerald. 2011. From Standard Scientific Realism and Structural Realism to Best Current Theory Realism. *Journal for General Philosophy of Science*, 42: 295–316.
Doppelt, Gerald. 2013. Explaining the Success of Science: Kuhn and Scientific Realists. *Topoi*, 32: 43–51.
Doppelt, Gerald. 2014. Best Theory Scientific Realism. *European Journal for Philosophy of Science*, 4: 271–291.
Dorato, Mauro. 2007. *Cosa c'entra l'anima con gli atomi*? Roma-Bari: Laterza.
Duhem, Pierre. 1906. *La théorie physique. Son objet et sa structure*. Paris: Rivière.
Fahrbach, Ludwig. 2011. Theory Change and Degrees of Success. *Philosophy of Science*, 78: 1283–1292.
Fano, Vincenzo. 2005. *Comprendere la scienza. Un'introduzione all'epistemologia delle scienze naturali*, Napoli: Liguori.
Fano, Vincenzo and Macchia, Giovanni. 2015. Scientific Progress. In Alai, Buzzoni, and Tarozzi, (eds.) 2015: 65–80. Cham: Springer.
Ferraris, Maurizio. 2012. *Manifesto del nuovo realismo*, Roma-Bari: Laterza.
Fine, Arthur. 1984. *The Natural Ontological Attitude*. In J. Leplin (ed.) 1984: 83–107.
Fine, Arthur. 1986. Unnatural Attitudes: Realist and Instrumentalist Attachments to Science", *Mind* 95: 149.
Fine, Arthur. 1991. Piecemeal Realism. *Philosophical Studies* 61: 79.
Foucault, Michel. 1966. *Les mots et les choses*. Paris: Gallimard 1966.
French, Steven. 2011. Metaphysical underdetermination: Why Worry? *Synthese* 180, 2: 205–221.
French, Steven. 2014. *The Structure of the World*. Oxford: University Press.

French, Steven and Ladyman, James. 2003a. Remodelling Structural Realism: Quantum Physics and the Metaphysics of Structure. *Synthese* 136: 31–56.
French, Steven and Ladyman, James. 2003b, The Dissolution of Objects: between Platonism and Phenomenalism. *Synthese* 136: 73–77.
French, Steven and Ladyman, James. 2011. In Defence of Ontic Structural Realism. In *Scientific Structuralism*, ed. Alisa Bokulich and Peter Bokulich: 25–42. Dordrecht: Springer.
Friedman. Michael. 1983. *Foundations of Space-Time Theories,* Chicago: University Press.
Frigg, Roman and Votsis, Ioannis. 2011. Everything you Always Wanted to Know about Structural Realism but Were Afraid to Ask. *European Journal for the Philosophy of Science*, 1 (2): 227–276.
Giere, Ronald N. 1988. *Explaining Science: A Cognitive Approach*. Chicago: University Press.
Giere, Ronald N. 2006. *Scientific Perspectivism*. Chicago: University Press.
Glymour, Clark 1980. *Theory and Evidence*. Princeton, NJ.: University Press.
Goodman, Nelson. 1978. *Ways of Worldmaking*. Indianapolis-Cambridge: Hackett.
Hacking, Ian. 1983. *Representing and Intervening*. Cambridge: University Press.
Hardin, Clyde L., and Rosenberg, Alexander. (1982). In Defence of Convergent Realism. *Philosophy of Science*, 49: 604–615.
Heisenberg, Werner. 1955. *Das Naturbild der heutigen Physik*. Hamburg: Rohwolt.
Hempel, Carl G. 1952. *Fundamentals of Concept Formation in Empirical Science*. Chicago: University of Chicago Press.
Hempel, Carl. G. 1958. The Theoretician's Dilemma. *Minnesota Studies in the Philosophy of Science* 2: 37–98.
Howson, Colin. 2000. *Hume's Problem: Induction and the Justification of Belief*. Oxford: Oxford University Press.
Hoyningen-Huene, P. 2011. Reconsidering the Miracle Argument on the Supposition of Transient Underdetermination. *Synthese* 180: 173–187.
Ivanova, Milena (ed.) Scientific Semi-Realism. *PhilPapers*. http://philpapers.org/browse/scientific-semirealism.
Kincaid, Harold, 2008. Structural Realism and the Social Sciences. *Philosophy of Science*, 75: 720–731.
Kosso, Peter. 1992. *Reading the Book of Nature. An Introduction to the Philosophy of Science*, Cambridge: University Press.
Kitcher, Philip. 1993, *The Advancement of Science,* Oxford: University Press.
Kuhn, Thomas S. 1962. *The Structure of Scientific Theories*. Chicago: The University Press.
Ladyman, James. 1998. What is Structural Realism? *Studies in History and Philosophy of Science* 29: 409–24.
Ladyman, James. 2002. *Understanding Philosophy of Science*, London, New York: Routledge.
Ladyman, James. 2011. Structural Realism versus Standard Scientific Realism: the Case of Phlogiston and Dephlogisticated Air. *Synthese* 180: 87–101.
Ladyman, James. 2014. Structural Realism. *Stanford Encyclopedia of Philosophy*. http://plato.stanford.edu/entries/structural-realism/.
Ladyman, James and Ross, Don, with Spurrett, David and Collier, John 2007. *Every Thing Must Go: Metaphysics Naturalised*. Oxford: University Press.
Lange, Marc. 2002. Baseball, Pessimistic Inductions and the Turnover Fallacy. *Analysis*, 62: 281–285.
Latour, Bruno. 1998. Ramses II est-il mort de la tubercolose? *La Recherche*.
Laudan, Larry. 1981. A Confutation of Convergent Realism. *Philosophy of Science* 48: 19–49. Repr. in D. Papineau (ed.) 1996: 107–138.
Laudan, Larry. 1984a. Explaining the Success of Science. In *Science and Reality: Recent Work in the Philosophy of Science*, ed. James T. Cushing, Cornelius F. Delaney and Gary M. Gutting: 83–105. Notre Dame: University Press.
Laudan, Larry. 1984b. Discussion: Realism without the Real. *Philosophy of Science*, 51: 156–162.
Laudan, Larry. 1996. *Beyond Positivism and Relativism*. Boulder, Co.: Westview Press.

Laudan, Larry, Leplin, Jarrett. 1991. Empirical Equivalence and Underdetermination. *Journal of Philosophy* 88: 449–472.
Laudan, Larry, Leplin, Jarrett. 1993. Determination Underdeterred: Reply to Kukla. *Analysis* 53: 8–16.
Leplin, Jarrett (ed.). 1984. *Scientific Realism,* Berkeley: University of California Press.
Leplin, Jarrett. 1987. Surrealism. *Mind* 96: 519.
Lewis, Peter. 2001. Why the Pessimistic Induction is a Fallacy. *Synthese,* 129: 371–380.
Lipton, Peter. 1991. *Inference to the Best Explanation.* London: Routledge (2004[2]).
Lyons, Timothy D. 2002. The Pessimistic Meta-Modus Tollens. In Clarke, Lyons (eds.): 63–90.
Lyons, Timothy D. 2006. Scientific Realism and the Stratagema de Divide et Impera *Brit. J. Phil. Sci.* 57: 537–560.
Lyre, Holger. 2011. Is Structural Underdetermination Possible? *Synthese* 180, 2: 235–247.
Magnus, P.D. and Callender, Craig. 2004. Realist Ennui and the Base Rate Fallacy. *Philosophy of Science* 71 (3): 320–338.
Massimi, Michela. 2012. Scientific Perspectivism and its Foes. *Philosophica* 84: 25–52.
Massimi, Michela. 2016. Four Kinds of Perspectival Truth. *Philosophy and Phenomenological Research.* doi:10.1111/phpr.12300.
Maxwell, Grover. 1970. Theories, Perception, and Structural Realism. In *Nature and Function of Scientific Theories*, ed. R. Colodny: 3–34. Pittsburgh: University Press.
McMullin, Eranan. 1984. A Case for Scientific Realism. In Leplin (ed.): 8–40.
Muller, Frederick A. 2011. Withering Away, Weakly. *Synthese* 180, 2: 223–233.
Musgrave, Alan. 1985. Realism vs. Constructive Empiricism. In Churchland and Hooker (eds.) 1985: 197–221.
Musgrave, Alan. 1988. The Ultimate Argument for Scientific Realism. In *Realism and Relativism in Science*, ed. R. Nola: 229–252. Dordrecht: Kluwer.
Musgrave, Alan. 2006–2007. The 'Miracle Argument' for Scientific Realism. *The Rutherford Journal* 2. http://www.rutherfordjournal.org/article020108.html.
Nanay, Bence. 2013. Singularist Semirealism. *British Journal for the Philosophy of Science* 64 (2): 371–394.
Nanay, Bence. 2016. Entity Realism and Singularist Semirealism. *Synthese* online. DOI 10.1007/s11229-016-1179-9.
Nickles, Thomas. Forthcoming[a]. Perspectivism versus Copernicanism. *Axiomathes.*
Nickles, Thomas. Forthcoming[b]. Strong Realism as Scientism: Are We at the End of History? In *Science Unlimited? The Challenges of Scientism*, ed. M. Boudry, M. Pigliucci. Chicago: University Press.
Nietzsche, Friedrich. 1886. *Jenseits von Gut und Böse. Vorspiel einer Philosophie der Zukunft.* Leipzig: C. G. Naumann.
Nietzsche, Friedrich. 1901. *Der Wille zur Macht. Versuch einer Umwerthung aller Werthe.* hrsg. v. Ernst Horneffer, August Horneffer und Peter Gast. Leipzig: C. G. Naumann.
Niiniluoto, Ilkka. 1987. *Truthlikeness*, Dordrecht: Reidel.
Niiniluoto, Ilkka. 1998. Verisimilitude: the Third Period. *The British Journal for the Philosophy of Science* 49.1: 1–29.
Niiniluoto, Ilkka. 1999. *Critical Scientific Realism*, Oxford: University Press.
Nola, Robert. 2002. Realism Through Manipulation. In Clarke, Lyons (eds.): 1–23.
Oddie, Graham. 1986, *Likeness to Truth,* Dordrecht: Reidel.
Papineau, David (ed.). 1996. *The Philosophy of Science.* Oxford: University Press.
Perrin, Jean. 1913. *Les atomes.* Paris: Alcan.
Peters, Dean. 2014. What Elements of Successful Scientific Theories are the Correct Targets for "Selective" Scientific Realism? *Philosophy of Science* 81: 377–397.
Poincaré, Henry. 1902. *La science et l'hypothèse.* Paris: Flammarion. Engl. transl. *Science and Hypothesis.* New York: Dover Publications 1952.
Popper, Karl R. 1963. *Conjectures and Refutations: The Growth of Scientific Knowledge.* London: Routledge and Kegan Paul.

Psillos, Stathis. 1999. *Scientific Realism: How Science* Tracks Truth. London-New York: Routledge.

Psillos, Stathis. 2011a. Choosing the Realist Framework. *Synthese* 190: 301–316.

Psillos, Stathis. 2011b. The Scope and Limits of the No-Miracles Argument. In *The Philosophy of Science in a European Perspective*, ed. Friedrich Stadler et al. Vol. II: 23–35. Cham: Springer.

Putnam, Hilary. 1975a. *Mathematics, Matter and Method. Philosophical Papers vol. 1.* Cambridge: University Press.

Putnam, Hilary. 1975b. *Philosophical Papers,* Vol. II: *Mind Language and Reality*. Cambridge: University Press.

Putnam, Hilary. 1975c. The Meaning of 'Meaning'. In Putnam 1975b: 215–271.

Putnam, Hilary. 1975d. Language and reality. In Putnam 1975b: 272–290.

Putnam, Hilary, 1978a. *Meaning and the Moral Sciences*, Boston-London-Henley: Routledge&Kegan Paul.

Putnam, Hilary, 1978b. Realism and Reason. In Putnam (1978a): 123–140.

Putnam, Hilary. 1978c. Equivalenza. In *Enciclopedia Einaudi*, vol. V: 547–564. Torino: Einaudi. Engl. transl.: Equivalence. In Putnam 1983: 26–45.

Putnam, Hilary. 1981. *Reason, Truth and History*. Cambridge: University Press.

Putnam, Hilary. 1994. Sense, Nonsense, and the Senses: an Inquiry into the Powers of the Human Mind. *The Journal of Philosophy* XCI, n. 9, 445–517.

Putnam, Hilary. 2012. *Philosophy in an Age of Science*. Cambridge, Mass.- London: Harvard University Press.

Quine, Willard V.O. 1951. Two Dogmas of Empiricism. *Philosophical Review* 60: 20–43.

Rees, Paul. 2012. *A Critique of The Arguments for Scientific Realism.* Reading: Cranmore Publications.

Rescher, Nicholas. 1987. *Scientific Realism: A Critical Reappraisal.* Dordrecht: Reidel.

Russell, Bertrand. 1927. *The Analysis of Matter.* London: Routledge Kegan Paul.

Saatsi, Juha. 2016. Replacing Recipe Realism. *Synthese*, First Online, 2016. DOI:10.1007/s11229-015-0962-3.

Sankey, Howard. 2008. *Scientific Realism and the Rationality of Science*. Aldershot and Burlington, VT: Ashgate.

Schlick, Moritz. 1938. *Form and Content. An Introduction to Philosophical Thinking*. In Schlick, *Gesammelte Aufsätze* 1926–1936, ed. F. Waisman. Wien: Gerold.

Schurz, Gerhard. 2011. Structural Correspondence, Indirect Reference, and Partial Truth: Phlogiston Theory and Newtonian Mechanics. *Synthese* 180: 79–85.

Smart, John J.C. 1963. *Philosophy and Scientific Realism,* London: Routledge&Kegan Paul.

Smart, John J.C. 1968. *Between Science and Philosophy*: New York: Random House.

Sosa, Ernest. 1991. *Knowledge in Perspective. Selected Essays in Epistemology.* Cambridge: University Press.

Stanford, Kyle P. 2000. An Antirealist Explanation of the Success of Science. *Philosophy of Science* 67: 266–284.

Stanford, Kyle P. 2006. *Exceeding Our Grasp. Science, History, and the Problem of Unconceived Alternatives*. Oxford: University Press.

Stanford, Kyle P. 2013. Underdetermination of Scientific Theory. *Stanford Encyclopedia of Philosophy*. http://plato.stanford.edu/entries/scientific-underdetermination/.

Suppe, Frederick. 1977. *The Structure of Scientific Theories*. Urbana, Ill.: University of Illinois Press.

Teller, Paul R. Pan-Perspectival Realism Explained and Defended. http://philosophy.ucdavis.edu/people/fzteller/site/truth-page.

Toraldo di Francia, Giuliano. 1976. *L' indagine del mondo fisico*. Torino: Einaudi.

Van Fraassen, Bas C. 1980. *The Scientific Image*. Oxford: Clarendon Press.

Van Fraassen, Bas C. 1985. Empiricism in the Philosophy of Science. In *Churchland* and *Hooker* (eds.) 1985: 245–307.

Van Fraassen, Bas C. 2002. *The Empirical Stance*, Yale: University Press.

Vattimo, Gianni and Rovatti, Pier A. (eds.). 1983. *Il pensiero debole*. Milano: Feltrinelli.

Votsis, Ioannis. 2011a. Saving the Intuitions: Polylithic Reference. *Synthese* 180: 121–137.
Votsis, Ioannis. 2011b. The Prospective Stance in Realism. *Philosophy of Science* 78:1223–1234.
Votsis, Ioannis. 2012. Putting Realism in Perspectivism. *Philosophica* 84: 85–122.
White, Roger. 2003. The Epistemic Advantage of Prediction over Accommodation. *Mind* 112 (448): 653.
Whorf, Benjamin Lee. 1956, *Language, Thought, and Reality*, Cambridge: MIT Press.
Wigner, Eugene P. 1967. Remarks on the Mind-Body Question. In Wigner, *Symmetries and Reflections*. Bloomington, In.: Indiana University Press.
Worrall, John. 1989. Structural Realism: The Best of Both Worlds? *Dialectica* 43:99–124.
Worrall, John. 1994. How to Remain (Reasonably) Optimistic: Scientific Realism and the 'Luminiferous Ether'. *PSA* 1994. vol. 1., ed. M. Forbes and D. Hull: 334–342. East Lansing, Mi.: Philosophy of Science Association.
Worrall, John. 1995. Il realismo scientifico e l'etere luminifero. In *Realismo/antirealismo*, ed. Alessandro Pagnini: 167–203. Firenze: La Nuova Italia.
Worrall, John. 2011. Underdetermination, Realism and Empirical Equivalence. *Synthese* 180, 2: 157–172.
Wright, John. 2002. Some Surprising Phenomena and Some Unsatisfactory Explanations of Them. In Clarke and Lyons (eds.): 139–153.
Wright, John. 2013. *Explaining Science's Success. Understanding How Scientific Knowledge Works*. Durham: Acumen.

The Truth of Theories and Scientific Realism

Evandro Agazzi

Abstract The thesis maintained in this paper is that scientific anti-realism was the consequence of having lost the confidence in the capability of science to attain truth, something that historically occurred at the end of the nineteenth century. Therefore, the requirement of truth was removed from science and replaced by the requirement of objectivity. This has a 'weak' sense, according to which scientific knowledge is 'independent of the single subjects' (intersubjectivity) In addition, however, every science is considered to investigate not reality in general, but only its specific objects ('strong' ontological sense of objectivity). These specific objects are 'clipped out' of the reality of common sense 'things' by considering them from a specific point of view focusing only on certain attributes of reality. In order to determine these clips, the scientific community elaborates certain standardized operational procedures for establishing whether certain statements regarding things are immediately true or false. In such a way these operational procedures are 'criteria of reference' and 'criteria of truth' for a given science, and moreover turn out to be the same used for securing objectivity in the weak sense. This amounts to recovering the characteristic of truth for scientific knowledge, and giving it a realist interpretation both ontologically and epistemologically, at least for its empirically testable statements. The contemporary struggle about realism, however, regards the unobservable entities introduced in scientific theories, and the strategy proposed in the present paper is that of suitably 'extending' to theories the notion of truth, which is immediately and directly defined for single declarative statements. From the referential nature of truth follows that if we have reasons for admitting the truth of a theory, we must also admit, for the same reasons, the existence of its referents, even when they are unobservable entities.

E. Agazzi (✉)
Interdisciplinary Center for Bioethics, Panamerican University,
Mexico City, Mexico
e-mail: evandro.agazzi@gmail.com

E. Agazzi (ed.), *Varieties of Scientific Realism*,
DOI 10.1007/978-3-319-51608-0_3

1 Introduction

There is quite a large display of characterizations of realism in general and of scientific realism in particular. For the limited purpose of the present paper we shall summarize them under only two headings that we shall call "ontological" and "epistemological". The first qualification has its historical roots in a distinction that was introduced in the philosophy of the eighteenth century, when "realist" was the opposite of "idealist". At that time the term "ideas" was used to denote our mental contents, our representations in general, and the question was debated whether *reality* has an existence independent of our cognitive activity or not. Realist were called those doctrines that assigned to reality such a mind-independent existence, and idealist those who reduced reality to the content of our cognition (the paradigmatic example of such an idealist position is the famous statement by Berkeley, *esseestpercipi*, that is, "to be is to be perceived"). Kant, for example, qualified himself as "empirical realist and transcendental idealist",[1] because he maintained that sense impressions are not produced by our mind but are passively received by our sense organs, whereas the objects of our knowledge are constructed by the transcendental forms of our intellect (categories). We call ontological this sense of realism since it has to do with the *existence* of reality. Realism in this first sense is often called "metaphysical realism" in contemporary literature.

The second meaning of the issue of realism, which we call 'epistemological', derived from a strange presupposition that modern philosophy adopted in epistemology: starting practically with Descartes, philosophers gave for granted that we know our representations (or ideas) and not reality but, admitting that our aim is to know reality, they asked whether we can be granted that our ideas correspond to reality. Are considered realist in this sense those thinkers who maintain that we can attain such an (indirect) knowledge of reality.

Before modernity philosophers were (with very few exceptions) realist in both senses: they admitted that reality exists independently of our knowledge, and that we know it as it is. This position, that we summarize here in a couple of words, was articulated and elaborated in details that we cannot examine here, and had produced the notion of *science* in a general sense, as full *knowledge* based on sense experience and rational deductive arguments. An obvious necessary requirement for science in that traditional sense (inaugurated by Plato and Aristotle and developed during the whole of Western "classical" philosophy) was that it is a *true* knowledge, but this requirement was not considered sufficient because this truth had to be also *.logically justified*. Such a foundation had to start from certain first principles whose *intellectual evidence* would provide universality, necessity and *certainty* to scientific knowledge. According to this view, philosophy was the highest science (the *scientia prima*, according to a terminology introduced by Aristotle), due to its universal scope, and for the particular sciences, the intellectual intuition of the

[1]*Critique of Pure Reason,* A 347.

essence of the investigated objects should constitute the starting point of the deductive justification.

But how must truth be understood? Also in this case we must note that the notion of truth has received a wide display of meanings along the history of Western philosophy that we are not certainly going to review here. Being specifically interested in the issue of *scientific realism,* we can restrict our attention to what we can call the "cognitive" sense of this notion, that is, to that sense which we commonly adopt in ordinary language when we qualify as true (or non-true) a given statement, or even a concept, a theory, a doctrine. Leaving aside for the moment the rather complex precisions needed for clarifying the different modalities with which it might be possible to speak of truth for those various forms of our cognition, we can note that there is a central core that all of them share when they are qualified as true, that is, the fact that they have a content that is *real*, that they attain *reality.* Indeed we find already in Plato and then in Aristotle several almost equivalent definitions of truth regarding declarative sentences, the most synthetic of which is the following: "To say of that which is, that it is not, or of that which is not, that it is, is false; to say of that which is, that it is, or of that which is not, that it is not, is true".[2] The force of this characterization of truth consists in its 'double directionality'. One direction is obvious: if something is the case (is *real*) and we describe it faithfully in a statement, this statement is *true*. But also the reverse holds: if a statement is *true*, then what it describes is *real* (since no true statement can tell "what is not the case").

2 Modern Natural Science

The natural science founded by Galileo at the beginning of the seventeenth century and developed by Newton in the second half of the same century did not radically differ from the classical paradigm. A decisive difference, however, consisted in the methodological decision to give up the frustrating illusion to grasp the *essence* of the "natural substances" (i.e. of the material bodies) by "speculation" (i.e. by an intellectual intuition), and restrict our attention to a few selected *accidents* (i.e. to certain mathematizable properties of the physical bodies). Instead of the speculative intellectual intuition, a new method of inquiry was invented (the experimental method), consisting in formulating a hypothesis regarding the phenomenon under consideration, and testing by an appropriate artificial experiment the consequences of the said hypothesis. This new way of making *natural philosophy* (as it continued to be called) remained *realist* in both the classical senses: (i) because the object of investigation was considered to be a reality independent of the human investigation (the new science was concerned with "real accidents" to use the literal phrase of Galileo), and (ii) because this investigation was considered to attain a *true*

[2] Aristotle, *Metaphysics* 1011b, 26–29.

knowledge of the said delimited domain of reality. Therefore, *truth* still was the fundamental characteristic attributed to this new form of knowledge.[3]

It is worth noting that working scientists, such as Galileo and Newton, were not affected by that strange "dualistic" presupposition according to which what we know are our ideas and not reality, and this has remained a constant attitude of scientists until the end of the nineteenth century. In particular, this realist consideration of natural science rapidly produced in Western culture the conviction not only that it was a genuine form of knowledge, but even the paradigm of knowledge as such (as is explicitly stated in Kant's Preface to the second edition of the *Critique of Pure Reason*). In the nineteenth century, positivism claimed that science was actually the only genuine form of knowledge, having overcome the illusionary pretentions of theological and metaphysical approaches to reality. Positivism, however, was affected by an inadequate appreciation of the role of reason in the construction of science. Galileo, who had so strongly underscored the role of concrete observations, measurements and experimental testing of hypotheses, had fully understood the indispensable role of reason that must even be ready to "do violence to sense" in order to uncover the real nature of phenomena by advancing a hypothesis that can finally be experimentally confirmed. Or, to give another example, when he formulated the principle of inertia, for which no empirical evidence exists, but must be admitted only through cogent reasoning. Positivism, on the contrary, was prisoner of a radical empiricism, reducing science to a diligent record of uninterpreted data that allegedly express sense perceptions (the only bearer of knowledge). Theoretical constructions were downplayed to useful tools for organizing perceptions for practical purposes, but without cognitive purport. This is notoriously the position of Ernst Mach in which we can see a clear forerunner of scientific anti-realism.

3 Contemporary Science

No absolute criteria exist for the "periodization" of historical events, and such more or less conventional criteria vary according to the particular domain considered, so that the periodizations of political history, history of literature, history of philosophy, history of science, history of fine arts—for example—are usually not overlapping. Owing to this, we propose to qualify as "contemporary" science that which begins in the last decades of the nineteenth century and continues up to now. Its defining characteristic is that of being *the science of the unobservable,* with a special reference to physics. Modern natural science had made abundant use of *idealizations*, that is, of concepts and statements that were an abstract representation of things and processes observable in ordinary experience, and were "visualizable". Therefore, a certain confidence was spontaneous in admitting the "real existence" of

[3]See for details Agazzi 1994.

such idealized entities when used in theories. For example, a corpuscular theory of light and a wave theory of light allowed scientists to conceive a light beam as consisting of a swarm of microscopic particles, analogous to grains of sand, travelling in the empty space or, respectively, in a wave propagating itself in an impalpable medium, analogous to the waves we see when a small stone falls in a pond of quiet water. This visualizability was the intrinsic force of such "mechanical models" which, in addition, constituted the intuitive basis for the elaboration of the rich mathematical apparatus of what was later called "classical mechanics". It was precisely the failure of these models in accounting for the second principle of thermodynamics and of the properties of the electromagnetic field that gradually led to phenomenalist and anti-realist positions such as that of Mach: unobservable entities may be introduced in a scientific theory as useful tools for organizing ideas and permitting more or less accurate predictions, but do not correspond to physically existing objects.

We can note that also in mathematics something similar was happening: the construction of the non-Euclidean geometries whose statements are often in contrast with the geometric intuitions, but are nevertheless on an equal footing with the Euclidean geometry as far as their internal non-contradiction is concerned, was opening the way to a purely formalistic and widely conventionalist conception of mathematics.

There is something puzzling in the situation we are considering. Contemporary natural science can be qualified in a broad sense as a science of the unobservables in multiple senses; because it has accomplished astonishing advancements in the investigation of the "microworld", as well as in the description of the almost unthinkably enormous spatiotemporal dimensions of the universe, or in the penetration of the hidden structure of living matter and living beings. These advancements have been possible thanks to a strict synergy of theoretical thinking and technological development that has permitted us to 'observe' instrumentally a great number of features that the pure sensory observation cannot attain. One must be aware, however, that we can rely on such 'observations' only because we accept the theories on whose grounds the sophisticated instruments have been designed and their outputs are interpreted (and even 'visualized'). Therefore, if we are not ready to accept that the elementary particles, or the DNA, or the extragalactic celestial bodies exist, we simply say that today's natural science is unable to know nature as it really is, contrary to what was capable to do classical science. More than simply puzzling, this position is paradoxical since is advocated in general by positivists, that is, by people who attribute to science the privilege of being the best form (if not even the only genuine form) of knowledge.

One can wonder how such a strange attitude could arise and become widespread. The answer is historical. The already mentioned difficulties of 'reducing' to classical mechanics the explanation of fundamental thermodynamic and electromagnetic phenomena was only the announcement of a crisis that had found especially in Mach an explicit forerunner due to his radical empiricism, but literally exploded at the beginning of the twentieth century with the creation of quantum mechanics and relativity theory. With these theories several laws and principles of classical mechanics were proved not to hold in the newly discovered domains, and even fundamental concepts seemed to have lost their original meaning. The whole of this well-known situation was (and still is) described by saying that classical physics had been proved to be *false*. And precisely because it had been found false despite its sophisticated mathematical formulation, the great number of empirical and experimental confirmations in different domains, and an incredible display of technological applications, it seemed wise the learn the lesson and say that no new scientific theory could advance the pretension of being finally the *true* one. No scientific theory is expected to be true and, therefore, the entities it introduces in its discourse are not expected to be *real*. In conclusion, the crisis of scientific realism was the historical consequence of having considered false classical mechanics and having therefore excluded truth from science. This is not an arbitrary reconstruction and is well reflected in the way in which Bas van Fraassen characterizes scientific realism in a celebrated work: realism is the position according to which "science aims to give us, in its theories, a literally true story of what the world is like; and acceptance of a scientific theory involves the belief that it is true".[4]

A consequence of this awareness is that, in order to vindicate a realist conception of science, one has to give back to science the traditional privilege of being a *true* discourse, obviously with all the precisions necessary for removing the obstacles that had led people to believe that even the best theories of classical mechanics were false.

This philosophical endeavor can profitably start by considering what has been proposed as an honorable replacement for truth in order to save the cognitive value of science, and then by a deepening of the very concept of truth.

4 Objectivity as a Replacement of Truth in Science[5]

The most immediate consequence of that generalized crisis has been that scientists no longer dared to call "true" even the best founded of their statements and tried to avoid the use of the notion of truth. But what will be then the reason for justifying the admission of a proposition or a scientific theory once we have dismissed the reference to its truth?

The solution to this problem that first occurs to mind is that the whole value of a theory or a single proposition should consist in the usefulness that they can show as instruments for an efficient conduct and also for an easier intellectual management of the external world and of our everyday experience (the view of Mach, as we have seen). This is the well-known conception (at the same time conventionalist and instrumentalist) that has dominated the first two decades of the twentieth century

[4]Van Fraasssen (1980), p. 8.

[5]The considerations that will follow constitute the core of my reflection on philosophy of science during several decades. They have been organically and systematically presented in Agazzi (2014).

and amounts to denying to science the capability and even the intention of actually producing genuine *knowledge*. It is certainly possible that such a view prevail during a short time of discomfort, but it is hardly possible that it lasts as a real conviction, especially with scientists, because a working scientist cannot be really persuaded that he *does not attain knowledge* when he is doing his investigation. Therefore, here is the impasse: how can we recover the confidence in the *cognitive* capability of science without falling back into the difficulties that surface when we attribute to science the capability of attaining truth? The way out that has been found was the invention of a kind of replacement of the notion of truth, by the introduction of the idea of *objectivity*. In such a way it became possible to say: well, we agree that science is not a 'true' knowledge, it is however, an 'objective' knowledge. At this junction, however, our problem becomes "What is objectivity?"

4.1 Objectivity as Intersubjectivity

If one consults the literature produced both by scientists and philosophers of science one finds a variety of implicit characterizations of objectivity, that we shall try to collect under two fundamental headings, that summarize the most salient proposals.

A first sense of objectivity is the following: objective means intersubjective. This meaning corresponds, in a first approximation, to the colloquial expression that science is a 'public' discourse. By this it is meant that it is a discourse open to everyone provided, obviously, that one puts oneself in the condition of entering this discourse. It is a discourse in which every statement is submitted to the control of anybody belonging to the scientific community, and this means someone that has undergone the standard training necessary for understanding and checking the statement at issue. It is a discourse in which what I have said does no longer belong to me in the first person, but also what my colleague has said does no longer belong to him: it is a discourse on which we ought to be in agreement. Now the difficulty resides precisely in this "ought to" or, said in a more precise way, the problem is that of answering the question "how can we found intersubjectivity?" since the problem is that of rendering public a knowledge that, as such, is always private (one always knows in the first person).

The difficulties that one encounters in looking for a solution to this problem are easy to imagine and also widely discussed in the literature. They concentrate especially on one crucial point, that is, on the impossibility of realizing a direct exchange of cognitive experience, at whatever level. I cannot 'have a look' in the thoughts of another person in order to see whether the notions she has are identical with mine. I cannot perceive another person's perceptions, I cannot be conscious of her states of consciousness, feel her emotions, and so forth. Taking all this into account, it seems that there is no ground for 'making public' whatever content of knowledge, and intersubjectivity appears as an illusory mirage.

Against such radical objections, however, stands an undeniable fact of life: humans, and also other animals, are able to communicate among themselves.

Therefore, what we need is not to know *whether* intersubjectivity is possible, but to understand *how* it is possible. And if the difficulties we have just mentioned are really insurmountable (as indeed they are), this means that this is not the price one has to pay in order to attain the intersubjective agreement.

Actually a little reflection shows us that what is needed in order to attain an intersubjective agreement regarding a certain notion is not the ascertainment of a conformity in the *way of conceiving* that notion, but the ascertainment of a conformity in *the way of using* it, and this ascertainment is generally possible, whereas the first one is never. We can easily confirm this fact by considering examples of concrete familiar notions as well as of complex abstract notions.

4.2 Objectivity as Reference to Objects

Something a little strange resides in the fact that we have defined objectivity as intersubjectivity, for actually the terms "objective" and "objectivity" contains in their own linguistic root a reference to the object much more than to the subject, as does, on the contrary, the term "intersubjective". Actually it is not difficult to recognize that the original sense of the notion of objectivity, the sense that we could call *strong*, entails the reference to the object: objective is—in this sense—a characteristic, a property, a judgment that concern "what is inherent in the object". From this strong sense follows the *weak* sense according to the following reasoning: if a property is intrinsic to the object, it must hold independently of the subjects who know the object, therefore, all the subjects should in principle recognize it in the same way. The inverse does not obviously hold. Nevertheless, for well-known historical reasons culminated with the philosophy of Kant, philosophers lost the confidence in the human capability of knowing the object such as it is *in itself* and, as a consequence, the strong sense of objectivity lost as well any concrete interest and was replaced by the objectivity in its weak sense. This last one—that fundamentally expresses the idea of an "independence with regard to the (individual) subject"—has received different formulations in philosophy, and its translation as intersubjectivity is its current version, especially regarding the sciences.

This I why, in the opinion of many philosophers and also scientists, there is no point in looking for an objectivity that would be stronger than the weak objectivity understood as intersubjectivity. In particular, every effort to attribute to objectivity an "ontological" sense, by conceiving it as a "reference to existing objects" would be the expression of an obsolete mentality. Statements like "in science we remain content with an objective description of phenomena without any pretension to know reality as it is in itself" seem to express this attitude very faithfully. Nevertheless, beside this kind of statements we find another one (even more widespread) which strongly stresses that every science is a "specialized" discourse dealing only with "its own specific objects". And it would be difficult to deny that such an expression contains the idea of a "reference to objects" with an implicit ontological understanding that requires to be investigated. Is this simply a colloquial 'way of speaking' or does this contain something deeper that must be well understood and made explicit?

In order to answer this question one must make a clear distinction between the "things" of ordinary experience and the "objects" of the different sciences, though recognizing that precise links exist between them. Now, while it would be wrong to say that every science specifically deals with a particular domain of "things" (because any "thing" can become the "object" of several sciences) one can say that every science deals with whatever thing "from its own point of view", and it is owing to this particular point of view that it makes this thing one of its proper "objects". Therefore, one could say that the objects of a science are the "clippings" obtained in things by considering them from the point of view of that science.

It may be useful to clarify this point by means of an example. Let us consider a watch that I hold in my hand and which as such can be considered a "thing" of ordinary experience that we find in the world. This thing can become an object of mechanics if, for instance, I ask some questions regarding its mass, the laws governing the motion of its internal gears; but it can also become an object of chemistry if I ask questions regarding the composition of the alloy of which its body is made, or the degree of purity of the rubies inside it; it can become the object of economics if I inquire about its price on the watch market; it can become a historical object if I ask the question whether or not this watch once belonged to Napoleon, or something of this kind. Therefore, one sees that whatever thing can be the object of whatever science, depending on the fact that it can be considered from the point of view of that science.

We cannot enter here the presentation of the details necessary for making precise the intuitive notion of "clipping" that we have used above but it is sufficient to point out that every science realizes its clipping by using in its language a limited number of specific *predicates* (whose sense is determined in a univocal and technical way) that it employs for speaking about things. These predicates are intended to correspond to certain *attributes* (that is, properties, relations and functions) that are present in things (though not necessarily all in whatever thing). So the use of predicates such as those of mass, length, duration and force determines the clipping (and hence the objects) of mechanics; the use of predicates such as those of metabolism, generation, etc. determines the objects of biology; whereas if we use predicates such as price, market value, supply and demand we are constructing the objects of economics.

We must now underscore that every science that we intend to qualify as "empirical" must rely upon certain means for 'touching' the things of ordinary experience. Therefore, it is indispensable that at least a part of the predicates constituting the language of an empirical science be of an *operational* nature in the sense of being directly linked with concrete standardized operations. These operations, on the one hand, enable us to 'manipulate' things and, on the other hand, to establish (and to establish in an intersubjectively ostensible way) whether propositions containing exclusively those operational predicates are *immediately true or false*.

The last affirmation entails two significant consequences. The first is that the operational conditions that constitute the foundations of intersubjectivity are at the same time the conditions that permit the construction of the scientific objects. In such a way we are justified in affirming that the two notions of objectivity (understood as intersubjectivity and as a reference to objects) practically coincide, though being conceptually distinct. The second consequence is that we can *recover the notion of truth* in the sciences, provided that we are aware that this truth is always "relative to the specific objects" about which the propositions are formulated. The crisis of the old notion of scientific truth depended on having conceived it as an *absolute and total* truth, that is, a truth regarding things in themselves. As a consequence this truth was seen to be ruined when new aspects of reality were discovered (that is, new "domains of objects") with which the old theories were unable to cope. The issue, however, appears under a completely different light if one is conscious that any theory has to be true only *about its own objects*.

5 A Deeper Notion of Scientific Object

Our discourse on scientific objectivity is still far from being concluded with the (important) acquisition consisting in recovering the legitimacy of scientific truth thanks to the referential purport of operations. This acquisition allows us to overcome a serious limitation of the radical empiricist epistemology for which *observations* are the only ground for meaning and reference: indeed observations are strictly *private* and, therefore, cannot grant neither intersubjetivity nor common reference to objects, whereas *operations* (as we have seen) are able to secure both conditions for objectivity. Nevertheless operations are still too close to empirical evidence and, if they were the unique criterion for scientific truth and the only ground for scientific realism, one would be obliged to reduce scientific truth to what science can tell about *things* such as they appear in everyday experience, and scientific realism would not go beyond commonsensical realism. This conclusion, however, would be in open conflict with the (fully justified) conviction that modern science has produced a great amount of *knowledge* going far beyond ordinary commonsensical knowledge, and concerning, in particular, an immense domain of sensibly *unobservable* objects.

In order to overcome the radical empiricist prejudice we must first reject the tempting idea that scientific objects are *things*, despite the fact that such an idea seems to be the best support for scientific realism. Indeed, in our preceding considerations we have pointed out: (i) that a thing can become the object of a given science *as far as* it is considered from the specific "point of view" of that science which determines a particular "clipping" within this thing; (ii) that *one single* thing can be the *object* of an *indefinite* number of scientific investigations. Therefore, equating a scientific object with a thing would amount to ignoring the distinction between thing and object that has been central to the whole of our analysis. The way out from this difficulty comes from a deepening of the above notions of "point of view" and "clipping": they are actually colloquial expressions for denoting the particular *structure of concepts* we can use to consider not only a single thing, but reality in general. From a genetic point of view (as we have noted) this conceptual structure is usually obtained by *abstraction* from the consideration of a domain of concrete things and, therefore, corresponds to a certain *structured set of attributes* (i.e. properties, relations, functions) of such things, and this is precisely the "clipping" that is operated in a thing through the application of the said conceptual structure.

But now it is precisely such conceptual structure that—after having been suitably refined and made totally explicit—becomes the effective *object* of the special science considered. Therefore, this is an *abstract object* that, in order to be investigated, must also be denoted by an appropriate structured set of *predicates* constituting the specific technical language of the science in question. Mass-point, uniform motion, frictionless motion, rigid body, perfect gas, adiabatic process, isolated system, insulating and conducting material are familiar examples of such abstract objects studied in physics, while chemical reaction, chemical equilibrium, metabolism, reproduction, homeostasis, perfect competition, marginal utility, per capita income, marginality, demographic transition are examples of abstract concepts elaborated respectively in chemistry, biology, economics, sociology. The theoretical dimension of a science consists indeed in the construction, articulation and development of such abstract objects that have the statute of intellectual entities, endowed with a quite precise *sense* and a *logical structure*. For this reason they are different from nothing, they are the content of thinking, they have a *kind of reality* that we could call mental or "noematic" (using a Husserlian terminology) and they are the *scientific objects* in a proper sense, because they are what a science *directly* investigates. They present the notable advantage of being endowed with universality and necessity (i.e. with the two fundamental features that were attributed to science during the Western tradition and were always put in connection with the powers of the intellect).

Empirical sciences, however, aim at investigating their own *domain of objects* (as we have noted) and this is not meant to be just a realm of abstract entities having a mental reality. How this can happen can be clarified by distinguishing *encoding* and *exemplifying*. An *abstract object* "encodes" a certain number of concepts that characterize it in an explicit and exact way, but it can be "exemplified" by many concrete *things* that are endowed with those attributes that are encoded in the abstract concept, and this normally occurs only within certain limits of approximation or tolerance, depending on several practical considerations. This happens because the same thing normally exemplifies many abstract objects (or concepts) and these simultaneous exemplifications inevitably entail only a partial satisfaction of any particular abstract concept. For instance, an iron bar is a good exemplification of the abstract concept of rigid body, however it is not 'perfectly' rigid and also presents some elasticity (something that may be very useful in certain concrete applications). We call r*eferents* the entities exemplifying an abstract object, and here we see that the objects of which we have spoken in the introductory part of our discourse when we have said that every science investigates only its own *domain of objects* are more properly called the referents of that science, that has its proper *domain of referents*. It is important to recognize that the two domains—that of the abstract objects and that of the referents—are deeply different and not just a kind of a reduplication of one other. For instance, on the one hand, no abstract concept enjoys the properties that it encodes (the concept of uniform motion is not in motion, the concept of a quadruped animal has no legs, etc.) and, on the other hand, no concrete thing encodes properties, because it is not completely characterized by any finite set of properties. The relation between encoding and exemplifying has represented a complex issue that has challenged the mind of great thinkers starting at least with Plato. Without entering such a difficult problem, however, we can propose a practical criterion for determining the referents of abstract objects: this is the use of *standardized operations* that, in the sciences, play the role of *criteria of referentiality*. For a correct understanding of this point it is useful to make a short digression regarding a fundamental issue of philosophy of language and semiotics.

6 Sense and Reference

We maintain that the *meaning* of a concept consists in the interrelation of two constitutive parts, *sense and reference*. These terms have been explicitly introduced in modern philosophical vocabulary by Frege,[6] but the corresponding notions have existed in the Western tradition since ancient and medieval philosophy under different denominations, and have been submitted to careful and subtle analyses. What was lost in certain trends of modern philosophy that were most directly influenced by empiricism was the awareness of the existence and specificity of an intellectual world of thoughts in which the *sense* of concepts has its place, and is not reducible to a dynamics of sensible perceptions (though having links with them). When the "linguistic turn" occurred in contemporary philosophy, a comfortable replacement for the embarrassing notion of a thought-content seemed to be offered: the sense of a term or a linguistic expression in general is determined by its *linguistic context* and in such a way is totally internal not only to language, but even to any particular language context. This "semantic holism" paradigmatically advocated by Quine was also a barrier regarding the access to a reality different from language and, therefore, did not leave room for *reference* in any proper sense. According to this position, meaning was reduced to sense, and sense was understood as the result of a linguistic context.

The genuine spirit of empiricism, however, pointed toward a different direction. Though sharing the elimination of the world of thought by simply considering language, the question regarding the *meaning* of a linguistic expression received a different answer: this meaning is what this linguistic expression is about, or *denotes*, that is, something that lies outside of the language itself and must be attained by sensible experience. In other words, in this case meaning was reduced to *reference*. The gigantic difficulty in this position regarded how to establish the proper link

[6]Cf. Frege (1892).

between a linguistic expression and "its" referent without resorting to an intellectual sense, since logical-linguistic machineries (starting with Carnap's "correspondence "rules") still remain of a linguistic nature and cannot provide the tools for over-stepping the linguistic barrier.

The weakness of both positions was the pretension to dispense with the specific dimension and role of thought: even admitting that language is in a way primary (because it is impossible to formulate, articulate and communicate thoughts without some kind of language), it remains clear that any language is in itself a pure set of material *signs*. This set becomes a language only if the signs are endowed with a *meaning*, that is, if they can be *understood,* and this very word directly calls into play the *understanding* (that is also called *intellect*), while we can call *concepts* in a very general sense the result of this understanding (since they are what we "conceive" when we understand the language). It is true that the linguistic context greatly contributes to the determination of meaning; not, however, because of the changing arrangement of the material signs of the language, but because of the intellectual interrelations among the various *conceptual* components of the context. By this we have recognized the merits of the 'contextualist' theories of meaning, but at the same time we have also recognized that they tacitly rely upon the admission of a specific realm of thoughts, which we can equate with the domain of *sense*, Yet, if we remain at this stage, we are unable to explain how the language can *speak about* something different from itself (or, to put it in colloquial terms, about "the world").

Equally frustrating is the effort of granting such a link between language and world by a direct assignment of constituents of the world to constituents of the language, since in this case one had to make an association of material entities (the signs of the language) with other material entities (the things of the world) without any reason for the choice. If we admit, instead, that the signs of the language have a meaning, and that this meaning has also certain recognizable relations with the world, it is possible also to assign *referents* to the signs of the language and to know "about what" it is speaking. Therefore, even this reduction of meaning to reference cannot function without admitting the 'intermediate' level of thought-contents, of concepts. As a consequence, if we accept to define *semantics* as the theory regarding the *meaning* of a language, we should advocate a *three-level semantics*, in which we envisage the level of *signs* (language), the level of *sense* (concepts) and the level of *reference* (the entities about which the language intends to speak).[7]

As we have already explained, *operations* play an essential role in this framework, because they provide the missing link between the level of sense and the level of reference, and they can do this because, on the one hand, they are *understood*, i.e. conceptualized as a part of the *sense* of a proposition, but on the other hand they belong to the "world", they consist in *doing something* and not just in speaking or thinking.

---

[7]An extensive treatment of this three-lever semantics is presented in Agazzi (2014), Chap. 4.

7 The Referential Nature of Truth

The difference and interrelation between sense and reference was recognized already in ancient philosophy in the efforts of defining *truth*, and was explicitly pointed out in the Aristotelian distinction between *semantic discourse* (i.e. the discourse that is endowed simply with meaning), and *apophantic discourse* (i.e. the discourse that affirms or denies something) because in this second case one has to consider "about what" the affirmation or negation is stated, and on this ground the discourse turns out to be true or false. To put it differently, it is not granted that whatever meaningful declarative discourse is true: in addition it must "say *of*" something what this something "really is". This is the familiar commonsensical notion of the truth of a statement that has also been accepted in philosophy until the twentieth century (and which underlies the different *criteria for truth* proposed by the different "theories of truth"). In the case of the sciences, the 'crisis' occurred especially at the beginning of the twentieth century (of which we have spoken in Sect. 3 of the present paper) had led many epistemologists to consider scientific theories essentially as formal systems whose global context provides at the same time the sense and the content of their statements, provided that they be internally consistent (i.e. free from contradiction). This was a kind of "syntactic conception of truth" or "coherence theory of truth", whose weakness became patent especially after Gödel's result concerning the impossibility that any formal system (satisfying certain minimal conditions) can prove its own consistency. This produced the challenge to reintroduce the traditional proper notion of truth also for the formalized languages, and this result was offered in the famous paper by Tarski (1933)[8] who explicitly wanted to qualify his doctrine as a "semantic conception of truth".[9] If one considers the core of Tarski's very complex and elaborated construction one recognizes two things: (a) that the "interpretation" of the formal language does not consist in connecting its signs with concepts or giving them a *sense,* but in connecting them directly with unqualified elements of a given set (i.e. with *referents*), and this was the reason for calling his conception "semantic"; (b) the necessary and sufficient condition for declaring the truth of a proposition is that the state of affairs described by the proposition actually holds, but no criteria are offered for checking whether or not such a condition is fulfilled. These can perhaps be considered weak points or at least limitations of the Tarskian definition of truth. Two fundamental acquisitions, however, must be recognized: (a) to have recovered the *referential* nature of truth, that is, the inadequacy of considering it as a purely internal property of a linguistic construction; (b) the 'bilateral' condition for the truth of a declarative sentence, that is: if a state of affairs (or a fact) obtains, then the proposition describing it is true, and if a proposition is true, then the state of affairs (the fact) it describes must obtain. One can discuss whether or not this Tarskian conception can be qualified as a "correspondence theory" of truth; this discussion is rather idle

[8]See Tarski (1933).

[9]See Tarski (1944).

because it is not univocally understood what such a correspondence should be. This is why we prefer to call it "referential". Not just for the sake of clarity, but also because we have already indicated how one can trace the referents of linguistic elements, since our operations constitute at the same time "criteria of reference" and "criteria of truth", as we have seen.

8 The Regional Ontologies

In ordinary discourse, as well as in scientific discourses, we use a great variety of declarative sentences that we qualify as true (for instance, "2 + 2 = 4", "Paris is the capital of France", "insects have six legs", "gold is more expensive than silver", "Hector is a Trojan warrior in the *Iliad*", "Napoleon was defeated at Waterloo", "Spanish is a neo-Latin language", "the Earth is a planet in the Solar system", the "Minotaur lived in Crete"). Since, as we have stressed above, a true sentence cannot be true "about nothing", the "fact" to which it refers must obtain, must be the case. It follows that also the entities mentioned in the declarative true sentence, as well as the properties and relations of these entities, must *exist*, though having *kinds of existence* very different in the different cases. So we can say that 2 is a mathematical entity, or has a mathematical existence; that Paris and France are geographical (or political) entities linked by the geographical (or political) relation of the first being the capital of the second; that Hector is a literary person or has a literary existence; that the Minotaur is a mythological entity or has a mythological existence. This way of speaking sounds very obvious, but it actually has a deep significance: it recovers the fundamental thesis of Parmenides that being simply is what is different from non-being, that is, from nothing, and at the same time retains the equally fundamental Aristotelian thesis of the *analogical sense of being*. Aristotle has presented this thesis especially regarding the "way of being" or existing (for example, the substance exists "in itself", whereas the accidents exist only "in a substance"; or something can exist "in potentiality" or "in actuality"). Our examples have to do with another aspect of the analogy of being, that is, with the different "kinds of existence", that were also considered in the philosophical traditions and have been revisited in the Husserlian doctrine of the *regional ontologies*. The temptation that must be resisted (at least in the present discourse) is that of trying to distinguish what "really exists" from what exists "only in a certain sense", a distinction that sometimes is presented as a difference between metaphysics (the discourse regarding what really exists) and ontology (the discourse regarding only thinkable or even fictitious entities). Very easily one may think that 'what really exists' is what exists in space and time, but how could we deny real existence to a deep sorrow that might push a person even to suicide, or to a bankruptcy that could suddenly reduce to poverty hundreds of people? Not to speak of the numberless people who believe in the existence of an immaterial god or a plurality of immaterial deities. The risk of such a pretension is to fall into a flat reductionism by dogmatically stating that a certain kind of reality is what "really exists" and then trying to reduce all the rest to a manifestation of that reality (which would be an unconscious form of uncritical metaphysics).

We can avoid such difficulties because we have explained how all these different "entities" about which a certain declarative sentence is true are its *referents* and, in addition, we have also explained how certain fundamental referents of a scientific discourse can be immediately attained by specific operations that play the role of criteria of referentiality and criteria of truth for the specific discipline adopting them.

Therefore, if I say that I saw a white horse in a dream, whereas I actually saw a black cat, my declaration is false and it would not be correct to say that it is neither true nor false, because these horse and cat "did not exist". Indeed they exist in the ontological region of the dreamed entities, that may have a not negligible place in a person's life, be the subject matter of literary works, and even become an important source of "data" for certain scientific theories such as psychoanalysis. Let us note, in addition, that the domain of referents, or ontological region, delimited by a certain group of concrete material operations is not necessarily material: for example, in order to ascertain whether it is true that Hector was a Trojan warrior in the *Iliad*, one needs to perform concrete operations like finding a book in a library, visually recognizing that its title is *Iliad*, reading it in a language that he knows, and finally finding out whether there is in that book the story of a certain personage named Hector playing the role of a Trojan warrior. This literary personage is not material and has just a literary existence.

This simple example helps us in clarifying that the operations used in an empirical science for determining its domain of reference are certainly material, and are the tools for recognizing *immediately* true or false statements. They provide the *data* of the science in question and for this reason are also the tools for *testing* the acceptability of other statements that are not fully expressed by means of operational predicates (by checking whether such statements logically entail operationally testable consequences). The construction of the most important sciences, however, goes far beyond the collection of such immediately true statements and, as we have seen, introduces new concepts by means of definitions and proposes *theories* in order to *explain* data. This amounts to introducing certain *abstract objects* that encode concepts not all of which are of an operational nature and which, therefore, cannot be exemplified by means of the concrete basic operations.

What is the ontological status of such objects? We can answer that their ontological region remains that determined by the fundamental operations: they are physical objects if they are introduced in a physical theory and do not convert themselves into literary, psychological or mathematical objects. Yet this answer is not totally clear. One can say: of course, as *abstract objects* they certainly exist and belong, for instance, to physics, but *do they really exist?* That is, are they also *exemplified* by physical *referents*? This is the core of the issue of *scientific realism* that, in its rough formulation, can be expressed as the question of the *existence of the unobservables* and remains intact also after having refined it by replacing the observations by the operations, as we have done.

9 Truth as Warrant for Existence

The solution of our problem is implicit in the 'bidirectionality' that we have already stressed as included in the referential nature of truth, that is: if a proposition is true (for whatever reason it could be stated as true), its referents must exist (for the same reason) in the corresponding ontological region. This, however, is just a hint, because for the moment we have conventionally restricted the notion of truth to single statements, and proposed a unique kind of criteria of truth and referentiality (i.e. the operations), that are appropriate only for *immediate truth*. Therefore, we should treat of the different methods for securing *indirect truth* to single propositions, that we could call synthetically *inferential methods*. They are widely examined in the standard literature of philosophy of science and need not be recalled here. Let us simply note that the majority of such methods is such that the conclusion of the inference is not endowed with *absolute certainty*. This, however, is only an *epistemic* (not an epistemological) condition that does not detract anything from the ontological commitment of the statements involved. This simply means that, if one is not certain about the truth of a statement, one is also not certain about the existence of the referents of this statement, but not that these referents do not exist *because* of this uncertainty. In addition, we know that such methods (especially those applied in the mature sciences) normally secure a significant degree of confidence in the truth of their conclusions, and this offers a (not absolute but very reasonable) warranty also for the existence of the corresponding referents.

A further step consists in defining truth for a *set* of propositions and this can be done only if a certain *structure* is introduced into this set; the most spontaneous idea is to consider this structure as consisting in *logical links* for which elementary formal logic offers us a suitable display of patterns and rules with direct impact on the truth of the statements involved. In particular, this is the conception underlying the *sentential view of scientific theories* that has dominated the analytical philosophy of science for most of the twentieth century and has promoted the *nomological-deductive model* of scientific *explanation*. The typical schematization of this approach was that of considering a scientific theory as the *logical conjunction* of all its hypotheses (commonly presented as "laws"), and the testing of a theory as a *logical deduction* from these hypotheses (in conjunction with some empirically testable premises) of a singular empirically testable conclusion: if this conclusion were found to be true, the theory would be confirmed, if it were found to be false, the theory would be "falsified". The trouble with this approach is that the confirmation is always not absolutely secured (for simple logical reasons), and that the falsification would concern the theory "as a whole" without indicating with certainty which one, among the various hypotheses, is false and makes false the logical conjunction making up the theory. These are the fundamental reasons that have shown the inadequacy of this approach to account for *scientific change* as it actually occurs and, especially after Kuhn's book on *The Structure of Scientific Revolutions*,[10] have started new trends in the philosophy of science.

We are not interested in discussing these trends but we point out that they have in common the rejection of the statement view of theories and of the nomological-deductive model of scientific explanation. The first aspect almost automatically prevents the extension of the notion of truth to theories: if truth has been defined for single statements and then extended to logically structured sets of statements, but theories are *not* logically structured sets of statements, it follows that scientific theories are not the kind of entities to which the property of truth or falsity applies. What are then theories? The answers to this question are quite similar: theories are images, representations, models, global perspectives or *Gelstalten* which are put forward in order to *understand* and *explain* the data ascertained in a certain domain. They perform this task by introducing *theoretical entities* that are the constituents of the representation or model but are not found in the data. It might be interesting to show how these characterizations tacitly rely upon that capability of *intellectual intuition* of which we have spoken above, but we want to consider other aspects of this issue.

It is right to point out that we do not usually qualify as "true" an image, a representation, a model, but we would rather qualify them as "adequate", "faithful", "correct", "appropriate", "useful", "reliable". Nevertheless it is no less obvious that such models, or images, or representations are of no use from a cognitive and practical point of view unless they are *expressed* and *made explicit* by means of certain *statements* which may be far from expressing all the representational content of the intellectually conceived model, but have the great advantage of being communicable, true or false, and testable. For example, the map of a city is certainly far from being considered "true" in a strict sense, but it enables one to extract from it a certain number of true statements, like "the railways station is located at London Square", "the distance between the Cathedral and your hotel is less than the distance from the City Park and the Modern Art Museum", "you must walk to the South if you want to go from your hotel to China Town", and so on. This simple example is sufficient to show that it is certainly wrong to maintain that a scientific theory is *just* a set of statements but it is equally right to maintain that it is *also* a set of *suitably* connected statements. This "suitably" has a complex meaning: from the one hand it indicates the *logical correctness* of the links, but on the other hand it alludes to the importance, to the "representational relevance" of the links.

Having (partially) redeemed the statement view of theories we are entitled to speak in an *analogical sense* also of the truth of a theory, considering the statements through which it is actually formulated. By this we do not pretend, as van Fraassen would put it, that a scientific theory tells a "literally true story" about the world, but that it offers us a testable cognitive representation in which certain *theoretical entities* occur, that are elements of certain *statements*. We can maintain that the presence of these elements contributes to the understanding and explanation of our

[10]Kuhn (1962).

data not *just because* from the statements where they occur we can *logically deduce* the statements describing the data, but because, in addition, these theoretical elements appear to be *causally related* with the data in virtue of what is proposed in the theory, and this is a condition impossible to capture from a purely logical point of view. Therefore, the nomological-deductive model also retains a minimal value as a necessary condition for scientific explanation (in the sense that logical consistency and compatibility with the data is a *necessary* condition for *fully* accepting a theory) but they are not sufficient for deciding which part of the theory is responsible for a "falsifying" result, or whether this failure would simply require a "readjustment" rather than a total rejection of the theory. The "perspectivist" or "gestaltic" view of scientific objectivity that we have advocated offers a ground for debating these issues.[11] For the limited purpose of the present paper it is sufficient to stress that, by affirming a correctly understandable notion of truth for scientific theories, and recognizing that this truth entails the existence of the referents of all the concepts occurring in the true statements of the theory, we are entitled to say that also the theoretical entities so introduced *exist* with the same kind of existence of the other referents belonging to that ontological region.

Let us note, in addition, that, usually, such theoretical entities are not practically accepted in mature sciences unless they are in a way "observed". One must be aware, however, that the term "observation" is used here in a sense very different from the radically empiricist one that reduces it to the content of unaided sense perceptions. Indeed, modern science relies upon *instrumental observation,* that avails itself of often very sophisticated instruments whose reliability and results (as we have already pointed out in Sect. 3) are granted by a robust application of scientific theories gradually accumulated up to the point of becoming reliable (within the scientific community) no less than sense perceptions in everyday life. Note that these instruments do not use theories, but the *applications* of theories, that is, the action of concretely existing things that exemplify, at different levels of complexity, the concepts of preexisting theories. For this reason technology is a very powerful warrant of scientific realism, both because it testifies of the capability to act of the referents of theoretical entities, and because it also enables scientists to "observe" even the unobservables.

References

Agazzi E. (1994), Was Galileo e realist? *Physis* 31/1, 273–296.

Agazzi E. (2014), *Scientific Objectivity and its Contexts*, Springer, Cham, Dordrecht, Heidelberg, New York.

Dilworth C. (2008), *Scientific Progress,* 4th edition, Springer, Dordrecht.

Frege G. (1892), Ueber Sinn und Bedeutung, *Zeitschrift für Philosophie und philosophische Kritik* 100, 25–50.

---

[11]This view is presented in details in Agazzi (2014) and also in Dilworth (2008).

Kuhn T. (1962), *The Structure of Scientific Revolutions*, Chicago University Press, Chicago.
Tarski A, (1933), Pojecieprawdy w jezykachnaukdedukcyjnych, in *Travaux de la.*
Société des Sciences et des Lettres de Varsovie, Classe III Sciences Mathémathiques et Physiques, no 34, Warsaw, 1933. Better known after the German translation: Der Wahrheitsbegriff in den formalisierten Sprachen, in *Studia Philosophica*, 1 (1935), pp. 261–405. Engl.trans. by J.H. Woodger in A. Tarski, *Logic, Semantics, Metamathematics. Papers from 1923 to 1938*, Clarendon Press, Oxford, 1956, pp. 152–278.
Tarski A. (1944), The Semantic Conception of Truth and the Foundations of Semantics, *Philosophy and Phenomenological Research,* 4. 341–375.
Van Fraassen B. (1980), *The Scientific Image,* Clarendon Press, Oxford.

Part I
Realism and Antirealism

Strict Empiricism Versus Explanation in Science

Alan Musgrave

Abstract It is reasonable to believe that the best (or only) explanation of certain phenomena is true. This is why we are entitled to believe both in successful scientific theories, and in scientific realism, which explains that theories are successful because they are true. This is denied by strict empiricists, according to whom only empirical evidence should determine theory-choice: for them there is no difference between a theory T, and its "surrealist transform" T*, the claim that all phenomena are as if T were true. But there is a fundamental difference, that T explains the phenomena while T* does not. Historical cases of surrealism were Bellarmine's instrumentalism, Berkeley's theological explanation of empirical regularities, and Gosse's creationist explanation of the geological evidence for evolution. Contemporary examples are van Fraassen's constructive empiricism and Stanford's explanation that T is successful because it is "predictively similar" to the true theory. Moreover, the successful predictions made by false theories are not a counter-example to the realist explanation of scientific success, as claimed by Laudan, because either they were not novel, or they can be explained by the partial truth of those theories (to be distinguished from verisimilitude).

1 Explanations of Phenomena and Strict Empiricism

As everybody knows, science aims to explains things—scientists propose theories to help them understand or explain observable phenomena. Since an explanation is not adequate unless it is true, scientists seek true theories. Or so scientific realists think. As everybody also knows, science is empirical—scientists use observation and experiment to try to decide between the competing theories that they propose. What is not so well known is this—that the dream of finding explanations of phenomena is at odds with strict or hardline empiricism. What is strict or hardline empiricism? It is the view that *only* empirical evidence should determine

A. Musgrave (✉)
University of Otago, Dunedin, New Zealand
e-mail: alan.musgrave@otago.ac.nz

E. Agazzi (ed.), *Varieties of Scientific Realism*,
DOI 10.1007/978-3-319-51608-0_4

theory-choice. There is nothing to choose between theories that empirical evidence cannot choose between.

So suppose we have two theories that say the same about the observable phenomena, two theories that are equivalent as far as the observable phenomena go, two theories that are 'empirically equivalent'. No observation or experiment can decide between these two theories. Observation or experiment may, of course, tell us that both of them are false—for they may both predict something that turns out to be wrong. But what if the two theories are 'empirically adequate' as well as empirically equivalent? What if they tell us the same things about observable phenomena and everything they tell us is true? Strict empiricists will say not only that there is nothing to choose between them, but also that they equally well serve the aim of science. For according to strict empiricism, also known these days as 'constructive empiricism', the name of the scientific game is 'saving the phenomena' (Van Fraassen 1980, p. 93). An empirically adequate theory that tells us nothing but truths about observable phenomena fulfils the aim of science perfectly well.

This means that there is nothing to choose between any scientific theory **T** and what I call its surrealist transform **T***: 'The observable phenomena are *as if* **T** were true'. The surrealist transform **T*** of **T** is by design empirically or observationally equivalent with **T**. They 'save the observable phenomena' equally well. So according to strict empiricism, there is nothing to choose between them.

You may think that surrealist transforms of scientific theories are a mere philosopher's plaything. Not so. They have a history, and served serious purposes. It was the ancient astronomers who discovered the first real example of empirically equivalent theories. They proved that two quite different theories about the sun's (apparent) motion made exactly the same predictions about the (apparent) positions of the sun. Both theories saved the phenomena equally well, yet both could not be true. However, astronomical observation could not tell the ancients which of them is true, and it is hard to see what else might have told them. Perhaps astronomers should forget about truth, perhaps the whole task of astronomical theory is just to save the phenomena.

2 Three Historical Cases of Surrealism

These ideas loomed large in the Scientific Revolution. The Copernican idea that the earth moves around the sun, and not the sun around the earth, clashed with certain passages in the Bible. Astute Church leaders saw a way to resolve the conflict. Cardinal Bellarmine, Chief of the Inquisition, later to become Saint Robert Bellarmine, told the Copernicans where to get off:

> For to say that assuming the earth moves and the sun stands still saves all the appearances … is to speak well. This has no danger in it, and it suffices for mathematicians. But to wish to affirm that the sun is really fixed in the centre of the heavens … and that the earth … revolves very swiftly about the sun, is a very dangerous thing, not only by irritating all the theologians

> and scholastic philosophers, but also by injuring our holy faith and making the sacred Scripture false... To demonstrate that the appearances are saved by assuming the sun at the centre and the earth in the heavens is not the same thing as to demonstrate that in fact the sun is in the centre and the earth in the heavens. I believe that the first demonstration may exist, but I have very grave doubts about the second; and in case of doubt one may not abandon the Holy Scriptures as expounded by the Holy Fathers.[Letter to Foscarine, April 1615.]

Following Bellarmine, Pope Urban VIII reinforced the point with what is now called the 'argument from the unthought-of alternative'. In an audience with Galileo, he argued as follows:

> Let Us grant you ... that it is entirely possible for things to stand as you [Copernicans] say. But now tell Us, do you really maintain that God could not have wished or known how to move the heavens and the stars in some other way? We suppose you will say 'Yes', because We do not see how you could answer otherwise. Very well then, if you still want to save your contention, you would have to prove to Us that, if the heavenly movements took place in another manner than the one you suggest, it would imply a logical contradiction at some point, since God in His infinite power can do anything that does not imply a contradiction. Are you prepared to prove as much? No? Then you will have to concede to Us that God can, conceivably, have arranged things in an entirely different manner, while yet bringing about the effects which we see. And if this possibility exists, which might still preserve in their literal truth the sayings of Scripture, it is not for us mortals to try to force these holy words to mean what to us, from here, might appear to be the situation... Have you got anything to object? We are glad to see that you are of Our opinion. Indeed, as a good Catholic, how could you hold any other? To speak otherwise than hypothetically on the subject would be tantamount to constraining the infinite power and wisdom of God within the limits of your personal ideas. You cannot say that this is the only way God could have brought it about, because there may be many, and perchance infinite, ways He could have thought of and which are inaccessible to our limited minds. We trust you see now what We meant by telling you to leave theology alone.
>
> [G. Santillana, *The Crime of Galileo*, p. 166]

Notice how the Pope has upped the ante. It was not enough for Galileo to show that Copernican theory saved the phenomena. Galileo had to show that they could be saved in no other way, show that an omnipotent God could not have fixed things so that it was merely as if the earth moved, show that the earth must move. This was a big ask. Galileo tried to answer it by proposing his famously mistaken theory of the tides, describing them as "physical effects whose causes can perhaps be assigned no other way" than by supposing the earth to move. But at the end of his *Dialogue on the Two Chief World Systems* he gives up. Simplicio, who has so far only been made fun of in Galileo's *Dialogue* as the spokesman for Aristotle, reiterates the Pope's argument that God could produce tides in the oceans some other way and that "it would be excessive boldness for anyone to limit and restrict the Divine power and wisdom to some particular fancy of his own". Galileo's spokesman Salviati is stumped by what he calls this "admirable and angelic doctrine" and the *Dialogue* ends. It is said that the Pope was cross that Galileo put his argument into the mouth of the idiot Simplicio. Perhaps. Galileo's earlier remarks should also have made him cross:

> Surely, God could have caused birds to fly with their bones made of solid gold, with their veins full of quicksilver, with their flesh heavier than lead, and with wings exceeding small. He did not, and that ought to show something. It is only in order to shield your ignorance that you put the Lord at every turn to the refuge of a miracle.

Nowadays secular philosophers of science turn Urban VIII's argument into the argument from the unthought-of alternative: we cannot infer the truth of any theory from the fact that it is the only empirically adequate theory we have, because there may be an alternative theory we have not thought of.

For a second example of surrealism with an apologetic purpose, let us jump forward to the middle of the nineteenth century. By then geologists had amassed much evidence of the great antiquity of the earth, and the fossils of extinct creatures in the rocks fuelled evolutionary speculations. Philip Gosse was a famous naturalist and explorer, author of 40 books, the David Attenborough of his age. He specialized in marine biology and invented the aquarium without which no Victorian drawing room was complete. He was also a Biblical fundamentalist who thought that the earth and all the creatures on it were specially created in about 4004 B.C. He agonized for years about the clash between science and his religion. Then he hit upon a brilliant solution. In 1857, two years before Darwin's *Origin of Species*, he published *Omphalos—Untying the Geological Knot.* 'Omphalos' is Greek for belly-button. Gosse's *Omphalos* begins with an erudite discussion of whether Adam and Eve had belly-buttons. Think about it—this is a good question for a creationist! Gosse soberly concludes that they probably did have belly-buttons. He soberly concludes, in other words, that God created them *as if* they had been born of woman and were not the first people. (Renaissance artists were not so sure: their paintings of Adam and Eve often had foliage tastefully arranged to hide not just their naughty bits but also whether or not they had belly-buttons!) It is the same, Gosse argued, with the growth-rings in the trees and the fossils in the rocks. God created them in 4004 B.C *as if* the teachings of biology and geology were true.

Gosse's son Edmund, in his brilliant *Father and Son*, describes what happened:

> Never was a book cast upon the waters with greater anticipations of success … My father lived in a fever of suspense, waiting for the tremendous issue. This 'Omphalos' of his, he thought, was to fling geology into the arms of Scripture, and make the lion eat grass with the lamb. … He offered it, with a glowing gesture, to atheists and Christians alike. … But, alas! atheists ad Christians alike looked at it, and laughed, and threw it away.
>
> [*Father and Son*, 1907, p. 105.]

But *why* did the geologists laugh at Gosse's hypothesis? They had not one scrap of geological evidence against it. "God created the earth in 4004 B.C. *as if* geology were true" is evidentially or empirically equivalent with geology. The geologists who laughed cannot have been strict empiricists, for whom only empirical evidence determines theory-choice. (By the way, when push comes to shove contemporary Creation Scientists also go Gossian, and say that God created the rocks a few thousand years ago *as if* theories of radio-carbon dating were true.)

For my third, and most radical example, I jump back in time to a great 18th century philosopher. George Berkeley, Anglican Bishop of Cloyne, was the scourge of 'materialism, scepticism and atheism'. He was a more radical thinker than Bellarmine or Gosse. He did not just dispute the truth of some particular scientific theory. He thought that all science was false. More radically still, he thought that our commonsensical belief in external or physical or material objects was also false. Common sense postulates external objects to explain our experience. I see a tree, shut my eyes, and when I open them I see the tree again. Why? Because there is a tree out there, existing independently of me, which somehow causes my experiences of it. Berkeley says this is wrong. According to Berkeley, the world consists entirely of minds or spirits or souls. There are the finite spirits—you and me—and there is the Infinite Spirit, God. There are no trees or rocks or rivers, no external material objects at all. Of course, we have tree-experiences. But our tree-experiences are not caused by trees. No, our tree-experiences are planted directly into our minds by God, as are all our perceptual experiences. It is just that God plants experiences in our minds *as if* our commonsense beliefs in independently-existing trees and the like were true.

Similarly with science. Science is all false. (More precisely, physical science—physics, chemistry, geology, biology, and so on—is all false. We have minds, so some psychology might be literally true.) But a scientific theory need not be true to be good. A good scientific theory is one that makes correct predictions, or 'saves the phenomena', or is empirically adequate. But a good theory is not true. It is just that God plonks experiences directly into our minds *as if* it were true. Strict empiricists cannot reject this doctrine, for there is no experience or evidence against it. It is empirically equivalent with common sense and science.

So, Bellarmine said that God produced astronomical phenomena *as if* Copernican theory were true, and Gosse said that God produced geological phenomena *as if* geology were true, and Berkeley went the whole hog and said that God produced all phenomena *as if* common sense and science were true. There is no disputing that God could do all these things—after all, She can do anything, can't She? And, to repeat, there is no way of refuting these views by appeal to experience or experiment.

It is no accident that in all three of these cases, God gets into the picture. The antirealist or surrealist or instrumentalist philosophy of science was invented, by Bellarmine, Berkeley, Gosse and Duhem, to dissolve clashes between science and religion. But the religious dimension is not essential. Instead of saying that God produces the phenomena *as if* some scientific theory were true, we can simply say that the phenomena just are *as if* that theory were true. This is secular surrealism. So, suppose we have some empirically successful scientific theory **T**. We form a *theological surrealist* version of it by saying "God produces the phenomena *as if* **T** were true" (call this $\mathbf{T}^{\mathbf{G}}$). And we form a *secular surrealist* version of it by saying "The phenomena are *as if* **T** were true" (call this **T***). These three theories, **T**, $\mathbf{T}^{\mathbf{G}}$ and **T***, are empirically or observationally equivalent. No observation or experiment can decide between them. From a strict empiricist point of view, they are equally good theories.

3 Constructive Empiricism and Other Contemporary Surrealisms

Secular surrealism is the same as Bas van Fraassen's constructive empiricism. Van Fraassen's slogan is "The name of the scientific game is saving the phenomena". A theory saves its phenomena if its predictions about observable states of affairs are true, if it is empirically adequate. Strict or constructive empiricists should never assert that a theory is true, only that it is empirically adequate. But to say that a theory is empirically adequate is just to say that the phenomena are *as if* it were true. Yet another version of secular surrealism is involved in Kyle Stanford's abortive attempt to provide an antirealist explanation of the success of science. Stanford says that we need not invoke a theory's truth to explain its success, we can simply say that it makes the same predictions as the (a?) true theory would make. But to say that a theory **T** makes the same predictions as a true theory would make, is just a fancy way of saying that the observable phenomena are *as if* **T** were itself true.

Sunday School precepts of scientific method help account for the popularity of surrealism. People say that what matters in science is predictive power. Well, **T**, $\mathbf{T}^{\mathbf{G}}$ and **T*** have exactly the same predictive power. People say that what matters in science is that its theories are falsifiable by observation and experiment. Well, **T**, $\mathbf{T}^{\mathbf{G}}$ and **T*** are equally falsifiable by observation and experiment. Vulgar American pragmatists say that the 'cash value' of any theory is what it tells us about experience. Well, **T**, $\mathbf{T}^{\mathbf{G}}$ and **T*** have exactly the same 'cash value'. Innocent victims of the verifiability theory of meaning say that theories that are verifiable by the same experiences are really the same theory expressed in different words. Well, on that view, **T**, $\mathbf{T}^{\mathbf{G}}$ and **T*** are really the same theory expressed in different words.

I think this is all wrong. Surrealism is not the same as realism. **T**, $\mathbf{T}^{\mathbf{G}}$ and **T*** are not the same theory, nor are they equally good theories. The difference between them has to do, not with predicting or saving the phenomena, but with *explaining* them. We may assume that the theory **T** with which we began explains its phenomena, shows how they are brought about, says what causes them. But **T***, the secular surrealist transform of **T**, does not explain its phenomena. I might explain why the streets are wet by saying that it is raining. But I do not explain why the streets are wet by saying that they are *as if* it were raining. The ancients explained the (apparent) daily motions of the fixed stars by saying that they are all fixed on an invisible celestial sphere that rotates once a day on its axis about the earth located at its centre. But you do not explain the motions of the stars by saying that they move *as if* they were all fixed on an invisible sphere. So, if we are interested in explaining phenomena, not just in 'saving' them, we should prefer the explanatory theory **T**, whatever it is, to its non-explanatory surrealist transform **T***. Realists who value explanation will prefer **T** to **T***. It follows that, for realists anyway, there are explanatory virtues that are not evidential virtues. It follows that realists are not strict empiricists, and do not accept that *only* empirical evidence should determine theory-choice.

Here, as my last example makes clear, a terminological point intrudes. Did the ancients really explain the motions of the stars, given that their theory is false and that the celestial sphere they postulated does not exist? Does it make sense to speak of a false explanation? As I use the words, this does make sense. This does not imply, of course, that explanation has nothing to do with truth. All it means is that truth is an adequacy condition upon explanation, not a defining condition of it. It makes sense to speak of a 'false explanation', but it does not make sense to speak of a 'false yet adequate explanation'. Nothing hinges upon this terminological preference of mine. If you think that 'false explanation' makes no sense, then speak instead of 'false *putative* explanation'. My philosophical point still stands, but must be differently expressed. I must now say that we should prefer the *putative* explanatory theory **T**, whatever it is, to its non-explanatory surrealist transform **T***.

What about theological surrealism? This does explain phenomena, after a fashion, by saying that God produces or causes them. Better this supernatural explanation than no explanation at all. Many of Berkeley's readers thought God the weak link in his system, and removed Her from it. The result is *phenomenalism*, a secular surrealist metaphysic that has minds and their experiences but no explanation of why the experiences are as they are. If I had to choose between phenomenalism (secular surrealism) and Berkeley (theological surrealism), I would prefer the latter from an explanatory point of view. But we do not have to choose between surrealisms. There is a third alternative, realism about both common sense and science. Is realism better than theological surrealism, from an explanatory point of view? Are natural explanations better than an all-purpose supernatural one? And if so, why?

This is not an easy question to answer. Strict empiricists often invoke simplicity as a virtue as well as empirical adequacy. Duhem's slogan was that science aims to save the phenomena in the simplest possible way. But simplicity is not much help to realists. What is simplicity, for a start? Occam said that it required us not to multiply entities without necessity. But nothing is simpler than theological surrealism: only one entity is postulated as the cause of all observed phenomena. Occam's Razor seems to favour theological surrealism over realism. Besides, realists seek truth, and what has simplicity to do with truth? Even if naturalistic explanations could be made out to be simpler in some sense than a supernatural one, what does this tell us about their truth? How do realists know that Nature is simple?

Perhaps the difference has to do, not so much with the number of entities that we postulate, but with our ability to understand how these entities work. This seems more promising. For it is admitted on all sides that we poor humans cannot comprehend how divine causality operates. "God said 'Let there be light'—and there was light". Gee, thanks! God puts tree-experiences into our minds as if there were trees. How does She do it? We have no idea, and are not meant to have any idea. We are just meant to comfort ourselves with the thought that She can do anything.

The trouble with this response is that some naturalistic explanations are also difficult for us to understand. How do celestial spheres work and what are they made of? How does gravity work, or magnetism? How can light behave both as a particle and as a wave? Does anybody understand Quantum Mechanics, the best predictive theory we have ever had? Down the ages, strict empiricists have grown impatient with this demand for 'understanding' and claimed that saving the phenomena is all that really matters. Duhem famously made fun of Lord Kelvin, who sought mechanical models of electromagnetic action: "We thought we were entering the tranquil and neatly ordered abode of reason, but we find ourselves in a factory" [Duhem 1954, pp. 70–72]. Nowadays those who despair of understanding Quantum Mechanics opt for surrealism about it—it is called 'the Copenhagen Interpretation of Quantum Mechanics'. But that brings us full circle. If the unintelligibility of electromagnetism or of Quantum Mechanics does not matter, neither does the unintelligibility of Divine causation. Indeed, better one big mystery than a host of little ones.

But is that really so? Reflection on the little mysteries that naturalistic explanations throw up has sometimes led to better and better explanations—even if 'better' is understood here in strict empiricist terms. Scientists seem to have found out more about how natural causes work, without perhaps ever understanding them fully. There is no finding out more about how divine causality works—that is beyond our ken. Such empirical progress as there is in theological surrealism is entirely parasitic on the progress of science. Gosse said that God fixed things *as if* the geological theories of his day were true. But Gosse would not have minded if the geologists refuted those theories and replaced them with better ones. He would simply incorporate the improved geology into a new version of his surrealism. And neither the original theory nor its amended version tells us anything about how divine causality works.

How are we to explain the fact that science has been empirically successful, that scientists have improved their naturalistic explanations of the phenomena? (Notice that what we seek to explain here is not a fact about the world, but rather a fact about our intellectual dealings with the world.) Suppose we have a theory that is completely successful, that saves all its phenomena, that is empirically adequate. How come? The obvious realist explanation is that the theory is true. The secular surrealist has no explanation or only a circular one—saying that the phenomena are *as if* the theory were true is just saying that the theory is empirically adequate. The theological surrealist avoids circularity by postulating miraculous divine intervention and saying that God fixes it that the phenomena are *as if* the theory were true. Which brings me to the so-called 'Miracle Argument' for realism. Hilary Putnam's slogan was "Realism is the only philosophy that does not make the success of science a miracle".

4 Inference to the Best Explanation

The Miracle Argument for realism is a special case of inference to the best explanation, hereafter IBE for short. IBE is a pattern of argument that is ubiquitous in everyday life as well as in science. The Miracle Argument for realism is a special case of it, a meta-level case. What we seek to explain here is not a fact about the world, but rather a fact about our intellectual dealings with the world. The Miracle

Argument claims that the best explanation of a theory's empirical success is that it is true.

Before turning directly to this, how does IBE work in general and what does it show? Philosophers assume that IBE, and its intellectual ancestor Peirce's abduction, are non-deductive or inductive or 'abductive' arguments which are supposed to show that some theory is true. I think the philosophers are wrong on both counts.

Peirce's abduction is supposed to go like this:

> The surprising fact, C, is observed.
> But if A were true, C would be a matter of course.
> Hence, … A is true.
> [C. S. Peirce, *Collected Papers*, 1931–1958, Vol. 5, p. 189.]

This is deductively invalid. We can validate it if we view it as a deductive *enthymeme* and supply its missing premise, "Any explanation of a surprising fact is true". But this missing premise is obviously false. Nor is any comfort to be derived from weakening the missing premise (and the conclusion) by replacing 'true' with 'probably true' or with 'approximately true'—though philosophers have cottage industries devoted to both these projects. It is a surprising fact that marine fossils are found on mountain-tops. One explanation of this is that Martians came and put them there to surprise us. But this explanation is not true, or probably true, or approximately true. Nor should philosophers derive any comfort from saying that although the argument is deductively invalid, it has some sort of merit in some special inductive or abductive logic.

IBE attempts to improve upon abduction by requiring that the explanation is the best explanation we have. It goes like this:

> F is a fact.
> Hypothesis H explains F.
> No available competing hypothesis explains F as well as H does.
> Therefore, H is true.
> [Lycan 1985, p. 138]

This is better than abduction, but not much better. It is also deductively invalid. We can validate it if we view it as a deductive *enthymeme* and supply its missing premise, "The best available explanation of a (surprising) fact is true". But this missing premise is also obviously false. Nor, again, will going for probable truth or approximate truth or some fancy explanationist logic help matters.

There is a way to rescue abduction and IBE. Peirce provided the clue to this. Peirce's original abductive scheme was not quite what we have considered so far. Peirce's original scheme went like this:

The surprising fact, C, is observed.
But if A were true, C would be a matter of course.
Hence, *there is reason to suspect that* A is true.
[C. S. Peirce, *Collected Papers*, 1931–1958, Vol. 5, p. 189.]

This is invalid, but to repair it we need the missing premise '*There is reason to suspect that* any explanation of a surprising fact is true'. This missing premise is, I suggest, true. After all, the epistemic modifier 'There is reason to suspect that …' weakens the claims considerably. In particular, 'There is reason to suspect that A is true' may be true even though A is false. If the missing premise is true, then instances of the abductive scheme may be deductively sound. We have not traded obvious invalidity for equally obvious unsoundness.

IBE can be rescued in a similar way. I even suggest a stronger epistemic modifier than 'There is reason to suspect that …', namely 'There is reason to believe (tentatively) that …' or 'It is reasonable to believe (tentatively) that …'. What results, with the missing premise spelled out, is:

It is reasonable to believe that the best available explanation of any fact is true.
F is a fact.
Hypothesis H explains F.
No available competing hypothesis explains F as well as H does.
Therefore, *it is reasonable to believe that* H is true.

This scheme is valid and instances it of might well be sound. Inferences of this kind are employed in the common affairs of life, in detective stories, and in the sciences.

Of course, to establish that any such inference is sound, the 'explanationist' owes us an account of when a hypothesis explains a fact, and of when one hypothesis explains a fact better than another hypothesis does. If one hypothesis yields only a circular explanation and another does not, the latter is better than the former. If one hypothesis has been tested and refuted and another has not, the latter is better than the former. These are controversial issues, but they are not the most controversial issue. That concerns the major premise. Most philosophers think that the scheme is unsound because this major premise is false, whatever account we can give of explanation and of when one explanation is better than another. So let me assume that the explanationist can deliver on the promises just mentioned, and focus on this major objection.

It is objected that the best available explanation might be false. Quite so—and so what? It goes without saying that any explanation might be false, in the sense that it is not necessarily true. It is absurd to suppose that the only things we can reasonably believe are necessary truths.

But what if the best explanation of a fact not only might be false, but actually is false. Can it ever be reasonable to believe a falsehood? Of course it can. Van Fraassen has a homely example:

> I hear scratching in the wall, the patter of little feet at midnight, my cheese disappears—and I infer that a mouse has come to live with me. Not merely that these apparent signs of mousely presence will continue, not merely that all the observable phenomena will be as if there is a mouse, but that there really is a mouse. [1980: 19–20.]

Now the mouse explanation might be false, a mouse might not be responsible for the scratching, the patter of little feet, and the disappearing cheese. Still, it is reasonable for us to believe it, given that it is the best explanation we have of those phenomena. Of course, if we *find out* that the mouse explanation is false, then it is no longer reasonable for us to believe it. But what we find out is that what we believed was wrong, not that it was wrong or unreasonable for us to have believed it.

But this means, it is objected further, that being the best available explanation does not prove a hypothesis to be true. Quite so—and again, so what? The explanationist principle—"It is reasonable to believe that the best available explanation of any fact is true"—means that it is reasonable to believe or think true things that have not been proved to be true. Philosophers who think that it can only be reasonable to believe what has been proved will reject the explanationist principle. Such philosophers accept what I have called the 'justificationist principle', according to which a reason for believing something must be a reason for what is believed.

Explanationism, or the explanationist principle, stands opposed to justificationism, or to the justificationist principle. As I see it, the rejection of justificationism lies at the heart of Karl Popper's critical rationalism. And explanationism, as I understand it, is part and parcel of critical rationalism. But most philosophers see things differently, accept justificationism, and go in for the usual formulations of IBE, where the conclusion is that the best explanation is true (or probably true, or approximately true). Critics rightly object that such conclusions do not follow. They can be made to follow, but only by adding to the premises an absurd metaphysical principle like "The best available explanation of any fact is true (or probably true, or approximately true)". Most philosophers, instead of making these absurd metaphysical principles explicit, reply that abduction or IBE has some special merit in some special abductive or non-deductive logic. Critical rationalists reject justificationism, and have no need of non-deductive reasoning, either. IBE is for them a perfectly valid deductive form of reasoning, with a non-justificationist epistemic principle as its major premise. Thus critical rationalism and explanationism go hand-in-hand.

But do explanationism and realism go hand-in-hand as well? Why cannot constructive empiricists also accept IBE and put their own gloss upon it? When we consider van Fraassen's mouse hypothesis, truth and empirical adequacy coincide, since the mouse is an observable thing. But when it comes to hypotheses about unobservables, truth and empirical adequacy come apart. Why cannot the constructive empiricist also accept IBE, but only as licensing acceptance of the best available explanation as empirically adequate, not as true? As Howard Sankey puts it: "The question is why it is reasonable to accept the best explanation *as true*. Might it not be equally reasonable to accept the best explanation as empirically adequate …?" (2006, p. 118).

My answer to this question is NO. Suppose that H is the best explanation we have of some phenomena. Remember the Truth-scheme: It is true that H if and only if H. Given the Truth-scheme, to believe that H and to believe that H is true are the same.

Given the Truth-scheme, to accept that H and to accept that H is true are the same. So what is it to accept that H is empirically adequate? It is not to accept H, for this is the same as accepting that H is true. Rather, it is to accept a meta-claim about H, namely the meta-claim "H is empirically adequate" or equivalently "The observable phenomena are as if H were true". Call this meta-claim H*. Now, and crucially, H* is *no explanation at all of the phenomena*. The hypothesis that it is raining explains why the streets are wet—but "The phenomena are as if it were raining" does not. *Ergo*, H* is not the best explanation—H is, or so we assumed. (Actually, all we need assume here is that H is a better explanation than H*.) So given IBE, H* should not be accepted as true. That is, given IBE, H should not be accepted as empirically adequate.

I wonder which part of this argument those who believe that there is a constructive empiricist version of IBE will reject. Not IBE—at least, they are pretending to accept it. Not, presumably, the Truth-scheme. Not, presumably, its consequence, that to accept H and to accept H as true are the same thing. Not, presumably, the equivalence of "H is empirically adequate" and "The observable phenomena are as if H were true". Not, presumably, the claim that H is a better explanation of the phenomena than "The phenomena are as if H were true". What this shows is that realism and explanation go hand-in-hand. If you try to recast IBE in terms of empirical adequacy rather than truth, you end up with something incoherent. You start off thinking that it is reasonable to accept the best explanation as empirically adequate. And you end up accepting something that is no explanation at all.

This does not refute constructive empiricism. It only refutes the idea that constructive empiricists can traffic in explanation, and in IBE, just as realists do. It is no accident that down the ages acute anti-realists have pooh-poohed the idea that science explains things. Van Fraassen should join Duhem in this, as he already has in most other things.

5 The Miracle Argument

So much for explanation—and for IBE—in science. The same considerations apply to IBE in meta-science, to the so-called Miracle Argument for scientific realism. Here what is to be explained is not a fact about the world, like the scratching in the wall or the disappearing cheese. What is to be explained is a fact about science, the fact that science is successful. The success in question is predictive success, the ability of a theory to yield true predictions about the observable, and the technological success that often depends upon this. The key claim is that the best explanation of a theory's predictive success is that it is true. Given this claim, IBE licenses reasonable belief in the truth of that theory.

It is only *consistent* empirical success that can be explained in terms of truth. You cannot explain the partial success of a falsified theory in terms of its truth. (This is an important point, to which I shall return.) So the *explanandum* is of the form "*All* T's predictions about observable phenomena are true", or (putting it in van Fraassen's terminology) "T is empirically adequate", or (putting it in surrealist terminology, following Leplin 1993) "The observable phenomena are as if T were true". The realist thought is, that T's actually being true is the best explanation of why all the observable phenomena are as if it were true.

6 Is Any Theory Empirically Adequate?

Can we know for sure that any theory is empirically adequate? To accept a theory as empirically adequate and set out to explain why is to generalise beyond the available evidence, to make an 'inductive leap'. But it is no different with explanation in science itself. Scientific *explananda* are typically general. Scientists typically seek to explain general statements, rather than statements of particular fact. For example, they seek to explain why sticks look bent when half-immersed in water, not why my walking-stick looked bent last Thursday when I dipped it in the Leith. Some radical inductive sceptics deny that any general statement is true. The best reply to them is to ask how they know this (self-refuting) statement to be true. Other radical inductive sceptics deny that any general statement can be known for certain to be true. The best reply to them is to agree, but to insist that some general statements can be rationally adopted as true. Scientists who set out to explain why sticks look bent in water obviously suppose that sticks *do* look bent in water … *always*. And it is *reasonable* for them to suppose this, despite the fact that it might turn out to be mistaken, as the inductive sceptics rightly point out. As in science, so also in metascience. Metascientists can reasonably suppose that a theory is empirically adequate, and set out to explain why. If the metascientific explanatory project is to be rejected because it involves an 'inductive leap', then science's typical explanatory projects must be rejected on the same ground.

Even so, to claim that a theory is empirically adequate is to make a very strong claim. It is to claim that *all* the empirical regularities predicted by a theory are true. There are both vertical and horizontal 'inductive leaps' involved in such a claim. To say that any particular predicted empirical generalisation is true involves the vertical inductive leap from examined cases to all cases. To say that all the predicted empirical generalisations are true involves the horizontal inductive leap from the generalisations we happen to have tested to all of the generalisations. So, what is to be explained, empirical adequacy, is already epistemically problematic. But let us set this aside. After all, we are dealing here with antirealists who think empirical adequacy an epistemically respectable category, but who baulk at truth.

So, supposing that it makes sense to try to explain empirical adequacy, how exactly does truth do it? Suppose the theory in question asserts the existence of unobservable or theoretical entities. The theory will not be true unless these existence claims are true, unless the theoretical entities really exist, unless the theoretical terms really do refer to things. So part of the realist story is that T is observationally adequate because the unobservables it postulates really do exist.

But this cannot be the whole realist story. Reference may be a necessary condition for success, but it cannot be a sufficient condition. A theory may be referential yet false and unsuccessful (more on this later). The other part of the realist story is that what the theory says about the unobservables it postulates is true.

The Miracle Argument says, not just that truth explains empirical adequacy, but that it is the only explanation, or at least the best explanation. To evaluate this claim, we need to pit the realist explanation of success, in terms of successful reference and truth, against other possible antirealist explanations. What might such antirealist explanations be like? van Fraassen replaces truth by empirical adequacy as an aim for science. But it is obvious that we cannot satisfactorily explain the empirical adequacy of a theory in terms of its empirical adequacy:

T is empirically adequate.
Therefore, T is empirically adequate.

This explanation is no good because it is blatantly circular.

Other antirealist explanations are also circular, but not so blatantly circular as this one. Laudan replaces truth by problem-solving ability as an aim for science. As he explains, an empirical problem is posed by a question of the form "Why G?", where G is some empirical generalisation. So to say that a theory is a good empirical problem-solver is just to say that it yields lots of true empirical generalisations. So, when we unpack the definitions, what we have once again is an explanation of empirical adequacy in terms of empirical adequacy.

Then there is Jarrett Leplin's *surrealism*, which is short for 'surrogate realism' (Leplin 1993). Surrealism arises by taking some theory T and forming its surrealist transform T*: "The observed phenomena are *as if* T were true". It is clear that "The observed phenomena are as if T were true" is merely a fancy way of saying that T is empirically adequate. That being so, we cannot satisfactorily explain the empirical adequacy of T by invoking the surrealist transform of T. For that is, once again, explaining empirical adequacy just by invoking empirical adequacy.

7 Stanford's Antirealist Explanation of the Success of Science

Kyle Stanford's 'Antirealist explanation of the success of science' (Stanford 2000) does no better either. Jack Smart suggested long ago that the Copernican astronomer can explain the predictive success of Ptolemaic astronomy by showing that it generates the same predictions as the Copernican theory does and by assuming the truth of the Copernican theory. Smart wrote:

> Consider a man (in the sixteenth century) who is a realist about the Copernican hypothesis but instrumentalist about the Ptolemaic one. He can explain the instrumental usefulness of the Ptolemaic system of epicycles because he can prove that the Ptolemaic system can produce almost the same predictions about the apparent motions of the planets as does the

> Copernican hypothesis. Hence the assumption of the realist truth of the Copernican hypothesis explains the instrumental usefulness of the Ptolemaic one. Such an explanation of the instrumental usefulness of certain theories would not be possible if *all* theories were regarded as merely instrumental. [Smart (1968), p. 151.]

Now by the 'instrumental usefulness' of Ptolemaic astronomy, Smart obviously means its predictive success. So, his suggestion is that realists can explain the predictive success of a false theory in terms of its predictive similarity to the true theory. Stanford considers Smart's suggestion, and says of it:

> Notice that the actual content of the Copernican hypothesis plays no role whatsoever in the explanation we get of the success of the Ptolemaic system: what matters is simply that there is some true theoretical account of the domain in question and that the predictions of the Ptolemaic system are sufficiently close to the predictions made by that true theoretical account. [Stanford (2000), p. 274.]

This is quite wrong. The detailed content of the Copernican theory, and the fact that some of the detail of the Ptolemaic theory is similar to it, is essential to the explanation of the success of Ptolemaic theory. There are many examples to illustrate this—I will give only one. The periodic retrograde motions of the superior planets are explained in Copernican astronomy by the fact that the earth overtakes those planets as it makes its annual journey around the sun. In Ptolemaic astronomy the earth is stationary at the centre of the universe, and makes no annual journey around the sun. Yet Ptolemaic astronomy also correctly yields the periodic retrograde motions of the superior planets. How? In Ptolemaic astronomy retrograde motions are explained by assigning each planet an epicycle-deferent system as it rotates around the stationary earth. It predicts the retrograde motions of the superior planets correctly because the period of the epicycle assigned to each superior planet is one year. The annual motion of the earth in Copernican astronomy 'corresponds' to the annual periods of the epicyclic motions of the superior planets in Ptolemaic astronomy. This is why the two theories make the same predictions in the case. The reason why the predictions are correct is that the Copernican theory is true, the earth does take a year to circle the sun.

Stanford suggests that "it is the fact that the Ptolemaic system is predictively similar to the *true* theoretical account of the relevant domain that explains its usefulness, not that it is predictively similar to the *Copernican* hypothesis as such." (ibid, 275). This, again, is quite wrong. No explanation, or no good explanation, of Ptolemy's usefulness is to be had simply by saying that it makes the same predictions as the true theory does. For this is just to say "It is predictively as if Ptolemy's theory were true".

Stanford generalises this example into an antirealist explanation of success in general. The predictive success of any theory is to be explained by saying that the theory makes the same predictions as the true theory (whatever that is). But this is explaining "T is predictively successful" by saying "It is predictively as if T were true", or for short, "T is predictively successful". It is incredible that earlier in his paper (ibid, 268-9) Stanford accepts that we cannot satisfactorily explain empirical adequacy in terms of empirical adequacy, nor can we adequately explain it in the surrealist way. Yet what he ends up with is just a variant of the surrealist explanation.

Stanford says, in defence of his proposal, that unlike the realist, constructive empiricist and surrealist proposals, which all appeal to some relationship between the theory and the world to explain its success, his proposal "does not appeal to a relationship between a theory and the world at all; instead it appeals to a relationship of predictive similarity *between two theories*" (ibid, 276). This seems to be a double joke. First, there are not two theories here at all, there is one theory and an existential claim that there is some true theory somewhere predictively similar to it. Second, you hardly explain the success of T by saying "T is predictively similar to some other theory T*", for T* might be false and issue in false predictions. The truth of T*, whether T* is spelled out or just asserted to exist (as here), is essential to the explanation of T's success. The relation of T* to the world is essential, in other words. [Note that Stanford's proposal collapses into the realist proposal if we allow T* to be identical with T. For then we explain the success of T by saying that it is predictively similar to some true T*, namely T itself. Nothing in Stanford's presentation rules this out. In particular, predictive similarity is a reflexive relation, which every theory bears to itself. However, Stanford obviously wants his proposal to be a rival to the realist proposal, so we ought in charity to assume that T and T* are distinct theories.]

Stanford counts it a virtue of his proposal that it does not involve asserting the truth of any particular theory—all that is asserted is that *there is* some true theory T* predictively similar to T. It might be thought that, simply by invoking the truth of some unspecified theory or other, Stanford's proposal remains a realist proposal. Not so. I can satisfy Stanford by invoking the truth of the surrealist transform of T. But then I end up saying that it is the truth of "The phenomena are as if T were true" that explains T's success. I can also satisfy Stanford by invoking the truth of Berkeley's surrealist philosophy. It is the truth of "God creates experiences in our minds as if science were true" that explains why science is successful. Surrealist transforms are by design structurally similar to what they are transforms of. There is nothing realist about them. And, to repeat, Stanford previously conceded that the explanations of success they offer are no good.

In the Ptolemy-Copernicus case, the empirical success of a false theory (Ptolemy) is explained by invoking its similarity to a true theory (Copernicus). The similarity explains why the two theories make the same predictions—the truth of the second theory explains why the predictions of the first theory are true even though the first theory is false. The surrealist transform of Ptolemy's theory —"Observed planetary motions are *as if* Ptolemy's theory were true"—follows from Ptolemy's theory *and* from Copernicus's. Realists about Copernicus become surrealists about Ptolemy, in order to explain the empirical adequacy of Ptolemy. But Copernican realism, not Ptolemaic surrealism, is doing the explaining here. Copernicus tells us *why* the phenomena are as if Ptolemy were true.

The key premise of the Miracle Argument was that the truth of a theory is the best explanation of the empirical adequacy of that theory. So far, at least, that key premise seems to be correct. From which it follows, provided we accept IBE, that it is reasonable to believe that an empirically adequate theory is true. (Of course, this argument assumes a realist theory of truth, which makes of truth something more than empirical adequacy. If we go in for an 'empirical adequacy theory of truth', which collapses truth into empirical adequacy, then the Miracle Argument also collapses.)

There are two worries about the argument so far. The first is that it concerns an extreme case, that of empirically adequate theories. How common in science are these? I shall come back to this worry in the next section. The second worry is more subtle. We have assumed that truth explains empirical adequacy better than empirical adequacy does, because the latter 'explanation' is completely circular. Now normally, when we go for explanatory depth as opposed to circularity, we would like some independent evidence that the explanation is true. But there can be no independent evidence favouring an explanation in terms of truth against a (circular) explanation in terms of empirical adequacy. The realist explanation may tell us more than the antirealist explanation, but in the nature of the case there can be no evidence that the more it tells us is correct. My response to this is to bite the bullet: there are explanatory virtues that do not go hand-in-hand with evidential virtues. How could the two go hand-in-hand, when the explanatory rival is *by design* evidentially equivalent?

It should really be obvious that explanatory virtues do not always go hand-in-hand with evidential virtues. The ancients explained the motions of the fixed stars by saying that they were fixed on the surface of an invisible celestial sphere which rotates once a day around the central earth. Compare that hypothesis with its surrealist transform, the hypothesis that the stars move *as if* they were fixed to such a sphere. The realist hypothesis is explanatory, the surrealist hypothesis is not, despite the fact that the latter is expressly designed to be evidentially equivalent with the former. Similarly with the nineteenth-century geological theory of fossil formation, G, and Philip Gosse's surrealist transform G*: God created the universe in 4004 BC *as if* G were true. There are quite different explanations here, but no geological evidence can decide between them—it was not on evidential grounds that nineteenth-century thinkers rejected G* out of hand. Finally, and most generally, consider the realist explanation of the course of our experience proffered by common sense and science, R, with its Berkeleyan surrealist transform R*: God causes our experiences *as if* R were true. Again, no experience can decide between R and R*, since R* is expressly designed to be experientially equivalent with R.

These examples are meant to show that realists should not be browbeaten by the fact that antirealists can come up with alternative hypotheses to the realist ones which empirical evidence cannot exclude. These alternatives can be excluded on explanatory grounds. Either they provide no explanations at all, or only incredible ones. It is the same with antirealist explanations of the success of scientific theories in terms of their empirical adequacy (however precisely formulated). Such explanations are either no explanations at all, or completely inadequate circular ones, and can be rejected as such. This, if accepted, only shows that the realist explanation of science's success is better than some antirealist ones. Perhaps there is another antirealist explanation that we have not yet considered? And in any case, how good an argument for realism is it, that the truth of a theory best explains its empirical adequacy?

8 Laudan's Historical Critique

Larry Laudan is the foremost critic of the miracle argument on historical grounds. He points out, first of all, that the global claim that science is successful is a hopeless exaggeration. Many scientific theories are spectacularly *un*successful. We must confine ourselves to successful theories, rather than to science as a whole. But even among successful theories, they are many that enjoy some success, but are not completely successful. What this means is that a theory yields some true observational consequences and some false ones, saves some regularities in the phenomena but gets others wrong. Now assuming that the scientists involved have made no logical or experimental error, and assuming that the false predictions have actually been tested, a partially successful theory of this kind has been *falsified.* No sensible realist can invoke the truth of a falsified theory to explain its partial success! As we saw, it is only *consistent* empirical success, or empirical adequacy, that can be explained in terms of truth.

Laudan's historical objection to scientific realism consists mainly in producing examples of theories that were successful yet neither referential nor true. But most of Laudan's historical counterexamples fall away, once we realise that they are examples only of *partially* successful theories, theories that were successful for a while but later turned out to be false (and in some cases, non-referential). No realist ever invoked truth to explain the partial success of a falsified theory. The Miracle Argument concerns only a very special case, the total predictive success of an empirically adequate theory. As we have already seen, the realist is right that truth (and reference) is the best explanation of empirical adequacy.

Actually, it is even worse. The chief target of Laudan's famous 'confutation of convergent realism' (Laudan 1981) is what we might call 'referential realism', the idea that "reference explains success". To be fair to Laudan, this was the view that one could glean from incautious formulations to be found chiefly in Putnam's writings. It is a view which has spawned what I have called 'entity realism', the idea that realists need not believe in the truth or near-truth of any theories, that it is enough just to believe in the theoretical entities postulated by those theories. It is not for nothing that Laudan attributes to the realist claims like "A theory whose central terms genuinely refer will be a successful theory". And he proceeds to refute this claim by giving examples of referential theories that were not successful, and of successful theories that were not referential.

Referential or entity realism is a hopeless form of realism. There is no getting away from truth, at least for realists. To believe in an entity, while believing nothing else about that entity, is to believe nothing or next to nothing. I tell you that I believe in hobgoblins. "So", you say, "You think there are little people who creep into houses at night and do the housework". To which I reply that I do not believe that, or anything else about what hobgoblins do or what they are like—I just believe *in* them. It is clear, I think, that the bare belief in hobgoblins—or equivalently, the bare belief that the term 'hobgoblin' genuinely refers—can explain nothing. It is equally clear, I think, that mere successful reference of its theoretical terms cannot explain the success of a theory. Laudan has an excellent argument to prove the point. Take a successful theory whose terms refer, and negate some of its claims, thereby producing a referential theory that will be unsuccessful. "George Bush is fat, blonde, eloquent and atheistic" refers to Bush all right, but would not be much good at predicting Bush-phenomena.

The ink spilled on reference is not wasted ink. That is because reference is (typically) a necessary condition for truth. A theory which asserts the existence of an entity will not be true unless that entity exists. But reference is not a sufficient condition for truth. A theory can be referential, yet false—and referential, yet quite unsuccessful. Laudan exploits the fact that truth requires reference—and adds that near-truth requires reference as well. He produces examples of non-referring theories that were successful, and argues that since they were non-referential theories they were neither true nor nearly true. What chance, then, of explaining success in terms of truth and reference?

But no sensible realist ever explained partial success in terms of truth and reference. Laudan produces no example of a consistently successful or empirically adequate theory that was (we think) neither true nor referential. The Miracle argument, as we have considered it so far, refers only to the special case of empirically adequate theories.

But can the realist take comfort from this rejoinder? Laudan might now object that the realist has jumped out of the frying pan into the fire. Empirical adequacy is an extreme case, rare, perhaps even non-existent, in the history of science. Most, perhaps all, theories in the history of science enjoy, at best, only partial success. It is the sum of these partial successes that phrases like "the success of science" refer to. Now if the realist is only going to invoke truth to explain empirical adequacy, partial success is left unexplained. And, since the "success of science" is a collection of partial successes, the success of science is left unexplained as well.

There is, moreover, a further, very obvious antirealist question. If partial success is explicable at all, it must be explicable in terms other than truth. So why can we not explain total success in terms other than truth as well? I shall defend an obvious realist response to this: just as total success is best explained in terms of truth, so also partial success is best explained in terms of partial truth.

9 Partial Truth Versus Verisimilitude

Return to the Ptolemy-Copernicus case. So far we have assumed that Ptolemaic astronomy was empirically adequate, and Copernican astronomy true. Of course, neither assumption is strictly correct. What is really the case is that Ptolemy's explanation of retrograde motions shared a true part with Copernican theory. That true part, common to both theories, sets out the relative motions of earth, sun and superior planets.

Partial truth is not the same as *verisimilitude*. Verisimilitude is closeness to the truth—the 'whole truth'—of a false theory taken as a whole. Partial truth is just truth of parts. A simple example will make the difference clear. "All swans are white" is false, because of the black swans in Australasia. (I had to get this baby example in—as some uncharitable soul once joked, having black swans in it is Australasia's chief contribution to the philosophy of science!) Despite its falsity, "All swans are white" is predictively successful in Europe, and bird-watchers find it useful to employ it there. I do not know how close to the (whole) truth "All swans are white" is, and none of the captains of the verisimilitude industry can tell me in less than 100 pages of complicated formulas. I do know that "All swans are white" has a true part (a true consequence) "All European swans are white", whose simple truth explains the success European bird-watchers have.

The simple example with the swans can be generalised. A false theory T might be successful (issue nothing but true predictions) *in a certain domain D*. Explain this, not by saying that T is close to the truth, but by saying that "In domain D, T" is true. A false theory T might be successful (issue nothing but true predictions) *when certain special conditions C are satisfied*. Explain this, not by saying that T is close to the truth, but by saying that "Under conditions C, T" is true. A false theory T might be successful *as a limiting case*. Explain this, not by saying that T is close to the truth, but by saying that "In the limit, T" is true. Notice that "In domain D, T" and "Under conditions C, T" and "In the limit, T" are all logical *parts* of T, that is, logical consequences of T. Of course, the conjunction S of the successes of T is also a logical consequence of T. But while S does not (satisfactorily) explain S itself, "In domain D, T" or "Under conditions C, T" or "In the limit, T" might explain S perfectly well. These restricted versions of T are not the same as its surrealist transform—restricted versions of T may be explanatory while its surrealist transform is not.

Of course, if we accept such an explanation, it immediately raises the question of *why* the restricted version of T is true while T is false. Typically, it is the successor theory to T that tells us that T is true in a certain domain, or under certain special conditions, or as a limiting case. Still, that this further question can be asked and answered does not alter the fact that a true restricted version of T can explain T's partial success while T's surrealist transform does not.

It is the same with *approximate truth*, as when we say that "It is 4 o'clock" or "John is 6 feet tall" are only *approximately true*. What we mean is that "It is *approximately* 4 o'clock" or "John is *approximately* 6 feet tall" are true. And if we want to be more precise, we can say that "It is 4 o'clock give or take 5 min" or "John is 6 feet tall give or take an inch" are true. Approximate truth is not to be explained by trying to measure the distance of a sentence from the (whole) truth. Approximate truth is truth of an approximation. Approximate truth is a species of partial truth, since the approximations in question are logical parts of what we began with. "It is 4 o'clock" logically implies "It is approximately 4 o'clock" as well as "It is 4 o'clock give or take 5 min", and "John is 6 feet tall" logically implies "John is approximately 6 feet tall" as well as "John is 6 feet tall give or take an inch".

I have come to believe that the entire verisimilitude project was a bad and unnecessary idea. Popper's definition of the notion of 'closeness to the (whole) truth' did not work. The plethora of alternative definitions of 'distance from the (whole) truth' that have taken its place are problematic in all kinds of ways. And what was the point of the verisimilitude project? Precisely to explain how a false theory can have partial success. Now it is obvious that a true theory will be successful—after all, true premises yield true conclusions. But it is not obvious that a theory which is close to the truth will be successful, since near-truths yield falsehoods as well as truths. We should eschew the near-truth of false wholes, in favour of the simple truth of their parts. We should explain partial success in terms of truth of parts. Whole truths are wholly successful, partial truths partially successful. Either way, it is simple truth, not verisimilitude, that is doing the explaining.

10 Novelty and Why Some Success Is not Surprising

I am not saying that partial success can always be explained by partial truth in this way, nor am I saying that it need be so explained. There is a kind of partial predictive success that needs no explanation at all, because it is no 'miracle' at all—it is not even mildly surprising! Here is a simple schematic example to illustrate what I mean. Suppose a scientist has the hunch that one measurable quantity P might depend linearly on another measurable quantity Q—or perhaps the scientist does not even have this hunch, but just wants to try a linear relationship first, to see if it will work. So she measures two pairs of values of the quantities P and Q. Suppose that when Q is 0, P is 3, and when Q is 1, P is 10. She then plots these as points on a graph, and draws a straight line through them representing the linear relationship. She has performed a trivial *deduction*:

$P = aQ + b$, for some a and b.
When Q is 0, P is 3 (so that $b = 3$).
When Q is 1, P is 10 (so that $a = 7$).
Therefore $P = 7Q + 3$.

Now, the point to notice is that the hypothesis $P = 7Q + 3$ successfully predicts, or 'postdicts', or at least entails that when Q is 0, P is 3, and that when Q is 1, P is 10. Are these successes miraculous, or even mildly surprising? Of course not. Those facts were used to construct the hypothesis (they were premises in the deductive argument that led to the hypothesis). It is no surprise or miracle that the hypothesis gets these things right—they were used to get the hypothesis in the first place. This trivial example illustrates a general point. Success in predicting, or post-dicting, or entailing facts used to construct a theory is no surprise. It is only *novel* predictive success that is surprising, where an observed fact is novel for a theory when it was not used to construct it.

Finally, a realist can say that accidents happen, some of them lucky accidents, in science as well as in everyday life. Even when a fact is not used to construct a theory, that theory might successfully predict that fact by lucky accident. It is not my claim that the correct explanation of predictive success is *always* in terms of truth or partial truth. My claim is that the best explanation of total predictive success is truth, and that the best explanation of partial predictive success (where it is not a lucky accident) is partial truth.

Nancy Cartwright argues that the predictive success of science is *always* a kind of lucky accident. It always arises from what Bishop Berkeley called the 'compensation of errors'. According to Cartwright, the laws or theories in science are always false (I shall come back to this). But scientists busy themselves to find other premises which, when combined with these false laws, will generate true predictions. And, scientists being clever folk, it is no wonder that they succeed. A trivial example may make the point clear. Suppose the 'phenomenological law' we want is "Humans are two-legged', and the false law of nature we have to work with is "Dogs are two-legged". What do we have to add to the false law to get the phenomenological law? Well, the auxiliary hypothesis "Humans are dogs" will do the trick. And two wrongs, carefully adjusted to each other, make a right.

Bishop Berkeley complained that the mathematicians of his day were only able to get correct results in their calculations because they systematically made mistakes that cancelled one another out. Berkeley observed that there was nothing so scandalous as this in the reasoning of theologians. Cartwright thinks the scandal is endemic in the reasoning of physicists: "Adjustments are made where literal correctness does not matter very much in order to get the correct effects where we want them; and very often … one distortion is put right by another" (Cartwright 1983, p. 140).

Now in a case like this, one would be crazy to suppose that the best explanation of the theory's predictive success is its truth. The success is accidental, from a logical point of view. Of course, the success is no accident at all from a heuristic point of view. It is, in fact, a variant of a case with which we are already familiar. We use a known fact ("Humans are two-legged" in the trivial example), and a false theory we have ("Dogs are two-legged", in the trivial example), to generate an auxiliary theory ("Humans are dogs", in the trivial example) that will get us back to the known fact. It is no miracle that we get out what we put in. And our success in getting it is no argument for the truth of what we get it from.

Why does Cartwright think that the laws of physics lie, that is, are always false? The laws lie, she thinks, because they idealize or simplify things—they are false because they do not tell the *whole* truth. This is a mistake. "Nancy Cartwright is clever" is not false, just because it does not tell the whole truth about Nancy Cartwright. Similarly, Newton's law of gravity is not false just because it does not tell the whole truth about the forces of nature.

Never mind this. The important point is that predictive success is no miracle if the predicted facts are used to construct the theory in the first place. What is miraculous is novel predictive success. And the best explanation of such 'miracles' is truth, either truth of wholes or truth of parts.

References

Cartwright, N. 1983. *How the Laws of Physics Lie*. Oxford University Press.

Duhem, P. 1954. *The Aim and Structure of Physical Theory*. Trans. E. Doland and C. Maschler. Chicago and London: University of Chicago Press.

Gosse, P. 1857. *Omphalos: An Attempt to Untie the Geological Knot*. London: J. Van Voorst.

Gosse, E. 1907. *Father and Son*. London: William Heinemann (Harmondsworth: Penguin Books 1963).

Laudan, L. 1981. 'A Confutation of Convergent Realism', *Philosophy of Science,* 48, 19–49.

Leplin, J. 1993: 'Surrealism'. *Mind* 97: 519–524.

Lycan, W. 1985. 'Epistemic Value'. *Synthese* 64: 137–164.

Pierce, C. S. 1931–58. *The Collected Papers of Charles Sanders Peirce*, ed. C. Hartshone and P. Weiss. Cambridge: Harvard UP, mass.

Sankey, H. 2006. Why is it Rational to Believe Scientific Theories are True? In *Rationality and Reality: Conversations with Alan Musgrave*, ed. C. Cheyne and J. Worrall, 109–132. Dordrecht: Springer.

Santilla, G. 1955. *The Crime of Galileo*. Mercury Books, London: William Heinemann, (1961).

Smart, J. 1968. *Between Science and Philosophy: An Introduction to the Philosophy of Science.* Random House.

Stanford, K. 2000. 'An Antirealist Explanation of the Success of Science', *Philosophy of Science,* 67: 226–284.

Van Fraassen, B.C. 1980. *The Scientific Image*. Oxford: Clarendon Press.

Misdirection and Misconception in the Scientific Realism Debates

Bas C. van Fraassen

Abstract The scientific realism debates have been plagued by misrepresentations of both realist and empiricist positions, sometimes by their adherents as well as by their critics. When positions are presented as contraries, there must be an isolatable question to which each gives its answer, in opposition to the other. Since philosophy does not provide a way to answer factual questions about the world, that common question must be about the character of science and scientific practice, rather than about what there is. Once what is at issue has been clarified, realists and empiricists can cooperate on an inquiry into what science is, what the criteria of adequacy are in scientific practice, and what epistemic or doxastic attitudes toward scientific theories are within the bounds of reason. In this inquiry, the writings of Weyl, Glymour, and Suppe provide an excellent guide. Just what is scientific realism, and what are its contraries? Despite, or perhaps because of, the many formulations of such positions that are found in the literature, it is not as easy to answer this as it might seem.

1 Do Electrons Exist? that is not the Question

Do electrons exist? Are atoms real? These are not philosophical questions.

Whether electrons exist is no more a philosophical question than whether Norwegians exist, or witches, or immaterial intelligences. Questions of existence are questions about matters of brute fact, if any are, and philosophy is no arbiter of fact.[1]

[1]In my usage of such terms as "exist" or "real" I follow Quine's seminal article "On what there. is". So 'Xs exist' and 'Xs are real' I understand as meaning simply that there are Xs.

B.C. van Fraassen (✉)
Department of Philosophy, San Francisco State University, California, USA
e-mail: fraassen@sfsu.edu

E. Agazzi (ed.), *Varieties of Scientific Realism*,
DOI 10.1007/978-3-319-51608-0_5

That is how I see it. But whether any such questions are within the scope of philosophy is itself, just as almost any other question about philosophy, something on which philosophers disagree.

Specifically, philosophers who classify themselves as scientific realists often present beliefs in the truth of certain scientific theories, theories that postulate the existence of unobservable objects or processes, as part of their philosophical position.[2] That is puzzling, and I shall give reasons to think that this rests in a confusion about what is at issue in the scientific realism debates.

Scientific realists' inclusion of beliefs in science, when they appear, tend to be conjoined with something still more troubling: the insinuation that empiricist alternatives to scientific realism approach science skepticism. That insinuation is completely at odds with the history of empiricism, which has at every stage begun with the conviction that scientific practice is a paradigm of rational inquiry. Empiricists' skepticism toward the claims made for science in other philosophical traditions, and especially for claims involved in various forms of metaphysics purporting to extend the sciences, is not skepticism toward the empirical sciences themselves. There is an admirable degree of passion in the scientific realism debates, as is to be expected in such fundamental philosophical controversies. It is disturbing, however, to have the impression, sometimes, that some of the passion displayed derives from a scarcely hidden conviction that only scientific realism respects the achievements of the sciences, or even that it is, so to speak, secularly impious not to profess emphatically one's belief in the reality of currently discussed unobservable, scientifically postulated entities.

To free the discussion from such bedevilment, let us consider anew the question what scientific realism is (although it has received so many answers), to see if we can clearly and distinctly separate it from any disputes about whether atoms, electrons, or other theoretically postulated entities are real.

2 What is Scientific Realism?

The upshot of my opening was meant to be this: whatever scientific realism is, if it is a philosophical position, it does not include such claims as that electrons, or other entities postulated in scientific theories, are real.

And similarly, whatever may be an empiricist position, contrary to scientific realism, if it is a philosophical position, it cannot include the claim that such entities do not exist. On questions of existence, as on all questions of brute fact, both sorts of philosophical positions must be mute.

My own view, constructive empiricism, may well have other eye-brow raising shortcomings, but it does not include any claim, whether positive or negative, about

---

[2] I owe this point and a list of references to substantiate it to Anjan Chakravartty. A dialogue between us on this topic will be forthcoming in the journal *Spontaneous Generations.*

whether the sciences are successful in their endeavors. I take it as common knowledge that the sciences are indeed successful by many important criteria, both practical and intellectual, and that science provides us with a paradigm example of rational inquiry. I also take it as commonly known among scientists that they are as yet unsuccessful in some respects. Scientists address questions about what there is in the natural world and what it is like, but these are matters aside from the basic question at issue in philosophy: *What Is Science?*

Remarkably, though, there are on the contrary scientific realists who present their position as centering on claims about what there is in the natural world and what it is like. These claims associate the philosophical position with concurrently accepted scientific theories, which the scientific avant-garde may well be questioning but the philosopher asserts to be true.[3]

Rather than quibbling with realists in general or paint them all with the same brush, let us focus on careful, thoughtful scientific realist philosophers who present this aspect of the position very guardedly, in a form so general that it is subject to less audacious interpretation. David Papineau begins his discussion with a characterization of realism, for any putative body of knowledge, as required

> to involve the conjunction of two theses: (1) an independence thesis: our judgments answer for their truth to a world which exists independently of our awareness of it; (2) a knowledge thesis: by and large, we can know which of these theses are true. (Papineau 1996, 2).

Taking this very strictly, this would mean that a scientific realist today would say, about the statement that electrons exist, (1) that it answers for its truth to a world existing independently of our awareness of it, and (2) that we can know whether or not it is true.

Together these do not imply that electrons exist, only that we can know, that it is possible to know, whether or not they do. So this position of scientific realism could be held by someone who also holds either (a) that electrons do not exist or (b) that we (can but) do not know whether or not they exist.

It is useful here to recall Peter Forrest's terms, scientific gnostic (someone who believes that our currently accepted scientific theories are true) and scientific agnostic (someone who suspends belief as well as disbelief with respect to currently accepted scientific theories). So here is a point of logic, which will not escape the careful reader: on Papineau's understanding at least, the scientific gnostic/agnostic distinction cuts across the scientific realist/anti-realist distinction. There can be philosophers of all four sorts: scientific realists who are scientific gnostics as well as ones who are scientific agnostics, and of course philosophers who are not scientific realists but are scientific gnostics as well as ones who are scientific agnostics.

Allow me to make a proposal, in two parts.

First of all, let us consider only philosophical positions that agree on Papineau's condition 1, his independence thesis. However broadly or narrowly this is construed,

[3]The assertion may be qualified, ranging from a minimal way by just inserting "approximately" to the extreme of limiting it to certain kinds of structural information. That does not affect the main point.

I take it that it is in accord with Michael Dummett's proposed usage for "realist": to take a realist position on some topic is to hold that a certain associated discourse has 'objective' truth conditions. In that sense (I would like to use the term "semantic realism" for this) all positions to be considered here agree, versions of scientific realism as well as their contraries.

Even adding Papineau's condition 2 we do not arrive at a substantive view of what science is. Papineau's conditions merely narrow the classification of science to an inquiry into objective fact, whose results can be known to be true–which applies equally to, for example, investigative journalism about politics.

So, what is to be included to formulate scientific realism, beyond this semantic realism? And what is to be included in an empiricist position, a rejection of scientific realism, beyond this semantic realism?

As second part, then, I would like to propose again, that it should be, in each case, a distinctive answer to the question What Is Science?, and that this should be an answer which purports to identify the aim of scientific inquiry, in the sense in which stating the aim amounts to specifying its most basic criterion of success.

Needless to say, that question needs spelling out before we will have enough clarity to see what will count as a relevant answer. I will return to this below, and to end I will lay out some recently discussed positive proposals for the investigation of what criteria are actually in force in scientific theory and practice. But I would like to first clear some more of the muddied waters that presently surround these scientific realism debates.

3 Straw Men and Windmills

There have certainly been ample well-targeted critiques of constructive empiricism, which had to be taken very seriously since its formulation in 1980. But there are also disconcertingly many attacks on various 'empiricisms', apparently designed for the purpose, whose humiliating defeats could insinuate much beyond their literal meaning.

For clear examples of critiques of empiricism biased by a tailor-made definition I can refer here to David Papineau and Alan Musgrave.

3.1 David Papineau's Empiricist

An example that may have reached many classrooms is in Papineau's introduction to his collection The Philosophy of Science:

> According to van Fraassen's 'constructive empiricism' … we ought never to believe in the truth of any theory which goes beyond the observable phenomena. (Papineau 1996, 8).

That "ought" may be true of some conceivable empiricist philosophy of science, but I don't know of any actually professed, and it is certainly not true of constructive empiricism.

Papineau's collection includes the article "To Save the Phenomena", in which the wording is very precise, and never gives leeway to such an implication about what we ought to believe:

> anti-realism is a position according to which the aims of science can well be served without giving … a literally true story, and acceptance of a theory may properly involve something less … than belief that it is true. (van Fraassen 1976, 623; Papineau 1996, 82)

This careful formulation respects the distinction between what is involved in acceptance of a theory and what someone who accepts a theory may believe apart from that. That is surely not a distinction difficult for a philosopher to appreciate. Whoever accepts a theory will have many opinions and beliefs that go beyond what is involved in that acceptance, and this philosophical position has no implications for what may or may not be included in those additional opinions and beliefs.

In The Scientific Image this distinction is emphatically observed as well:

> Science aims to give us theories which are empirically adequate; and acceptance of a theory involves as belief only that it is empirically adequate. This is the statement of the anti-realist position I advocate; I shall call it constructive empiricism. (van Fraassen 1980, 12).

Much is deliberately left open, including that it is not by any lights irrational to believe in the truth of a scientific theory. The italicized assertion implies only that the criterion of success in science, and hence the belief involved in acceptance of a theory as successful, does not demand so much:

> I do not advocate agnosticism about the unobservable, but claim that belief is supererogatory as far as science is concerned; you may if you like, but there is no need. (van Fraassen 2007, 343).

Empiricist leanings will incline one to believe less rather than more, that is true. But if anyone were to have the position that "we ought never to believe in the truth of any theory which goes beyond the observable phenomena", as Papineau puts it, what could be the reason for that? The only reason there could be is that the constraints of rationality require that disbelief! But what constraint could that be, in the case of a theory consistent with all that we know about the phenomena? The idea is entirely at odds with the liberal form of epistemology that I espouse: rationality is but bridled irrationality.[4]

---

[4]The first time I argued this was in a reply to Hilary Putnam in Florence 1978: "secondo me, *la razionalità è solo irrazionalità imbrigliata*" (Piatelli Palmerini 1984, 110). I elaborated on this later in analogy with Justice Oliver Wendel Holmes' distinction between Prussian law and Anglo-Saxon law (van Fraassen 1989, 171–172).

3.2 *Alan Musgrave's Empiricist*

Leaving epistemology aside now, for a moment, there are also questions about what is involved in theory acceptance other than belief. For any empiricist philosophy of science that must be a crucial question for scientific practice, and the empirical inquiry that it involves, is clearly theory-driven.

Alan Musgrave is much more careful than Papineau, and in his paper "Strict Empiricism versus Explanation in Science" he is careful to define his target 'strict or hardline empiricism':

> What is strict or hardline empiricism? It is the view that only empirical evidence should determine theory-choice. There is nothing to choose between theories that empirical evidence cannot choose between. [...] Strict empiricists will say not only that there is nothing to choose between them, but also that they equally well serve the aim of science. For according to strict empiricism, ..., "the name of the scientific game is saving the phenomena" ....[5]

What I have omitted in this paragraph includes "also known these days as 'constructive empiricism'". Rather than cavil about this again, let me address Musgrave's implication that on an empiricist view, theory-acceptance is reduced to a reliance on the evidence at hand. As I see it, Musgrave is quite right that any such view would be woefully inadequate.

Science aims to give us theories which are empirically adequate; and acceptance of a theory involves as belief only that it is empirically adequate (van Fraassen 1980, 12). But what does acceptance involve beyond belief? A practicing scientist who accepts a theory (what Musgrave refers to as theory choice) involves himself in a research program, s/he is committed to meeting the phenomena within the theoretical framework in question, and this program could be much different if an alternative, though empirically equivalent, theory were accepted (ibid.).[6] So clearly, it is not only empirical evidence that determines theory-choice, and it is hard to imagine any philosopher naïve enough to think so.

Musgrave's strict empiricist is not a good philosopher of science because s/he ignores the pragmatic dimension of theory acceptance. Dennis Dieks' contribution to the same conference provides us with examples of current interest.[7] Anyone who had quantum mechanics in their formative education years was immersed in a plebeian form of the Copenhagen interpretation. That includes those who, for

---

[5]Musgrave, this volume pp. 79–94. Originally presented at the Conference on "Scientific Realism: Objectivity and Truth in Science", A Coruña, September 2015. The phrase in quotation marks is a quotation from van Fraassen 1980, 93, which explains why "even the anti-realist ... will counsel the search for explanation".

[6]Given how much Musgrave values explanation, the remainder of that page may interest him as well. The conceptual resources of the theory determine "the terms in which we shall seek explanations. If the acceptance is at all strong, it is exhibited in the person's assumption of the role of explainer ....".

[7]Dennis Dieks, "Realism and Objectivity in Quantum Mechanics", presented at the Conference on "Scientific Realism: Objectivity and Truth in Science", A Coruña, September 2015 pp. 295–314.

example, turned later to actively developing Bohmian mechanics or the GRW version of quantum mechanics, and have thus immersed themselves in a new conceptual framework in which to engage the phenomena–new phenomena that the experimentalists are meanwhile creating in fabulous new ways.

Much as Musgrave values explanation, he does not, in the context of this critique of empiricism, explore the pragmatic dimensions of our explanatory practices, nor ask just what explanation is good for, or what purposes it may serve. It seems to me that this might well be an area in which scientific realists and constructive empiricists could cooperate, in order to reach a better understanding of science in practice.[8]

There can clearly be conceptual and pragmatic differences between theories that are empirically equivalent, and these can play an important role in theory choice. When a practicing scientist accepts a theory, the stakes (personal, communal, intellectual) can be great, there is a gamble here on the future of the scientific discipline. So it is not surprising if Musgrave finds that there will be differences, important to science and in practice, between a given theory T and a theory T* which is true if and only if T is empirically adequate. That will typically be so. (Only typically![9]).

Musgrave's strict empiricist would be embarrassed by this. That is probably why there aren't any strict empiricists.

4 Realists' Retreat on Inference to the Best Explanation

As I understand it, what is at issue in the scientific realism debates is not a question about rational management of opinion and belief—the main topic of epistemology today—but the question what science is.

However, it is not surprising that so much in those debates has focused on issues in epistemology. For the debate to start at all, it has to make sense to think that in response to the empirical evidence one could have a choice between believing that the theory is true or believing only that the theory is empirically adequate. Any position in epistemology which entails that, given the evidence, there is a unique rationally compelled conclusion, would leave nothing more to be said.

The main entry of epistemology into the scientific realism debates was the rule of Inference to the Best Explanation (IBE). There were various formulations of course, but they had in common precisely that under suitable conditions, the question of what to believe on the basis of available evidence would have a unique answer. That answer would be 'the best explanation of the facts adduced in that evidence'.

---

[8]What explanation is good for has hardly been touched by philosophers; in contrast there is an illuminating psychological study by Gopnik (1998).

[9]If T is only about observable phenomena and does not include postulation of unobservable entities, then T is true if and only if T is empirically adequate. Musgrave would presumably agree that the phenomena can sometimes be explained without postulating unobservable entities.

I place this in scare quotes, because each of the main terms in this phrase beg for explication (and thereby hangs a tale …).

Not only in philosophy of science but elsewhere this rule received much attention. Is it a good and rational policy to apply this rule when managing your beliefs and opinions? David Armstrong gave short shrift to this question, offering the following truly irrefutable argument:

> If making such an inference is not rational, what is? (Armstrong 1983, 53).To infer to the best explanation is part of what it is to be rational. If that is not rational, what is? (ibid. p. 59).

This rule, IBE, was the subject of much criticism, and here I think it is appropriate to defer to Alan Musgrave, and his acute, insightful critique of what various scientific realists have written about the 'rule' of IBE. He agrees with the empiricist critique that it cannot be a matter of inferring truth from explanatory success, but he offers in effect the following moderate form:

It is reasonable to believe that the best available explanation of any fact is true.
F is a fact.

Hypothesis H explains F.
No available competing hypothesis explains F as well as H does.
Therefore, *it is reasonable to believe that* H is true.[10]

So let us call this the Musgrave version of IBE, "MIBE" for short. While Musgrave feels that this version of IBE needs a defense, namely a defense of its first premise, I have no such worry. The premise seems to me eminently reasonable. (Since "reasonable" is a normative term, the question of truth does not arise.)

But this form of argument has no bite unless it can be accompanied by another conclusion:

(*) Therefore, it is unreasonable not to believe that H is true.

To mention just the application that Musgrave obviously has in mind, there is no empiricist objection to be made to a full belief that our best scientific theories are true. (At least in principle, perhaps not currently in practice: arguably, we are currently in a situation where our best theories in fundamental physics do not form a coherent whole.) But for the constructive empiricist position, such a belief is reasonable enough, but supererogatory. Against this, MIBE without (*) is powerless.

That Musgrave did not notice the lacuna is clear from the argument in which he ostensibly applies his rule. I'll quote the relevant part:

> Suppose that H is the best explanation we have of some phenomena. […]So what is it to accept that H is empirically adequate? It is … to accept a meta-claim about H, namely the meta-claim "H is empirically adequate" or equivalently "The observable phenomena are as if H were true". Call this meta-claim H*. Now, and crucially, H* is *no explanation at all of the phenomena*. The hypothesis H that it is raining explains why the streets are wet–but "The phenomena are as if it were raining" does not. *Ergo*, H* is not the best explanation–H is, or so we assumed. (Actually, all we need assume here is that H is a better explanation

[10]Musgrave, this volume, Sect. 3.2.1. Italics in the original.

than H*.) So given IBE, H* should not be accepted as true. That is, given IBE, H should not be accepted as empirically adequate.[11]

Why conclude that H should be accepted as true?[12] Because it is the best explanation, and H* does not dislodge it from that status. But MIBE allows only to conclude that it is reasonable to believe that H is true. Without the strengthening by (*), MIBE does not rule out that it is also reasonable to suspend belief about H, and to believe only the logically weaker H*.

Is there an argument, perhaps along lines pointed to by Musgrave, to strengthen MIBE with (*)? I see no reason to think so: at this point the only sort of consideration that could help would have to be ones to establish Lipton's (2004) contention that the lovelier is likelier to be true. But Lipton's admirable efforts were ultimately unsuccessful (cf. van Fraassen 2005).

5 How Should We Approach the Question *What is Science?*

There are activities whose criteria of success are public and explicit. That is so for games with standard rules, and with building projects under contract, and of course for political campaigns which need to end with a clear winner. But when we turn from specific, local, clearly specified activities of this sort what counts as success tends to be vague and subject to ambiguity or controversy, often enough at odds with what is publically avowed as the aim (think of the Iraq war!).

What about science? That is a very broad, and not clearly defined, area of human enterprise. Specifically, attempts at demarcation, at drawing lines between what is (genuine) science and what is not have been famously unsuccessful. Yet we have on the whole a fairly confident sense of our ability to refer to, discuss, admire, and argue over science and the sciences. Even if the concept of science is a cluster concept, it seems to be accepted that its central activities include the construction of theories and models, and that questions of success and failure are by and large settled within the relevant scientific communities themselves.

But this is just where philosophical battle lines are drawn: what are the criteria of success and failure? Is there a 'bottom line', a basic criterion of success?

Scientific realism, in its various versions, tends to focus on truth, or on criteria that involve truth, such as true explanation, or accurate representation. Empiricism, in its various versions, tends to focus on truth 'within our ken', empirical adequacy.

[11]Musgrave, this volume, Sect. 3.2.1. Italics in the original.

[12]In Musgrave's terminology, as he emphasizes in this section, that "to accept H and to accept H as true are the same thing". In my terminology, the word "accept" does not have this meaning (cf. van Fraassen 1980, 8; 12). What Musgrave calls the Truth-scheme I take to imply only that the hypothesis H and the hypothesis that H is true imply each other, that is, each is true if and only if the other is true.

These are contrasting assertions about what counts, at bottom, as success in science. It is not assertions, and variations thereon, that are lacking, but ways to judge these assertions.

There is a serious, relevant inquiry that has not been taken on to any great extent as yet:

How can we determine what are the criteria of adequacy in a given human activity?

And specifically, how can we assess such a claim as that in (current, or Western, or …) scientific practice the 'bottom line' criterion of adequacy is empirical success rather than truth overall, or the other way around?

Careful examination and analysis of science in practice is what is needed here, but the analysis needs to pursue a question that is not peculiar to any specific episode in scientific practice.

While I hope that this problem will be taken by the horns in new research, I can for the moment point to two inquiries which seem relevant.

The first is Frederick Suppe's paper on credentialing, and the second is the attempt by Clark Glymour to follow Hermann Weyl's version of 'co-ordination', which I have tried to re-codify as empirical grounding.

5.1 Suppe on Credentialing

Fredrick Suppe launched a sustained effort to separate two senses of "confirmation" that had been conflated in the philosophical literature: belief formation and evidential support.

The topic of belief formation, based on evidence and reasons of all sorts, is the traditional subject of the supposed logic of induction or abduction and Bayesian confirmation theory. Evidential support is what we find in actual scientific publications when data, and auxiliary information, are submitted to back up scientific claims. To the traditional eye, the two must be the same, and Suppe's radical assertion, on the basis of a thorough examination of the form of articles in scientific technical journals, is that they are not the same at all.

What is significant especially in Suppe's writing on this topic is that it presents a quite new characterization of what is sought and pursued as success in the scientific reports of data, experimental and observational findings, and the conclusions based thereon. The criterion for new scientific claims to satisfy is that they should come with satisfactory credentials to enter the body of scientifically accepted information and theory.

It is unfortunate that Suppe presents his case enmeshed in much extraneous pleading, argument, and debate (for example, with social constructivists). But the details of his case can be gleaned from three lengthy papers (Suppe 1993, 1997, 1998). Specific analyses of technical literature and episodes in the recent history of science are presented there, to show how the actual narrative and argumentative structure differs from any of the schemata presented in traditional and formal epistemology.

The process of credentialing actually found is basically a 'mine sweeping' operation to remove obstacles and possible objections or doubts that could reasonably be raised against the claims submitted by the study's authors. The typical paper reporting experimental studies:

- presents the reduced data or results of the experiment
- details the relevance of the experiment and its results for the target scientific community
- provides details concerning the experimental set-up, the apparatus, and the circumstances of the experiment that would be needed to replicate or evaluate the study
- provides an interpretation of the reduced data which yields the specific experimental claims
- marshals evidence to anticipate and remove specific doubts that the data could have been due to interference, inadequate statistical analysis, or be 'artifacts of measurement'
- marshals further arguments to rebut possible alternative interpretations of the data.

The refutation of envisaged alternatives does not, of course, imply the truth of the claim unless the range of alternatives is logically exhaustive, which is not possible in practice. But it does establish the credentials that the claims in question must have to be candidates for acceptance within the target scientific community. Suppes remarks:

> To the extent that there are no unrebutted objections or competing interpretations, the typical form of the interpretative argument is "Φ because not-A_1 and… and not-A_n." …. This feature of science proves troublesome for most epistemologies and accounts of scientific knowledge. It is most readily accommodated by divorcing belief-formation processes from the evidential basis for knowledge, denying the probative force of such arguments in scientific articles, and relegating them instead to belief formation processes. (Suppe 1997, 392–393)

Note well that a major part of this credentialing consists in specifying the empirical details needed for actual physical replication of the experiment, with the reduced data deriving directly from the physical end-state of the measuring apparatus used in the experiment. As Suppe points out, this "credentialing is always done against the background of a discipline's shared domain, background beliefs, presuppositions, and evidential standards or canons of reasoning" (Suppe 1993, 161). That background includes a good deal of theory accepted prior to the study in question, so the credentialing process is performed within a theoretical context. But that replication of the experiment, itself an observable phenomenon, will generate the same concrete data as outcome, is not therefore a self-fulfilling prophecy.

5.2 Weyl, Glymour, and Empirical Grounding

Hermann Weyl detailed, in a different way, quite clearly what the link between a scientific theory and relevant measurement is meant to be, if that theory is to be both testable and applicable in practice. What he outlined made concrete the relation of theory to experience that Schlick and Reichenbach discussed under the heading of "coordination".[13]

When first introduced, a model or theory may involve theoretically postulated physical quantities for which there is as yet no measurement procedure available. To that extent then they are as yet uncoordinated with any possible deliverances of experience, and there is no way to apply calculations involving those quantities to observable phenomena.

This possibility is well illustrated by the advent of the atomic theory in the early 19th century, for example. The masses of the atoms or molecules, or their mass ratios, played a significant part in the models offered for chemical processes, but could not be determined from the measurement data. During that century the theory was developed, various hypotheses were added beginning with Avogadro's, and slowly it became possible to connect theoretical quantities to measurable ones. Such development, simultaneously strengthening the theory and introducing new measurement procedures, is not adventitious or optional: it is a fundamental demand on the empirical sciences.[14]

To make that demand concrete, Weyl laid down two conditions to be met by a theory with respect to relevant measurement procedures.[15] There are three parts to it, the first two laid out by Weyl and the third by Glymour (Weyl 1927/1963, 121–22; Glymour 1975, 1980). They are:

Determinability: any theoretically significant parameter must be such that there are conditions under which its value can be determined on the basis of measurement.

---

[13]Schlick's concept of truth as unique coordination influenced how Einstein as well as Reichenbach wrote about theory and experience. That concept is, in retrospect, not easy to understand, but we may tentatively take it as a version of what later came to be called the verification principle. Objections to that principle do not apply to Weyl's conditions on an empirical theory, which are not presented as truth conditions (cf. Howard1984; Howard 1996, 126—127; van Fraassen 2008, 115—124.)

[14]This point has often appeared in the scientific and philosophical literature as demands to "operationalize" theoretical concepts, sometimes in polemics against rival theoretical approaches to a common domain–e.g. between advocates of the atomic theory and those advocating energetics, or between behaviorist and cognitive psychology. Such demands fell into disrepute among philosophers because they typically included the presumption that perfectly theory-neutral evidence could be had, or even that theoretical concepts could be reduced to operational ones. But at heart, and however imperfectly, those demands reflect norms operative in scientific practice.

[15]It should be noted here that what counts as a measurement is itself to a significant extent theory relative (cf. van Fraassen 2012, 2014).

Concordance, which has two aspects:

Theory-Relativity: this determination can, may, and generally must be made on the basis of the theoretically posited connections
Uniqueness: the quantities must be 'uniquely coordinated', there needs to be concordance in the values thus determined by different means.

Refutability, which is also relative to the theory itself:

there must be an alternative possible outcome for the same measurements that would have refuted the hypothesis on the basis of the same theoretically posited connections.

6 Conclusion

Attention to misdirection and misconception can at best be a prolegomenon to what is important: to find a positive and constructive way forward. Polemics are amusing, but the work to be done challenges all of us, on every side of the scientific realism debates, equally.

Philosophers who enter upon a reflection on the sciences come with a great diversity of worldviews, ontologies, temperaments, and epistemic as well as other values. Perhaps all of the discussants today are thoroughly immersed in an intellectual context significantly shaped by the sciences themselves. In principle at least they have in the theory and practice of science a meeting ground, a place where cooperation is possible. The analysis of the criteria and norms that are in force in scientific practice, and through these an understanding of what that practice is, and of the place it can have in a rational appreciation of ourselves in our world, can thus be a cooperative effort.

References

Armstrong, David M. (1983) *What Is A Law Of Nature?* Cambridge: Cambridge University Press.
Dieks, Dennis (2015) "Realism and Objectivity in Quantum Mechanics" pp. 295–314.
Glymour, Clark. 1975. "Relevant Evidence." *Journal of Philosophy* 72, 403–26.
Glymour, Clark. 1980. *Theory and Evidence*. Princeton, NJ: Princeton University Press.
Gopnik, Alison (1998) "Explanation as Orgasm". *Minds and Machines* 8, 101–118.
Howard, Don (1996) "Relativity, *Eindeutigkeit*, and Monomorphism: Rudolf Carnap and the Development of the Categoricity Concept in Formal Semantics", pp. 115–164 in Giere, Ronald N. and Alan W. Richardson (eds.) *Origins of Logical Empiricism*. Minnesota Studies in the Philosophy of Science XVI. Minneapolis: University of Minneapolis Press, 1996.
Howard, Don (1984). "Realism and Conventionalism in Einstein's Philosophy of Science: The Einstein-Schlick Correspondence." *Philosophia Naturalis* 21: 618–629.

Lipton, Peter (2004) *Inference to the Best Explanation.* Second Edition. London: Routledge.
Musgrave, Alan "Strict Empiricism versus Explanation in Science" pp. 71–94.
Papineau, David (1996) *The Philosophy of Science.* Oxford: Oxford University Press.
Piatelli Palmerini, Massimo (1984) *Livelli di Realtà.* Milan: Feltrinelli Editore.
Suppe, Frederick (1993) "Credentialing Scientific Claims", *Perspectives on Science* 1, 153–203.
Suppe, Frederick (1997) "Science Without Induction", pp. 386–429 in John Earman and John D. Norton, *Cosmos of Science: Philosophical Problems of the Internal and External Worlds.* Pittsburgh: University of Pittsburgh Press, 1997.
Suppe, Frederick (1998) "The Structure of a Scientific Paper", *Philosophy of Science* 65, 381–405.
van Fraassen, Bas C. (1976) "To Save The Phenomena", *Journal of Philosophy* 73: 623–632.
van Fraassen, Bas C. (1980) *The Scientific Image.* Oxford: Oxford University Press.
van Fraassen, Bas C. (1989) *Laws and Symmetry.* Oxford: Oxford University Press.
van Fraassen, Bas C. (2005) *"Wouldn't It Be Lovely? Explanation and Scientific Realism", Review Symposium for Lipton, Inference to the Best Explanation*, 2nd ed. *Metascience*14: 344–352.
van Fraassen, Bas C. (2007) "Reply: From a view of science to a new empiricism", pp. 337–383 in Monton, Bradley (ed.), *Images of Empiricism: Essays on Science and Stances, with a Reply from Bas van Fraassen.* Oxford: Oxford University Press, 2007.
van Fraassen, Bas C. (2008) *Scientific Representation: Paradoxes of Perspective.* Oxford: Oxford University Press.
van Fraassen, Bas C. (2012) "Modeling and measurement: the criterion of empirical grounding", *Philosophy of Science* 79: 773–784.
van Fraassen, Bas C. (2014) "The Criterion of Empirical Grounding in the Sciences", pp. 79–100 in Gonzalez, W. J. (ed), *Bas van Fraassen's Approach to Representation and Models in Science.* Dordrecht: Springer Verlag.
Weyl, Hermann (1927/1963) *Philosophy of Mathematics and Natural Science.* NewYork: Atheneum.

Selective Scientific Realism: Representation, Objectivity and Truth

Michel Ghins

Abstract In this paper I advocate a version of selective epistemological realism. I begin with analyzing the conditions in which a scientific model successfully and correctly represents an identified target. I stress that the realistic import of models rests on the truth of some predicative statements. I then examine the notions of objectivity and truth in order to be able to assess the reasons to believe in the existence of some unobserved objects posited by our best scientific theories and in the truth of some assertions about the properties that these objects possess. I distinguish between the properties that are observable in principle by means of instruments which enhance our perceptive capacities (the OP properties) and the properties that are beyond any possible observation by us, namely the properties which are purely theoretical (the PT properties), such as charm and strangeness in elementary particle physics. The OP properties are identical or similar to the observed properties of ordinary perceived things, such as velocity, volume and, admittedly more controversially, charge and mass. I propose four stringent requirements for rationally believing that an unobserved object posited by a theory possesses a specific property. Firstly, this property must be an OP property. Secondly, it must be measurable. Thirdly, it must play a causal role in producing the observed data. Fourthly, distinct independent methods for measuring this property must deliver concordant results. I then show that the generality and acceptability of these four criteria is grounded on a parallelism with the reasons we adduce for (rightly) believing in the existence of ordinary observable things which we don't immediately perceive such as mice, in some circumstances. However, an agnostic attitude is to be recommended with respect to the possession of PT properties by an object posited by an—even successful—scientific theory.

M. Ghins (✉)
Université Catholique de Louvain, Louvain-la-Neuve, Belgium
e-mail: michel.ghins@uclouvain.be

E. Agazzi (ed.), *Varieties of Scientific Realism*,
DOI 10.1007/978-3-319-51608-0_6

1 Representation

According to a still popular belief, science represents the world and is an image of reality. Certainly, the model-theoretic view of theories, according to which theories are foremost classes of models, makes such a view more plausible than the statement view, for which theories are sets of sentences or better, propositions (which are the semantic contents of sentences) which are asserted, i.e. statements or assertions. Indeed, models are structures, namely sets of elements organized by some relations, just as images are. An image is a visual two-dimensional object composed of colored patches—its elements—which stand in spatial relations. Thus, a scientific model is a structured set which can be meant to *represent* something else —a target—which I will call its *referent*.[1]

In the last decades, representation has become a trendy topic in the philosophy of science, due to the increasing influence of philosophers such as Bas van Fraassen who promoted the model-theoretic view of theories. Such a view is the heir of the *representational conception of knowledge* which accompanied the birth of modern science. We all know that both 17th century rationalists like Descartes and empiricists such as Locke construed knowledge as a faithful correspondence between our mental representations—our ideas—with reality. However, the representational conception of knowledge is confronted with a serious problem, called the "bridge problem" (Agazzi 2014, 29). Surely, we are immediately present to our ideas and we can safely investigate the properties of the idea of triangle without worrying about its possible correspondence with real entities. But how can we justify the belief that some of our ideas or representations faithfully mirror—at least in some respects—something external existing independently of our minds? This is the *idealistic predicament* which plagues the representational conception of knowledge. No fully satisfactory solution has been offered to this ill-posed problem (Agazzi 2014, 245) so far and it is doubtful that such a solution will ever be found. Simply because I'm unable to ascent to an overarching or divine point of view from which I could contemplate my representation and at the same time enjoy a direct non-representational—access to a real external entity in order to verify that my representation correctly represents such reality.

The anti-psychological turn at the beginning of the 20th century did little towards the solution of the bridge problem. Extracting ideas[2] out of the human mind and endowing them with flesh and blood in the external world does little to solve the problem. Granted, doing so localizes representations in representing artifacts—such as scientific models—which are accessible to a community of human subjects just as physical, as opposed to mental, images are. By this move,

[1]For more details on this see van Fraassen (2008) and Ghins (2010, 2016b).

[2]Mental ideas, whatever their status, exist as mental entities and can be the target of scientific investigation.

representations escape the privacy of our minds and do acquire an objective status, in the sense of intersubjectivity. Human subjects can immediately see a colored two-dimensional entity, and even come to an agreement in admitting that it is a portrait of someone in an appropriate context. But to resolve the bridge problem, we have first to muster arguments in favor of the belief that the person we aim at representing by the portrait is unambiguously identified and exists. In other words, we have to argue in favor of the existence of the particular target or the referent of the portrait. Furthermore, one must provide good reasons to believe that the portrait is faithful (or not) to its referent, always in some respects only (because a portrait is not a mere duplicate of its target). Such arguments and reasons cannot be gained from an examination of the internal properties of a possibly representing artifact, because the success in representing a target by means of a specific artifact depends upon factors which are *external* to it. No entity possesses internal properties that make it automatically the representor of a particular target, essentially because representing always implies the appropriation of an artifact by a user.

As van Fraassen rightly stresses, to represent is to act. To represent involves someone who represents (a user or subject S), a representing artifact (representor) R, something that is represented (a target T) in some context C. As in any action, we can succeed (or fail) in representing the target that we intend to represent. Success (or failure) can obtain with respect to three distinct aims: *identifying the target, representing it as such and such, and partially representing it in a correct way.*[3]

In representing, we must beforehand unambiguously identify our intended target, provisionally bracketing at this first stage the issue of its existence. The entity which we decide to employ as a representational artifact has some properties that we can ascertain. But this fact alone does not provide any clue for identifying the target. Anything, whether natural or artificial, can be used by someone to represent any target. The user must initially stipulate which properties of the representor are relevant for his purpose. These are the properties which are meant to convey some information about the intended target. In general, the cultural context suffices to determine the relevant properties since it contains some specifications or conventions implicitly agreed upon by a community. Cultural contexts are of course external to the artifact itself. Moreover, these contexts widely vary in space and time. The physical characteristics of maritime maps for example considerably differ among cultures. To find their way at sea, Micronesians use sticks and shells bounded with ropes whereas we use sheets of papers covered with patches and lines of different colors, forms and widths. When an entity is used to represent, it becomes a construct, our artificially fabricated representor. But this does not prevent our construct from actually possessing some definite properties, which are real and internal to them, whether they have been selected as relevant or even added by us. Only some, not all, of the properties of the representor are considered to be *relevant* for our representing action. In fact, only a few properties of the representor

---

[3]More on this in Ghins (2016b).

play a representational role. These are the properties which the user makes to correspond to the properties of interest in the intended target.

Identification of a target thus depends on the correspondence, established by the user, between some properties of the representor, which are considered to be pertinent, and certain properties of the intended target. Such a correspondence fixes a *code*. The stipulated correspondence or the code is a matter of convention. However, the actual possession of specific properties by the representor and the intended target, especially if the latter exists, is *not* conventional.

Arrived at this point, I wish to introduce a distinction between two ways of proceeding when we perform representing actions. First, we could start from a given perceived entity, which is immediately and unambiguously identified, such as a pipe, and attempt to represent it in a certain way. Or, and this is the second way, we can construct a representor, which functions as an unambiguous representation of an intended target, such as a unicorn, and inquire about the existence of unicorns. Surely, representation as such is directed: we use representors to represent targets, and not the other way around. In fact, within the representational conception of knowledge the users proceed by starting from their representors. Then, they ponder the reasons to believe that they represent some entities, which may or may not exist. Such a manner of proceeding is characteristic of modern times. Descartes asked himself if some ideas in his mind had internal characteristics which immediately pointed to a definite target and at the same time warranted its existence as well as the conformity of the idea with its target. He believed that the idea of an entity having all perfections implied the existence of God and also the conformity of this idea to what it represents. No external argument was needed to ground the success and adequacy in representing God by the mental idea of a perfect entity. Kant (rightly) disagreed and criticized this a priori proof of the existence of God. But he also embraced the representational view of knowledge and forcefully defended what he called the Copernican revolution according to which the phenomenal content given in space and time, which are the forms of human sensibility, together with the categories innate to the human subject constitute the objects of knowledge. According to this view, the represented object and its representor are made to coincide and the issue whether the constituted object corresponds to something external—a "thing in itself"–becomes at the very least insoluble and perhaps even meaningless.

After representors have been expelled from human minds and became entities in the world, the success and adequacy of a representing action not only acquired the status of meaningful (and interesting) issues, but must be assessed on the basis of considerations which do not solely belong to the representors and their intended targets. In scientific theorizing about unobserved objects, at the beginning of the representational *demarche,* I (the user) have to start from a potential representor to identify its possible target, and not the other way around, in the same way as the modern thinkers started from their internal representations and not from things in the world. Thus, the identification of a target of a representation can only be achieved by taking my representor as an initial point of departure. Then, I select in it some relevant properties corresponding to some properties of the intended target, irrespective of its existence, as in the example of unicorns. After having tacitly or explicitly specified the code, in order to convince myself of the existence of the target, I must obviously somehow verify that there is something in the world which possesses some relevant properties. For example, to revisit one of van Fraassen's examples, I must have reasons to believe that Bismarck did have a moustache if this property has conventionally been associated with a patch of a certain form and color in a picture meant to be a representation of *him*, and not of someone else.[4]

Once the target has been identified, the second aim pursued when we perform representations is to represent the target *as* having certain properties, which are distinct from the properties used for its identification. For clarity, it is convenient to distinguish in the representing artifact the properties A relevant for the *identification* of its target from the properties B that we consider to be relevant to *convey interesting information* about the target. Again, success in this respect can be achieved whether the intended target exists or not and whether it possesses the properties B or not. Some codified internal properties of the representor, such as distortion of bodily features in a portrait, can be used to succeed in representing Bismarck *as* vainglorious. In Spott's caricature, its referent is identified by means of pictorial elements internal to the representor associated with some properties of the intended target such as having a moustache, being bald etc. Additionally, Bismarck is represented *as* vainglorious since in the caricature his body is partially distorted in such a way that some of its parts resemble certain features possessed by peacocks. In our culture, peacocks are conventional symbols of vanity. Success in representing Bismarck as vainglorious is thus achieved independently of his being actually so (of course, he was…).

So far, we remained confined to the domain of our representations. Evidently, some ingredients *external* to our representors, such as conventions and selection of relevant properties, had to be mobilized in order to achieve success in identifying an intended target and in representing it as such and such. But the goal of *correctly* representing a target has not been sufficiently addressed yet. Obviously, the issue of correctness only arises for existing targets which may have properties corresponding to some properties of our representors according to the established code. Thus, we must have good reasons to believe that the intended target exists before raising the faithfulness issue. Such reasons cannot be extracted from the representor and the conventions adopted by the users only.

In the case of perceptual entities, actual perception can do the trick. We see a pipe which we compare with its representor, which could be Magritte's famous painting *La trahison des images* http://collections.lacma.org/node/239578. The sentence written on the canvas *Ceci n'est pas une pipe* makes clear that the existence of the pipe does not follow from the existence of the painting, since the word "ceci" refers to the painted pipe and not to a "real" pipe. But we can perceptually ascertain the existence of a pipe and verify that the brown area in the painting

[4]See van Fraassen (2008, 14) and Ghins (2010, 525) for a discussion of this particular example.

corresponds to a feature possessed by the real pipe which we see, e.g. the property of being brown.

In so doing we attribute a property to the pipe in an *act of predication*. In fact, acts of predication have been present, although tacitly, throughout the entire process which initiated with the identification of the referent and ended with the verification of the partial (i.e. in some respects) correctness of the representor to its target. To identify a target, I had to select or insert in the representing artifact some characteristics and assume that they do belong to the artifact. Further, I had to select or insert in the artifact elements which allow me to employ the artifact as a representor of the target in some of its respects. Given this, I (or another user) is able to represent the target *as* having certain properties. Irrespectively of its possible existence, those properties are predicated of the target or referent. Thereafter, an existing perceived entity is identified as a pipe, that is, as having the property of being a pipe. Finally, the faithfulness of the painting with respect to a specific feature of the target, its color for instance, is confirmed by observation and expressed in the act of predication of the property of being brown to both the real pipe and the painted pipe, provided we conventionally assume that the brown expanse in the painting corresponds to the property of being brown for the pipe (we could have decided that the brown expanse corresponds to being pink in reality…).

An act of predication consists in attributing a property to an entity. Such an act is equivalent to asserting that an entity possesses a property, by means of a judgement, a statement or an assertion. The word "judgement" has traditionally been associated with a judging subject. Nowadays, the terms "statement" and "assertion" are taken to be more neutral, or impersonal, in the sense that the truth of a statement or assertion transcends the particular idiosyncrasies of the judging subject. We will come back to the problem of objectivity below. As for now, it must be emphasized that the entire business of representing depends on the success of acts of predication and further, on the truth of statements.

As we saw, the user must specify at least implicitly a correspondence between some properties of the representor and some properties of the target. The faithfulness of a representor to its target in some respects cannot be conveyed by some feature intrinsic to the representor. Suppose a specific mark, such as a cross or a green dot, in the representing artifact is to signify that it is faithful to its target in some respect. Here, we inserted a property of the representor which does not correspond to a property of its target, and which consequently does not play a representational role. Success in representing presupposes a correspondence, a matching, between the representor and its target. What is called a *representative function* (Da Costa and French 2003, 49) has to be constructed by the user. Such a function must be an isomorphism or a homomorphism[5] which preserves the form,

[5]A homomorphism is a function which preserves the form or the structure but is not necessarily bijective. As a particular case of homomorphism, an isomorphism is a one-one correspondence between two sets. See Suppes (2002, 56).

or the structure, common to the representor and its target. An established *structural similarity* is a *necessary* (not sufficient) condition for the success of representation. If we add a mark to the painting of the pipe to indicate that it correctly represents existing pipes in some respects, we have constructed *another* representing artifact and the problem of correctness has been pushed one step back (van Fraassen 2008, 31; Ghins 2010, 527) .

In the example of the painting of a pipe (without Magritte's sentence), it seems that the bridge problem has been solved. On the one hand, we have a representing entity, the painting, and on the other, a represented entity, the pipe. Both are immediately perceptible and we can compare them to assess the partial correctness of the painting to its referent. The idealistic predicament has been dissolved by objectifying mental ideas into real perceivable entities and thereby making representors and their targets inhabitants of the same perceptual world in which the relevant comparisons can be performed.

This is an illusion. The initial modern defenders of the representational conception of knowledge claimed that we have direct cognitive grasp of our *mental* ideas or representations *only*. Thus, they deny that representing artifacts and perceived targets can be immediately known. The replacement of mental ideas by worldly representing artifacts has not been achieved at all. The comparison between the representor and its target is nothing else than a comparison between two mental representations. There is no warrant whatsoever that we have hit upon external real entities.

Even if we avoid any recourse to mental representations and defend some version of direct realism about paintings, pipes etc., we must concede that the cognitive content does not belong to representing artifacts per se but pertains to facts which are extrinsic to them among which the truth of some predicative assertions certainly is the most important. Representors are not the primary vectors of knowledge. What we originally know is that some perceived entities, representors and targets, possess some properties because we are immediately perceptually present to them. Such knowledge provides the ground for the success of our representing activity. I insist that the predicative statements which express this knowledge do *not* trade on representation. To assert that an entity possesses a specific property is not to say that the property (adequately?) represents the entity *albeit* partially and much less that the predicate term represents that entity as having such property (Ghins 2010 533; van Fraassen 2010, 553).

In science we often deal with theoretical models which are meant to represent targets which lie beyond our possibility of perceiving them. In such instances, the perceptual comparison between the model and its purported target cannot be carried out. The epistemological debate of scientific realism hinges on the cogency of the reasons which the realist can adduce in favor of belief in the existence of unobservable referents or targets. But before addressing this issue, a discussion of objectivity is in order.

2 Objectivity

In his latest book Agazzi (2014) introduces an illuminating distinction between *things* and *objects*. So far, I mostly used the generic and encompassing term "entity" to refer to things, objects, events, processes etc. A *thing* is an entity in the world which possesses a large, perhaps infinite, number of properties. An *object*, according to Agazzi, is a structured set of properties (Agazzi 2014, 284). Properties are *encoded*—a term borrowed from Zalta (1988), *albeit* in a different sense[6]—by the object. By abstraction, we select some properties of *things* and thereby construct *objects* which are "clipped out of things" (Agazzi 2014, 89). An object can be an abstract object, or an object of thought only, a *noema*. But, in a predicative statement, we attribute, correctly or wrongly, some property to a thing. For Agazzi, when an object is instantiated by a thing, the thing does possess all the properties which make up the concrete object.

"Scientific objects exist as abstract objects (i.e. as intellectual constructions) that encode certain properties, while not being purely abstract since they are exemplified (within certain margins of accuracy) by concretely existing objects" (Agazzi 2014, p. 104, n. 48)

But a thing cannot be reduced to an object, since a thing also has properties which do not belong to the object it instantiates. Moreover, the instantiation of an object by a thing need not be exact in the sense that a property of an object might only approximate a property of a thing. An object is thus constructed by adopting a certain point of view, some perspective on a thing. An object cannot encode all the properties of a thing. The predicative statements that attribute the properties of an object to a thing only convey a *partial* knowledge of the latter.

When we attempt to represent some observed targets, we similarly clip out some relevant aspects of those targets and connect them with some properties of our representors. In doing so, *what we represent are objects* and *not* things.[7] In what follows, to clarify the notion of objectivity, my starting point will be our perceptual presence to things. Thus, I will not conduct the discussion by beginning from our representing artifacts as I mostly did above. Contrary to Kant and the exponents of the representational conception of knowledge, I adopt a direct realism with respect to ordinary perceived things. Such a direct or common sense realism is also embraced by an empiricist such as van Fraassen (2008, 3) although he significantly refers to "observable phenomena" instead of things.[8] From these phenomena,

[6]Zalta's abstract objects are individuals which both encode *and* exemplify properties. They aren't Agazzi's abstract objects. For Zalta, when a concrete object has all the properties encoded by the abstract object, it is a physical "correlate" of the abstract object (See Ghins 2016a).

[7]Although the targets of our representations are things, we represent only some aspects of them, that is, structures or objects (see Ghins 2010).

[8]The word «phenomenon» inevitably suggests something that appears to me or us, whereas the word "thing" immediately makes us think of something in the external world.

we extract what van Fraassen calls "appearances" (2008, 8). Appearances present some analogy with Agazzi's objects since they are constructed by adopting a certain *perspective* on observable phenomena.

In what sense can my representations be objective? When users agree that a representor adequately or correctly represents its target in some respects, we get objectivity in the 'weak' sense of intersubjectivity (Agazzi 2014, 51ff.). But such agreement does not guarantee per se that there is something in the world that instantiates the relevant properties, i.e. the aspects selected when constructing the object of the representation. Intersubjective agreement fails to substantiate the claim that there is a thing "out there" possessing properties which conform to those encoded by the object of the representation. Such conformity is what Agazzi calls 'strong' objectivity. Surely, I concede that we can collectively be wrong in attributing properties to things. I thus endorse fallibilism.

Now, van Fraassen is quite aware of the importance of reaching strong objectivity. However, he subscribes to a representational conception of knowledge, albeit not in the classical, psychological, sense which took representations to be mental ideas. Then, he must propose a solution of the bridge problem, which he baptizes "the loss of reality objection" according to which our representors, our scientific models, do not hit on something external.

"*How can an abstract entity, such as a mathematical structure, represent something that is not abstract*, something in nature?" (van Fraassen 2008, 240)

The answer he gives is *pragmatic* and amounts to a "dissolution" of the loss of reality objection. When I claim that my representation is adequate to the phenomenon I also claim that my representation is adequate to the phenomenon *as represented by me* (van Fraassen 2008, 253–261). I cannot assert that my graph is adequate to the evolution of the deer population in Princeton as represented by me and at the same time deny that it represents such evolution in a correct way. If I do this, I fall into pragmatic inconsistency.

I'm happy to concede that the indexical ingredient in the activity of representing, which is performed by me or us, prevents from positioning myself at a God's eye point of view and enjoying an overarching view of my graph and the deer population. Yet, such a difficulty arises only when knowing is identified with representing. Such epistemological posture entails that the loss of reality objection must be overcome by resorting to a pragmatic move.

Moreover, even if we grant, as I do, that *within the representational conception of knowledge*, the loss of reality objection is pragmatically dissolved, doubts can be raised about our capacity of knowing real things by means of our representors, above all if we adopt what I will call a *predicative conception of knowledge*. According to the latter conception which I favor, knowing is attributing to things properties which they exemplify and to have good reasons for doing so. Such acts of predication are expressed in statements such as "At t_0, the deer population in Princeton is n_0". Within the representational conception of knowledge, strong objectivity collapses into weak objectivity and the bridge problem vanishes or "dissolves" as van Fraassen aptly says. Since we can't ascend to a godlike pedestal, we are forever confined to the domain of intersubjectivity or weak objectivity. Yet, such confinement is a consequence of the erroneous philosophical thesis according to which to know is to represent.

Notice that adequacy in van Fraassen's sense is not to be confused with truth, and certainly not with truth in a correspondence sense. Adequacy for him amounts to weak objectivity, that is, the invariance of some aspects of targets with respect to the particular points of view or perspectives of the various users of the same representors. For van Fraassen a representation is adequate or objective if it correctly represents its referent, in the sense that the usually agreed upon verification procedures have been implemented and consensus has been reached on their outcomes. Contrary to van Fraassen, I insist that the correctness of a representation rests on the truth of some predicative statements which attribute properties to things.

3 Truth

I maintain that truth is a property of statements[9] and I defend a correspondence *view* of truth. Now, I wish to stress the distinction I make between a *theory* of truth and a *view* of truth.[10] A theory of truth attempts to articulate in a precise way the correspondence between the statement and what makes it true, its truth maker (as in Wittgenstein's *Tractatus*). Granted, no fully satisfactory *theory* of truth as correspondence has been proposed so far. Yet, this is no sufficient reason to reject a correspondence *view* of truth according to which the truth maker of a statement is a *real fact*. In the case of a predicative statement, its truth maker is the fact that a given entity possesses a specific property referred to in the assertion. Such correspondence view of truth can be briefly advocated by pointing to the possibility of error when a statement "clashes", as Popper said, with reality. Then, we are constrained to admit that we are mistaken.

Thus, I sustain that truth is a relational property of statements. However, I submit that truth is absolute and not relative. Surely the truth of a statement is relative to the existence of the fact that makes it true (Agazzi 2014, 229). But truth is not relative to what we may believe or agree upon, nor does it depend on the arguments adduced in support of belief. This latter view of truth is epistemic, and thus relative to what we might judge to be a good justification. On the contrary, the existence of facts is independent of arguments or reasons. It would be wrong to object that the absoluteness of truth contradicts the fallibilist attitude. We may believe that a given

[9]See Agazzi (2014, 194).

[10]See also Ghins (2010, 527).

statement is true even when there is no fact which makes it true, and therefore the statement is false. Absolute truth is not to be confused with absolute certainty. The absolute character of truth is part and parcel of a correspondence view of truth. Absolute certainty on the other hand reflects the state of mind of a dogmatic person who pronounces that some beliefs are irrefutable. But he is not necessarily unable to provide arguments in favor of their truth. Famously, Descartes attempted to specify criteria which would guarantee that some of our ideas would be faithful representations beyond any possibility of doubt. It is significant that the criteria proposed by Descartes, namely clarity and distinctiveness, are *internal* to the ideas themselves. Indeed, dogmatism can only be supported by internal arguments.

On the contrary, a correspondence view of truth makes dogmatism quite difficult to maintain since resorting to any kind of external definitive authority would lead to an unstoppable regress. How can we justify that it is true, in a correspondence sense, that a given authority is an incontrovertible warrant of truth?

The correspondence view of truth may be considered to be a *realist view* of truth, because what makes a statement true must be real. Moreover, it is an essential component of any realist position. Any genuine realist must claim that there are things which exist independently of our language, our minds etc. and that some such things are cognitively accessible to us. As a consequence, a realist must be both an entity realist, who believes in the existence of external independent entities, and a statement realist, who believes that the predicative statements about things are true or false in virtue of the properties actually possessed by those external things.

4 Scientific Realism

As I conceive it, selective scientific realism maintains that some non-observational statements belonging to scientific theories are true or false in virtue of the existence of external things and further that we are at times in a position to provide good reasons to believe in the truth of these statements even if they speak of things whose properties are accessible only by means of instruments. Thus defined, scientific realism involves two main claims. At the ontological level, the scientific realist is committed to the existence of external things, i.e. things whose existence is independent of our language and internal states of mind such as thoughts, beliefs, representations, wishes and so on. At the epistemological level, the scientific realist contends we have good reasons to believe in the reality of some unobservable objects which actually possess some of the properties attributed to them by our theories. The challenge for the realist is to formulate criteria for selecting the objects posited by our theories which deserve to be called "real".

I broadly construe theories as having two major components: models and statements. Scientific models are artifacts. They are possible representors of concrete objects. Typically, models are mathematical structures composed of a set of elements which make up the domain of the structure and the relations which stand between the elements of the domain. Given some conventions, a two-dimensional curve functions as a representation of a specific object. Its elements are points which are meant to correspond to couples of properties specified by the coordinates, such as the number of deer in some area at some time, instantiated by the target, like the deer population in a defined area around Princeton. Here, the relations between the points of the curve are spatial and its shape provides some information on the variation of the deer population in time.

In order to construct the graph, we have to measure some selected properties of the object clipped out of the targeted thing. Generally, we manage to construct data models by means of measuring operations with the help of instruments. These instruments deliver data, that is, observable items such as digits on a screen, coincidences of a needle with black lines associated with numbers on a screen, colored surfaces (e.g. microscope images) etc. These data must be interpreted such that they might provide information on the properties possessed by the targeted thing. In the process of measurement, we can distinguish four steps, at least: the identification of the target, the observational interaction with the target, the production of a data model and finally the information the data model delivers about the target. Let us briefly discuss these four steps through the example of the determination of Avogadro's number by Jean Perrin.

In Perrin's experiments, the targeted things are emulsions, which are mixtures of liquids and small grains in Brownian motion. The realist at this stage presupposes that the grains and the liquid exist. The grains are identified by—indirect—observation by means of an instrument: the microscope. Once the target has been identified, we collect the data, which are the microscope images at different heights of the emulsion. Here, the interaction obviously is optical. Then, and this is the third step, we must select the properties which will be measured. We aim to give an answer the following question: is the liquid continuous or is it composed of discrete particles, namely molecules? (Molecules are considered to be indivisible, i.e. atoms in this context). More precisely, we want to measure the concentration of those grains in function of their height in the recipient. The distribution of the grains and their height are properties which are both observable by us (through a microscope) and measurable. We are then in a position to produce a graph (as in the case of the deer population) which represents the variation of the concentration of the grains in function of height. This graph is an exponential function which expresses in a precise way how the concentration of the grains decreases with height (the figure below (Perrin 1913, 144) shows the decrease in concentration of the grains with height). At this stage, we have a data model, namely a set of measurement results organized by an exponential function.

What does this data model teach us about the emulsion, and crucially about the nature of the liquid? Making a long story short (see Psillos' papers (2011) and (2014) for a detailed reconstruction of Perrin's reasoning), the data model permits to determine the value of Avogadro's number, which is a *finite* number. Consequently, the liquid is composed of molecules which can be counted. Matter is not continuous and the atomic hypothesis has been vindicated.

But what is so special about Perrin's experiments and arguments? What is it that makes them so convincing? Does Perrin rely on the inference to the best explanation (IBE) which is the favorite lever used by realists to boost our confidence in the existence of unobservable concrete objects? Yes, but not only, as I'm going to argue. The existence of molecules certainly *explains*, together with additional other hypotheses, why the data are organized by an exponential function. But what is more, the explanation is *causal*. The distribution (and the motion) of the grains is *caused* by the collisions with the molecules of the liquid. The presence of such a causal link is certified by showing that there is no other reasonable explanation of the distribution of the grains, because the alternative explanations have a very low degree of probability. Why? Because the value of N obtained by Perrin *concords* with the values obtained by different, independent, methods of measuring Avogadro's number. Perrin speaks of the "miracle of concordance": it is very unlikely that these methods would give roughly the same result if matter were continuous.

However, objections have been addressed by antirealists and realists alike against the force of IBE (of which the "no-miracle argument" NMA is a particular instance) even in its stronger—causal—version.[11] Some of these objections are a direct consequence of a staunch empiricist position, such as van Fraassen's, which eliminates from the domain of knowledge any statement about entities unobservable by our unaided senses. Perrin was quite aware of this kind of objection:

"Sensation is the only reality provided all *possible* sensations are adjoined to the actual sensations" (1903, p. X)

Since the word "sensation" habitually refers to internal mental items, I will rephrase Perrin's advice thus: we are entitled to believe in the existence of objects only if they are in principle accessible to observation or perception. Such constraint is in line with rather strict empiricist requirements. It implies that we can rationally believe in the existence of an entity only if it possesses at least one perceptible property: *it is not enough that the existence of this entity is an indispensable ingredient of the only possible explanation of some data*. Even if we have somehow managed to prove that a proposed explanation is the only acceptable one (and *a fortiori* the best available one…), this is not sufficient to ground our belief in the existence of unobservable entities and true statements about them. Those entities must possess properties which are the same as—or at least similar to—the properties possessed by ordinarily observed things. Otherwise, it would be impossible to make, as Perrin contends, molecular agitation visible. Visible, but not actually seen

[11]For a discussion of the strengths and weaknesses on the NMA, see for example Alai (2012).

through the microscopes available at Perrin's time. As he says, the behaviour of grains observed through the microscope makes molecular agitation visible in the same way as a floating cork follows the motion of sea waves better than a ship (1909, 353).

To avoid confusion I introduce a distinction between two categories of unobserved properties. The first category contains properties which are identical or similar to the perceptible properties of everyday things, but which cannot be observed because of the unavailability of appropriate instruments or other obstacles. Let's call them "observable in principle" (OP). Indeed, molecules are supposed to have properties such as size, velocity, acceleration etc. which were in principle perceptible at Perrin's time and also possessed by observable things such as billiard balls. Such properties are observationally accessible and measurable nowadays by means of more powerful scientific apparatuses. The second class of properties, which I will call "purely theoretical" (PT)[12] is composed of properties which are beyond the reach of any possible perception, even with the aid of the most powerful instruments, such as the properties of strangeness and charm in elementary particle physics. They might be measurable, but they are too remote from our ordinary perceptual experience.

Such version of scientific realism seems quite restrictive. The existence of unobservable objects can be defended only if they possess some (though not necessarily only) properties which are of the same *type* as the properties of ordinary observable things, namely the OP properties. Scientific realism is defensible with respect to objects which are unobservable only if their unobservable character results from the fact that the reach of our direct perception is limited *and* that we don't have today apparatuses sufficiently powerful to bring these properties within perceptual access. I'm aware that such a position casts a doubt upon the existence of properties such as strangeness, charm, baryonic number, internal spin etc. since those properties have no analogue in ordinary experience. Yet, the scientific realist is still allowed to rationally believe in the existence of such objects to the extent that they possess some OP properties. Thus, the existence of quarks with a specific mass can be defended since (as I will argue) mass is an OP property. But one may remain sceptical about their having some exotic properties such as charm, strangeness etc. which are PT properties. If this is correct, the version of scientific realism I advocate is not that restrictive after all. In physics, which is the discipline that postulates explanatory objects endowed with properties most distant from ordinary experience, the vast majority of posited objects do have one or more OP properties.

To conclude this section, let's summarize the philosophical lessons drawn from Perrin's experiments and reasonings by stating four conditions or requirements that a solid argumentation in favour of the existence of unobservable objects must satisfy:

---

[12]The OP properties attributed to molecules certainly are theoretical, since they belong to objects posited by a theory. This is why I use the adjective "purely" to refer to theoretical properties which are beyond any possibility of indirect observation.

1. *Requirement of observability in principle*: objects that are not directly perceptible must possess some OP properties (at least one…) which are identical or similar to the immediately observable properties of commonly perceived things (such as the velocity of a billiard ball) and as a consequence are in principle amenable to observation with the help of suitable scientific instruments.
2. *Requirement of measurability*: the OP properties must be quantitatively measurable by means of adequate instruments (such as the microscope).[13]
3. *Requirement of causality*: the OP properties of these objects are causally responsible, and thus provide an explanation, of the observed data.[14]
4. *Requirement of concordance*: distinct and independent experimental methods of measuring these properties must deliver results that are concordant (within acceptable approximation).

Taken together, these four requirements are a necessary and sufficient condition for the justification of the belief that a scientific object possesses a given property. They are jointly satisfied by Perrin's argumentation in favor of the existence of molecules and the truth of the atomic hypothesis. One might first object that some properties taken to be observable in principle, such as having a mass, cannot be considered to be so. True, the history of science teaches us that scientists came to agree on a precise definition of mass with much difficulty. However, once mass has been defined (and also before that…), it is easy to empirically ascertain that it is more difficult to set in motion a ball made of lead than a ball made of wood. Such difference is manifest because a ball made of lead has a larger *inertial* mass than a ball made of wood (with equal volumes). Such immediate observations allow to defend that the property of inertial mass is a OP property. Moreover, we observe that a ball exerts a stronger pressure on the hand if it is made of lead rather than wood. This shows that they have different *gravitational* masses. From these simple experiences, we can immediately conclude that there is a relation between the inertial and the gravitational mass without getting into an elaborate theorizing about gravitation.

The four requirements mentioned above provide grounds for a defensible version of selective scientific realism. How can we go beyond Perrin's experiments and establish the generality of these four requirements? The answer is straightforward: on the basis of our ordinary perceptual experience. Our next task is to show that these four requirements form the backbone of any argumentation in favor of the reality of commonly observable entities, but which have not been observed yet.

---

[13]Mario Alai (2010, 672) mentions a method of measurement used by Perrin to determine the upper limit for the size of molecules, which relies solely on an inductive inference.

[14]Admittedly, an OP property often plays a causal role together with some PT properties. Each requirement is necessary but none of them is sufficient in isolation.

5 The Parallelism with Ordinary Experience

Empiricists philosophers are quite right in stressing the epistemic role of observation or perception as warrant of our assertions of existence. In order to justify the belief in the existence of objects inaccessible to our immediate perception, we are only allowed to resort to arguments analogous to the ones used to comfort our beliefs in the existence of things immediately perceptible in the context of our everyday experience. In ordinary practice, we increase our degree of confidence in the reality of perceived things by augmenting the number of our observations, by repeating them, and also by relying not only on our sense of vision but on the other modalities of perception. We judge that our belief in the existence of a thing such as a rose is more assured when we have managed to look at it in satisfactory perceptual conditions and when we have witnessed an agreement between our perceptions, that is, when we have noticed that some properties remain constant, invariant, e.g. that the stem of the rose is covered with thorns. We increase our degree of belief in the existence of the rose by touching it, smelling its perfume and perhaps by tasting its petals as the ancient Romans did before mixing them with wine. By trusting our five senses, or at least several of them, that is, by resorting to various modularities of perception, we enhance the degree of confidence in the truth of an assertion of existence (Ghins 1992), without however reaching absolute certainty. We understand now why scientists give much importance to the concordance between measurement results obtained by means of *independent* methods, which corresponds to the fourth requirement mentioned above.

In the case of direct perception, it is evident that the first requirement is also satisfied, because the relevant properties are observable and perceptible. The second requirement is not, since in ordinary experience, it is not necessary to perform quantitative measurements to confirm the concordance between perceived properties in different circumstances, in accordance with the fourth requirement. The second requirement is a prerequisite for the satisfaction of the fourth. In most ordinary contexts there is no need to be in a position to determine if a rose is crimson, scarlet, purple or amaranth to truly state that it is red. Observations of this kind can be taken to be rough measurements, which are sufficient to establish the presence of a property in most cases. In science however, precise measurements are required to verify the accuracy of the predictions of a theory as well as to comply with the requirement of concordance.

The third requirement is also satisfied since it is easy to convince oneself that causal processes are at work in perception by verifying the systematic variation of some perceived properties in function of the changes in the observing conditions. Also, we can see that these properties vanish when the thing is removed. This is just a crude application of Mill's methods. Full knowledge of the laws and the causal mechanisms at work in perception is not indispensable because the immediate perceptual contact doesn't necessitate to resort to causal laws to guarantee that the perceived thing exists.

How do we infer the existence of a perceptible thing, which is not actually present in perception? In a by now famous example, van Fraassen (1980, 19–20) discusses the clues which permit to conclude that a—unobserved—mouse has come to live with me. Nobody has seen the mouse (which is observable…) yet, but some gray hair lies on the ground, cheese disappeared, tiny noises have been heard etc. Such evidence makes me believe that a mouse is living in the house. How is such belief justified? Well, we have seen mice before and we know that they have grey hair etc. We suspect that several causal links obtain between the existence of a mouse and the observed clues. Each causal relation points toward a particular property, such has loosing grey hair or eating cheese. The set of these properties and the awareness of the context allow to identify the thing which is causally responsible for their presence. Past experience is sufficient to make sure that the specific causal relations are in place without having knowledge of the causal laws. It is not mandatory either to perform precise measuring operations on the clues. (In other instances, a detailed quantitative inquiry is sometimes required, when for example the identification of a murderer is at stake.)

At this point, we come to grips with the justification of the belief in the existence of *unobserved* objects which are typically postulated by a theory to causally explain why some phenomena occur. The existence of molecules was asserted by Boltzmann because he wanted to causally explain the properties of gazes, such as pressure, temperature etc. I distinguished above between two kinds of properties that can fulfill an explanatory role, the OP properties which are observable in principle and the PT properties which are purely theoretical and escape de dicto any possibility of observation, even by means of instruments. With respect to objects which possess PT properties only, an agnostic attitude is to be recommended until perhaps some theoretical investigations lead us to attribute OP properties to them.

Thus, the selective realism I defend can give credence to observable properties only, but these properties are not limited to directly observable ones: they include OP properties, that is properties which could be observed through adequate instruments. OP properties are eligible candidates for membership in our ontology. As regards the sense of sight, which enjoys a privileged status in the sciences, observation instruments are extremely varied. First, we have the banal glasses commonly used by short-sighted, farsighted and presbyopic people whose eyesight is mildly deficient. Who will dare to claim that the observations made by someone who uses appropriate glasses are less reliable because they were not performed directly with the naked eye? Next,[15] we have the different kinds of microscopes and telescopes which make visible entities such as viruses, remote stars and galaxies. Notice again that it is not necessary to know the causal laws or the underlying mechanisms for these instruments to trust them. Polished glasses were already dependably used by the ancient Romans even though they didn't know anything

[15]See Maxwell (1962, 7).

about the laws of refraction and electromagnetism. Galileo and his contemporaries ignored the working of the rudimentary telescope they used to explore the sky. Yet, Galileo's contemporaries rapidly came to agree on what was seen through the telescope. (Although they disagreed about the impact of telescopic observations on issues such as the immutability of celestial bodies…)

Moving further away from ordinary perceptual experience, we can mention the cause of the light rays directly observed in the cathode ray tubes. These light rays are produced by the interaction of unobservable particles—the electrons—emitted by the cathode, which interact with the molecules of the rarefied gas filling the tube.[16] J.J. Thomson succeeded in determining the values of the charge and the mass of an electron without knowing how the cathode rays were produced. However, knowing the laws permit to manufacture instruments, such as wire and drift chambers to detect and measure the properties of charged particles. These devices make visible the trajectories of charged elementary particles, such as protons. In order to be justified in claiming that the lines immediately observed on photographs or computer screens represent the trajectories of protons, we must know the causal mechanisms, i.e. the causal laws[17] which describe the interactions between the protons and the molecules of a gas such as argon. Typically, the protons remove electrons from the molecules of the gas. These electrons are instantly captured by the wires in their vicinity, which detect them and make visible the trajectories of protons.

This case is analogous to the emulsions studied by Perrin. The molecular agitation is rendered visible by the movements of the grains of mastic which collide with the molecules of the liquid, in the same way as the trajectory of a proton is made visible by the electric discharges resulting from the interactions of the protons with the molecules of the gas. These discharges then produce luminous spots on a screen. For molecules as well as for protons, the knowledge of the laws which govern their interactions is necessary to comply with the requirement of causality and justify our existence assertions.

Whether we look at directly observable entities, like a rose or a mouse, or indirectly observable, such as the grains of an emulsion or galaxies, or at entities whose existence can only be inferred, such as atoms, electrons, protons or gravitational waves, our reasoning is grounded on the attested presence of causal relations (how we attest such presence is examined in Sect. 6). The working of causal mechanisms, the laws of which can be known or not, is crucial to justify our belief in the existence of entities, whether directly or indirectly observable, or inferred only. Surely, actual perceptual presence in adequate conditions is what guarantees the truth of the assertion that an entity possesses directly observable properties. But perception is supported by causal mechanisms that we have reasons to believe to be present even if we might still ignore the details of their action, if only because when

---

[16]See Nola (2008).

[17]A causal law is a mathematical law containing a time derivative which refers to the effect, whereas the other terms refer to the cause(s). (Blondeau and Ghins 2012).

we intervene on an object in a certain way, systematic variations in the perception of its properties occur. In the same way, if we have a complete knowledge of the functioning of a measuring instrument, we have reasons to believe that the causes of certain observed effects are objects with specific OP properties. As a consequence, nothing prevents that these objects might be (indirectly) observed someday. Molecules can now be seen through instruments which were not at Perrin's disposal: the electronic microscopes.

My vindication of selective—local—realism is grounded on a parallelism between the reasons to believe in the existence of an ordinary thing and the reasons why we are rationally entitled to assert the reality of some concrete unobservable object and some true statements about it. Most scientists resort to such arguments to sustain the existence of objects which have properties which are de facto—but not de dicto—unobservable, namely the OP properties. We resort every day to inferences to the best explanation, for example when we infer the presence of a mouse from some directly observed clues. But such a way of reasoning rests on previous observations which warrant the reality of a causal link between the presence of a clue and the existence of a thing which has some specific properties. We can thus rely on *several* inferences to the best causal explanation. Each inference justifies the belief in the existence of a concrete object with a particular property such as having gray hair or eating cheese and so on. Thus, my vindication of selective scientific realism is explanationist (Psillos 1999, 78). But causal explanation is a particular—very restrictive—case of IBE. Moreover, at the end of the day, the existence of the mouse is established by means of a variety of IBE. In fact, the mouse is the only animal which has all the properties causally responsible for the observed clues.

The concordance of the results obtained by the various independent methods of measuring Avogadro's number is comparable to the concordance of the various ways of observing the *same* property of a mouse. One could say that the finiteness of Avogadro's number is the best explanation of an obtained value. The same holds for the other *finite* values obtained by other methods. It is this concordance that certifies that the measured values are finite and approximately correct. In an analogous manner, several perceptual accesses to a property of a thing, whether seen, touched etc., can attest that a perceived thing such as a tabletop is rectangular. The presence of the rectangular form explains, via various causal processes, the agreement between the various perceptions. But we know that each individual perception is immediate evidence that the tabletop is rectangular, irrespective of a possible knowledge the underlying causal processes. In the same way, when we deal with OP properties, whether of mice or electrons, we must have reasons to suppose that causal processes are at work and that they furnish the basis for the causal explanations of our direct or indirect observations.

For each theory and each posited object, we must examine if causal chains connect specific phenomena to some specific OP properties and if independent measuring operations generate concordant results for the same property. Every existence assertion deserves a special justification, as we saw for the determination of the Avogadro number and the masses and dimensions of molecules. Yet, the philosopher aims at universality. For this reason, I tried to show that each particular argumentation articulated by the realist must obey the four requirements that I formulated and motivated above.

6 Causal Link and Causal Law

It remains to be shown how we can certify that causal links obtain or, eventually, that some causal laws are true. This is an important task. The justification of our assertions of existence hinges on the reliability of the supposed causal connections between what we observe and what exists. When perception is immediate, no causal link need to be invoked. The actual presence of a glass of beer immediately evidences its existence beyond reasonable doubt in a normal context of perception. Direct or immediate realism is the indispensable ground for epistemological realism about external things.

By reflecting on our perceptions, we are in position to ascertain the invariance as well as the orderly variation of some properties of the perceived thing, e.g. when our spatial points of view are modified. If we alternatively shut and open our eyes, we perceive the same properties, provided the perceptual environment remains unchanged. These banal observations provide sound reasons to believe that some causal connections are in place between the perceived concrete objects and the perceptions of their properties.

To discover and mathematically formulate causal laws, thorough empirical investigations and in-depth theoretical work are necessary. As a first stab, we can fall back on the well-known methods of agreement—and above all of difference—of John S. Mill (1911). Think of Newtonian dynamics. If we have reasons to believe that forces exist, it is in the first place because of our bodily experience. Weighty things exert a force on our arms when we carry them. An impressed force also produces variations of velocity. A body at rest can be put in motion by exerting pressure on it. When no net force acts, there is no acceleration. In science we must denote a force by a mathematical symbol, namely a vector. Moreover, we must be able to measure forces with accuracy. In order to measure the gravitational force, for example, we can use a scale. When it is horizontally in equilibrium, we can conclude that the forces, called "weights", exerted on the arms of the scale are equal. This static way to measure certain forces can be generalized to measure the values of different forces in some unit.

Accelerations can be measured by means of various experimental devices, such as the Atwood machine. In numerous cases, we can measure the forces and the accelerations they cause in order to verify that accelerations are proportional to forces, in conformity with the fundamental law of Newtonian mechanics. The proportionality factor—the inertial mass—can be measured by the method of

Huygens' collisions, independently of Newton's second law. When the vectorial sum of the forces is (approximately) nil, the velocity of the mobile is (approximately) constant in magnitude and direction. This is the inertial motion. Numerous observations and measurements give excellent reasons to believe that the law of inertia holds for all bodies.

Perrin's argumentation for the existence of the molecules can then rely on the fundamental laws of mechanics, which can furthermore offer a causal explanation of the property of gas pressure. We must therefore distinguish several levels in a causal explanation. The first hinges on the mechanisms described by a theory when Perrin relies on the kinetic theory and the laws of collision to explain what is going on in an emulsion. The second level consists in the causal interactions between our apparatuses, such as a microscope, and some observed properties, like the distribution of grains in an emulsion. Ii is not necessary to know the laws which govern the interactions of the second level to be entitled to believe that some objects having not directly observable (OP) properties exist.

7 Conclusion

A defensible version of scientific realism can only be fallibilist and selective. In order to be in a position to claim that a concrete object with particular properties exists, a specific argumentation in favor of the presence of each property must be provided. I have attempted to show that any such convincing argumentation must jointly satisfy four requirements: the requirement of observability in principle, the requirement of measurability, the requirement of causality and the requirement of concordance. The justification of these requirements is based on the fact that they are satisfied by everyday arguments in favor of the existence of ordinary objects, whether actually perceived or not. The requirement of measurability however, doesn't always have to be satisfied as far as ordinary objects are concerned. But in the sciences, the requirement of measurability is a prerequisite for the satisfaction of the all-important requirement of concordance. I submit that this version of epistemological selective realism on top of being forceful is also in compliance with a moderate empirical philosophy.

References

Agazzi, E. (2014), *Scientific objectivity and its contexts*. Berlin: Springer.

Alai, M. (2010), "van Fraassen, Observability and Belief", in D'Agostino, Giorello, Laudisa, Pievani & Sinigaglia (eds.) *New Essays in Logic and Philosophy of Science*, 663–675. London: College Publications.

Alai, M. (2012), "Levin and Ghins on the No-Miracle Argument and Naturalism". *European Journal for Philosophy of Science* 2, 85–110.

Blondeau, J. and Ghins, M. (2012), "Is There an Intrinsic Criterion for Causal Law like Statements?" *International Studies in the Philosophy of Science* **26**, 381–401.

Da Costa, N. and French, S. (2003), *Science and Partial Truth. A Unitary Approach to Models and Scientific Reasoning.* Oxford: Oxford University Press.

Ghins, M. (1992), «Scientific Realism and Invariance». *Proceedings of the Third SOFIA Conference on Epistemology. Campinas. July 30–August 1, 1990. Philosophical Issues (Vol. 2: Rationality in Epistemology).* 249–62. California: Ridgeview.

Ghins, M. (2002), «Putnam's No-Miracle Argument: a Critique» in Clarke & Lyons (eds.) *Recent Themes in the Philosophy of Science: Scientific Realism and Commonsense. Australasian Studies in History and Philosophy of Science Vol. 17,* 121–138. Dordrecht: Kluwer Academic Publishers.

Ghins, M. (2010), «Bas van Fraassen on Scientific Representation», *Analysis* 70(3): 524–536. http://analysis.oxfordjournals.org/cgi/reprint/anq043?ijkey=8y8OZR6C4lhmJCy&keytype=ref.

Ghins, M. (2016a), "Evandro Agazzi on Scientific Objectivity". Essay review of Evandro Agazzi's *Scientific Objectivity and its Contexts.* Berlin: Springer. 2014. *Metascience* 25(1), 151–157, 2016. http://link.springer.com/article/10.1007/s11016-015-0036-7?wt_mc=internal.event.1.SEM.ArticleAuthorAssignedToIssue.

Ghins, M. (2016b), "Bas van Fraassen on Success and Adequacy in Representing and Modelling" in Lorenzo Magnani & Claudia Casadio (ed.), *Model-Based Reasoning in Science and Technology. Theoretical and Cognitive Issues,* Studies in Applied Philosophy, Epistemology and Rational Ethics (SAPERE) 27, ch. 2, 21–42. Berlin: Springer. http://www.springer.com/gp/book/9783319389820.

Maxwell, G. (1962), "The Ontological Status of Theoretical Entities", *Minnesota Studies in the Philosophy of Science, Vol. III.*

Mill, J. S. (1911), *A System of Logic: Ratiocinative and Inductive.* London: Longmans, Green and Co. First published in 1843.

Nola, R. (2008), "The Optimistic Meta-Induction and Ontological Continuity: the Case of the Electron". In L. Soler, H. Sankey and P. Hoyningen-Huene (eds.) *Rethinking Scientific Change and Theory Comparison: Stabilities, Ruptures, Incommensurabilities.* Berlin: Springer. 159–202.

Perrin, J. (1903), *Traité de chimie physique. Les principes.* Paris: Gauthier-Villars. http://iris.univ-lille1.fr/handle/1908/3160.

Perrin, J. (1909). «Mouvement brownien et grandeurs moléculaires». Paris: Radium 6 (12), 353–360. https://hal.archives-ouvertes.fr/jpa-00242381/document.

Perrin, J. (1913), *Les atomes.* Paris: Alcan.

Psillos, S. (1999), *Scientific Realism. How Science Tracks Truth.* London: Routledge.

Psillos, S. (2011), "Moving Molecules Above the Scientific Horizon: On Perrin's Case for Realism", *J Gen Philos Sci* 42: 339–363.

Psillos, S. (2014), «The view *from Within* and the View *from Above*: Looking at van Fraassen's Perrin», in Gonzalez, W. (ed.) (2014), 143–166. M. Ghins, Université Catholique de Louvain, Belgium.- michel.ghins@uclouvain.be.

Suppes, P. (2002), *Representation and Invariance of Scientific Structures.* Stanford: CLSI.

Van Fraassen, B. (2008), *Scientific Representation. Paradoxes of Perspective.* Oxford: Oxford University Press.

Van Fraassen, B. (2010), "Reply to Contessa, Ghins, Healey". *Analysis* 70(3): 547–556. http://analysis.oxfordjournals.org/cgi/reprint/anq043?ijkey=8y8OZR6C4lhmJCy&keytype=ref.

Zalta, E. (1988), *Intensional logic and the metaphysics of intentionality.* Cambridge, MA: MIT Press.

Robustness, Intersubjective Reproducibility, and Scientific Realism

Marco Buzzoni

Abstract It is common to distinguish three main senses of the term "robustness": (1) Robustness of models; (2) Robustness as stability or insensitivity of output as against variations in parameter values; (3) Robustness as consilience of results from different and independent hypotheses, procedures or sources of evidence. The purpose of this paper is to discuss the last two meanings of robustness, in order to cope with some difficulties with which robustness as consilience is confronted and which have indirect consequences for the problem of scientific realism. On the one hand, robustness regarded as reproducible stability as against perturbations and variations in parameter values (robustness-as-stability) and robustness as consilience of results from different and independent pieces of evidence (robustness-as-consilience) are conceptually distinct. On the other hand, however, robustness-as-stability is a condition of robustness-as-consilience; and the converse holds also: robustness-as-consilience is an essential ingredient of robustness-as-stability. There is no vicious circle here, but a technical-practical synergy, which is at the heart of the experimental method, and which can help us out of the two main problems for robustness-as-consilience.

1 Introduction

Three main senses of the term "robustness" are often distinguished: (1) Robustness of models; (2) Robustness as stability or insensitivity of output as against variations in parameter values; (3) Robustness as consilience of results from different and independent hypotheses, procedures or sources of evidence. The purpose of this paper is to discuss the last two meanings of robustness, in order to cope with some difficulties with which robustness as consilience is confronted and which have indirect consequences for the problem of scientific realism.

M. Buzzoni (✉)
University of Macerata, Macerata, Italy
e-mail: marco.buzzoni@unimc.it

E. Agazzi (ed.), *Varieties of Scientific Realism*,
DOI 10.1007/978-3-319-51608-0_7

In Sect. 2, I shall characterize the two types of robustness that are most important for understanding experimental sciences and for the issue of scientific realism. Section 3 examines the important step towards answering both objections that was taken by Ian Hacking, who implicitly referred to robustness in his criticism of the no miracle argument. As we shall briefly see, Hacking only went half way, and finally fell into a more traditional view of robustness, which cannot avoid sceptical conclusions, akin in some respects to that of Kuhn's incommensurability thesis.

This will lead, in Sect. 4 to think in terms of the two kinds of robustness working together in such a way that we are able to cope with some difficulties concerning robustness-as-consilience that have indirect consequences for the scientific realism debate. According to one of the central points emphasized throughout this paper, the two senses of the term "robustness" have to be distinguished in order to avoid confusion, but, on the other hand, they are intimately connected with one another. Robustness-as-stability and robustness-as-consilience are two specifications of *the same experimental and intersubjective reproducibility, which, in this sense, may be plausibly presented as the substantial common core of robustness in the empirical sciences.*

As we shall see in Sect. 5 there is no vicious circle here, but a technical-practical synergy (or, if you prefer, a robustness of higher order)—which is at the heart of the experimental method and which can help us out of the two main problems for robustness-as-consilience. If the different notions of robustness-as-stability and robustness-as-consilience are not only distinguished, but also connected with one another, and if their intimate relationship is recognised, we find a notion of robustness of higher order. As we shall see, on the basis of this notion and by drawing on Hacking's hint to improve the no miracle argument, the spiral synergy of robustness-as-stability and robustness-as-consilience provides some support for scientific realism, at least in the minimal senses that (1) it furnishes an epistemic warrant for a reality which is independent of us, and (2) it puts a great number of reliable findings at our disposal, which we are not willing, at least provisionally, to put in question. Thus, from a technical-operational point of view, we may express differently one of the best known pragmatic theories of truth, by saying that the truth is something like 'what shows itself most robust' in the widest context and the longest run.

2 Two Senses of Robustness and Hacking's Scientific Realism

As already mentioned, it is common to distinguish different senses of the term "robustness". For example, drawing on Wimsatt (2007), Calcott (2010) distinguishes the three following meanings:

(1) Robustness of models. "Robustness" was originally introduced for statistical purposes and for model-based simulations (cf. above all Box 1953 and Levins 1966).

(2) Robustness as stability or insensitivity of output. In this sense, a phenomenon, mechanism or system may be said to be robust, if it shows a relative *stability or insensitivity of output* as against variations in parameter values. In this sense, robustness is important both in engineered systems and in living systems because it provides resilience against internal or external perturbations.
(3) Robustness as consilience of results coming from different and independent sources of evidence. These latter are said to be robust if they are supported in a *variety* of independent ways. It is this sense of robustness that is emphasized by those who espouse the no miracles argument.

I do not wish to discuss meaning (1) in this paper. I shall say something about the last two meanings of robustness, and I shall call them robustness-as-stability and robustness-as-consilience.[1]

According to the first sense that we shall discuss, robustness is usually defined as "a property that allows a system to maintain its functions against internal and external perturbations." (Kitano 2004, 826; cf. also Stelling et al. 2004, and Clausing 2004, 25). In this sense, robustness is very important in engineered systems, as it makes them more resistant to events that cannot de facto (and perhaps in principle) be predicted in the early stages of their development: think of a robot that must work in still unexplored portions of nature, for example. This kind of robustness is usually obtained by building redundancy into a mechanism: if one element fails, another element can play its role (cf. Calcott 2010; Clausing 2004).

In this same sense, robustness is also a very important property in living systems, as resilience against internal or external perturbations. For example, the genetic code can be described as a robust encoding of amino acids into codons, or we may say that proteins, developmental pathways, metabolic networks, and tumours are robust against, respectively, translation errors, environmental or genetic disturbances (e.g. "gene knockout experiments"), changes in enzyme efficiency, and various chemotherapies (cf. Wilke 2006, 695; Strand and Oftedal 2009). This meaning of the term is also exemplified by mechanisms by which living beings increase their survival rate, such as the mechanism of anhydrobiosis by which, under extreme dehydratation, tardigrades suspend metabolism almost completely

[1]The main reason for this limitation is that this paper is mainly concerned with the role robustness plays in the experimental sciences, while a discussion of the first meaning would require a discussion of the role that computer simulation plays in real world experiments, which is beyond the scope of the present paper. Many authors found interesting analogies between model-based simulations and real-world experiments (cf. for example Norton and Suppe 2001; Parker 2009; Morrison 2009). However, the analogy between real world experiments and computer simulations, though correct within certain limits, fails to the extent that computer simulations may give us information about the actual world only because we have independent and empirical-experimental evidence of the model's meaning (for a similar objection, see above all Hughes (1999 [2010]): 203; see also Buzzoni 2015, where further references will be found). For this reason, what will be said below may be extended to the first meaning of "robustness" in the measure in which model-based simulations can be considered as falling under the more general concept of experiment.

by extensive production of trehalose and become active again upon rehydration (cf. Crowe and Crowe 2000; Singer and Lindquist 1998).

The second meaning of robustness, that is, robustness as consilience or coincidence of a variety of different (independent) pieces of evidence, is both a method which is frequently used in everyday life and a venerable concept in the philosophy of science. Apart from some partial anticipation (on this point, cf. especially Wimsatt 2012; Stegenga 2009; Hudson 2014), it is well-known that William Whewell was the first important author to be fully aware of the importance of this concept, which he called "*Consilience of Inductions*" (cf. Whewell [1840] 1847, Vol. 2, 65–66). On the contrary, Bridgman (Bridgman 1927, 56–60, who employed the notion—though not the term–of robustness-as-consilience as a criterion for proving the physical reality of theoretical entities), Popper (for the importance of independent tests: 1963[1972] and 1972), and Glymour (for his "logical pincer movement" or "bootstrap testing": 1974 and 1980) are to be especially mentioned among the most important authors that have been too much neglected in the robustness debate hitherto.

Newton's theory of universal gravitation is one of the best examples. No relation between Kepler's first, second, and third laws concerning the motion of the planets around the sun was present until Newton's theory of universal gravitation explained all of them at once. Moreover, Newton's theory of universal gravitation did not explain only the perturbations of the moon and planets by the sun and by each other, from which it was originally inferred, but also the apparently independent fact of the precession of the equinoxes (cf. Whewell [1840] 1847, Vol. 2: 65–66). But the most discussed case in the literature is perhaps the measurement of Avogadro's number, that is, the number of molecules per gram-mole of any gas, which Jean-Baptiste Perrin computed by quite different methods (among which the most important was perhaps that based on Brownian motion) and found to be the same, with a relatively good—but not so good as it is sometimes supposed (cf. Perrin [1916], 87)—approximation to today's accepted value, that is, $6.02,214,179 \times 10^{23}\ mol^{-1}$ (cf. Psillos 2011). The fact that all of these measurements essentially agreed with the experimental data within the accuracy of the observations would be an unexplainable coincidence, it would be "miraculous", if each of the measurements referred to an artefact and matter were not composed of molecules and atoms.[2]

There are two main objections to robustness as used in arguments of this kind for scientific realism. The first is that robustness-as-consilience reasoning does not provide any *autonomous* epistemic warrant if one lacks minimally reliable observational data or procedures (cf. Hudson 2014, 200). For this reason, the robustness requirement must be integrated by a requirement of "minimal reliability" (Hudson 2014, 18; cf. also Stegenga 2009, 653).

[2]The no miracles argument was already present, at least in its essential points, in Poincaré's (1905, 154) and Duhem's ([1906]1914, 36–39) works, and it is well-known that it was taken up by, among others, Smart (1963), Sellars (1963, Chap. 4), Putnam (1975, 60–78), Salmon (1984, 206–207), and Kosso (1989, 247 and 1992).

The second objection goes farther. Even if it were true that any single piece of evidence was highly reliable, it still would be very difficult, if not impossible, to ascertain the independence condition for robustness. In fact, it is widely recognized that robustness-as-consilience in science (and the "no-miracle" argument for scientific realism) can be credited with epistemic warrant only if the multiple kinds of available and consilient evidence are independent. Wimsatt, for example, recognized that "one of the most critical and important problems in the study of robustness analysis". (Wimsatt [1981] 2012, 83) is "the failure of the different supposedly independent tests, means of detection, models, or derivations to be truly independent" (Wimsatt [1981] 2012, 82).

3 Robustness and Hacking's Scientific Realism

An important step towards answering both objections was taken by the experimentalist turn, and notably by Ian Hacking, who implicitly referred to robustness in his criticism of the no miracle argument. As we shall very briefly see, however, Hacking only went half way, and finally fell into a more traditional view of robustness-as-consilience, which cannot avoid sceptical conclusions, akin in some respects to that of Kuhn's incommensurability thesis.

On the one hand, Hacking claims that, in its most common form, the "cosmic accident argument" (which is another name for the no miracle or, to use Smart's expression, the "cosmic coincidence" argument: Smart 1963: 39) is circular:

> The cosmic accident argument notes that often in the growth of knowledge a good theory will explain diverse phenomena which had not hitherto been thought of as connected. […] Once again, this seems to me to beg that realist/anti-realist issue. The anti-realist agrees that the account, due to Einstein and others, of the mean free path of molecules is a triumph. It is empirically adequate—wonderfully so. The realist asks why is it empirically adequate—is that not because there just are molecules? The anti-realist retorts that explanation is no hallmark of truth, and that all our evidence points only to empirical adequacy. In short the argument goes around in circles (as, I contend, do all arguments conducted at this level of discussion of theories). (Hacking 1983: 54–55)

Hacking does not develop this objection as fully as could be wished. But his point is that scientific realism may be defended using experimental practice instead of traditional representational approaches:

Hacking's "Grid Argument" is perhaps the best interpretation of the no miracle argument (or, to use Hacking's expression, of the "cosmic accident argument") in terms of his 'interventionist' point of view. This argument starts from the fact that we manufacture grids to identify dense bodies in blood by now standard techniques. When we look the tiny metal grid so obtained through almost any kind of microscope, we

> see exactly the same shapes and letters as were originally drawn on a large scale. It is impossible seriously to entertain the thought that the minute disc, which I am holding by a pair of tweezers, does not in fact have the structure of a labelled grid. I know that what I see

> through the microscope is veridical because we *made* the grid to be just that way. [...] Moreover we can check the results with any kind of microscope, using any of a dozen unrelated physical processes to produce an image. [...] Is it a gigantic conspiracy of totally unrelated physical processes that the large scale grid was shrunk into some non-grid which when viewed using different kinds of micro-scopes still looks like a grid? (Hacking 1983, 201).

The "Grid Argument" supports realism about entities by insisting on the robustness of something manipulated, and not just 'observed', using different and independent techniques. In my opinion, Hacking's main point is that intersubjective reproducibility, considered from the point of view of a pragmatist conception of knowledge and in relation to our interactions with the surrounding world, may be understood as the most valuable clue to the independent existence of all empirical contents, including theoretical entities. If Hacking is interpreted in this way, I think he is right in advocating a technical-operational 'criterion' for the reality of theoretical entities: true ideas 'agree' with reality, in the sense and to the extent that they make it possible to master it.[3]

However, Hacking did not coherently apply this idea to both senses of robustness. Owing to an excessive and one-sided emphasis on the independence of experiment from theory,[4] in 1983 he is led to attach more importance to robustness-as-stability than to robustness-as-consilience, where robustness-as-stability is understood as consisting primarily in the stability of particular or 'local' technical-operational interventions on reality.[5] His later work, on the contrary, seems to attach very much more importance to robustness-as-consilience than to robustness-as-stability: the cause of the "stability" in science is found by him in a particular process of reciprocal adaptation between "thoughts, acts and manufactures" leading to a body of "types of theory and types of apparatus and types of analysis that are mutually adjusted to each other" (Hacking 1992: 30) . He also explicitly rejects any approximation to a unique reality and upholds a consilience due to a progressive and increasingly coherent synthesis between heterogeneous elements such as ideas, apparatuses, observations (Hacking 1992: 58). Since this consilience is only the result of the attempt to integrate the various factors (that he painstakingly classifies) that might influence scientist's decisions, he openly admits that these systems of theories, apparatuses and conceptual tools are "closed

[3]There is no question of here entering into any serious discussion of an experimental-technical theory of truth, which I have briefly outlined elsewhere. For the most general features of such a theory, I must refer the reader to (Buzzoni 2008).

[4]There is, unfortunately, no space available to argue for this point, but it seems hard to deny that this already follows from Hacking's thesis that experimentation 'has a life of its own' (Hacking 1983: xiii). For more details on this point, see (Buzzoni 2008), Chap. 1.

[5]This is indirectly confirmed by the fact that Hudson 2014 neglects Hacking's idea of combining robustness-as-stability with robustness-as-consilience—which is our starting point in this paper—, and essentially only finds in this author robustness-as-stability as "a key part" of his critique of robustness-as-consilience (Hudson 2014: 6).

systems", "self-vindicating" and "essentially irrefutable" (Hacking 1992: 30. Hacking 1999 only slightly mitigates, but does not change significantly this idea of robustness: see for example pp. 71–74, 85–86, 231).

It is clear that this perspective, rejecting the truth of very general theories and making that of "local" laboratory statements relative to a system of self-justifying theoretical presuppositions, leads Hacking to embrace, even though in a somewhat particular sense, the incommensurability thesis.

As a matter of fact, he claims quite explicitly that in science there can be stabilisations of theories that

> would not even be comparable, because they would be true to different and quite literally incommensurable classes of phenomena and instrumentation. I say incommensurable in the straightforward sense that there would be no body of instruments to make common measurements, because the instruments are peculiar to each stable science (Hacking 1992: 31).

According to Hacking, Newton's and Einstein's theories are incommensurable not because the statements of one could not be expressed in the other, but because one is true "to one body of measurements given by one class of instruments, while the other is true to another" (Hacking 1992: 54).

All these statements could come from *The Structure of Scientific Revolutions*, including the last one, since Kuhn (even though Hacking seems unaware of this) was explicitly concerned with the problem of the incommensurability of instruments relative to different paradigms (cf. Kuhn [1962] 1970: 126).

To sum up these considerations concerning Hacking, it may be said that, on the one hand, he deserves credit for having made evident the *epistemological* importance of experimenting for robustness. On the other hand, Hacking only goes half way. Instead of reaching a deeper comprehension of both notions of robustness, he reaches only a superficial schema concerning robustness as consilience (which is a merely analogical extension of the statistical meaning to the experimental sciences), to which, in the later work, he does accord a privileged epistemic status.[6]

We have now to consider how to develop more coherently Hacking's idea, by applying it to both notions of robustness.

4 Robustness, Intersubjective Reproducibility, and Scientific Realism

Although the distinction between different notions of robustness is necessary in order to avoid confusion (cf. Woodward 2006), we should not run into the opposite error of neglecting important similarities between these different kinds of

[6]Boon (2012) has stressed the importance of this last concept of robustness in Hacking. But precisely because I essentially agree with Boon that robustness should be considered in connection with practical-technical applications (and not merely with statistics, as in Hacking's case: cf. Hacking 1999: 231, Footnote 6), I think that Hacking's contribution to robustness-as-consilience is much less important than that to robustness-as-stability and technical reproducibility.

robustness. For this purpose, so far as the epistemological importance of robustness-as-stability is concerned, the most important point to notice here is that the robustness of a mechanism (or of an organism, if regarded in a mechanistic perspective) is intimately connected with the notion of intersubjective reproducibility, which is perhaps the main pillar of scientific experimentation.

As many authors already emphasized–from Frege to Poincaré, from Wittgenstein to Popper-, a particular or single event of perceptual awareness (for instance, *my* perception of a blank sheet of paper lying before me) is not only absolutely certain, but also unavoidably subjective and private, because it is not accessible to any other person. As such, it has no right of citizenship either in science or in empirical knowledge in general. Such a perception belongs to a subject, and not to an object, and for this reason it is not intersubjectively testable in principle. As for example Popper rightly noticed, we do not take even our own observations seriously, if they are not in principle intersubjectively testable (Popper 1959[2002], 45). On reflection, to regard such a perception as the property of some empirical object is much like conceiving it as a mere hallucination. If someone sees a lion in a room of the house, s/he would perhaps look again and/or ask someone to test whether s/he is seeing the same thing because there is something decidedly strange, if not impossible, in this perception. If no one could find any trace of the lion later on, s/he would know that what s/he had seen was a hallucination, no matter how frightening it was.

If, according to the interpretation of Hacking's realism presented in the preceding section, we consider intersubjective reproducibility from the point of view of a technical-experimental conception of knowledge, the present point may be formulated by saying that *we do not believe that our own observations refer to something real, if they are not in principle reproducible and therefore intersubjectively testable, in relation to the interactions of human beings with the surrounding world.*[7] From this point of view, what makes robustness-as-stability so important in science is the fact that it is intimately connected with intersubjectively testable reproducibility: intersubjectively constant and stable reproducibility is, in the last analysis, the most valuable clue to the independent existence of particular empirical objects (and their properties) and therefore of the truth of propositions which refer to them.

But what about robustness-as-consilience? Is it also connected with intersubjective reproducibility as an essential pillar of scientific experimentation?

There is at least an important sense in which robustness-as-consilience is a valuable clue to reality as intimately connected with intersubjective reproducibility, a sense that will put us in the position to appreciate the limits of the objections mentioned at the end of Sect. 2. In this section I will focus on the first objection; I shall postpone the discussion of the second one until the next section.

[7]This point is certainly broadly pragmatist or Deweyan (cf. for example Dewey 1938: 438 ff.), but many authors considered it as the more general characteristics of the notion of objectivity: cf. for example Janich (1997): 315, and Agazzi (2014): 76.

The first objection was that robustness-as-consilience reasoning does not provide any *autonomous* epistemic warrant if one lacks minimally reliable observational data or procedures; for this reason, the robustness requirement must be integrated by a requirement of "minimal reliability" (Hudson 2014, 18).

It must be conceded to Hudson that reproducibility of robustness-as-consilience is reliable only if it is already present, at least to some extent, at the level of robustness-as-stability. Therefore, we must admit that there is an important sense in which robustness-as-consilience presupposes robustness-as-stability. To underwrite the reliability of a single experimental procedure is the first step one has to take before that of examining whether or not different experimental procedures converge. To Hudson's arguments for the importance of robustness-as-stability I would even add the following one: it does not matter how many purported pieces of evidence we might have, they were, *ceteris paribus*, less reliable than the same pieces of evidence of which one or more elements have already been submitted to new and perhaps more severe tests. But it is apparent, on a little reflection, that the converse is also true.

Take for example sense perceptions in everyday life. On the one hand, robustness-as-stability must be already present at the level of everyday life, when for example we see again the same thing or hear again the same noise in order to increase the reliability of the judgments we have made. If a certain intersubjective reproducibility of what is perceived by the sense organs, taken separately, were not presupposed from the beginning, adding a second reproducibility to increase the reliability of the first one would be pointless.

On the other hand, however, if a certain intersubjective reproducibility of robustness-as-consilience—that is, of the consistency with which different pieces of evidence point to the same conclusion—, were not given, improving the degree of reproducibility of each piece of evidence would be equally pointless. We feel confident in the reproducibility of the results of any of our interventions on reality (robustness-as-stability) if the different interventions are *stably* consilient, that is, if robustness-as-consilience is itself intersubjectively reproducible to a sufficient degree—sufficient for our purposes. The fact that household objects such as flour, sugar, milk, eggs and currants can be handled with high reproducibility would not be a 'fact' (strictly speaking, they would be absolutely useless to us!) if they were not regarded as stably consilient, that is, as coherently placed into the whole of our (in this case domestic) life. For this reason, pace Hudson, robustness-as-consilience is as fundamental as robustness-as-stability.

This point may also be aptly illustrated by a brief analysis of the notion of scientific experiment, which somewhat reflects the unity and distinction of the two senses of robustness for which we have been arguing. For this purpose, we may adopt Mach's definition of experiment. It would lead me too far away from my present purpose to discuss Mach's definition. I shall simply adopt it as one which seems to me to provide a viable definition of experiment.

As Mach wrote, a scientific experiment is based on the 'method of variation' (*Methode der Variation*), whereby some variables are systematically modified to establish which relation of dependence, if any, holds between them:

> The basic method of experiment is the method of *variation*. If every element could be varied by itself alone, it would be a relatively easy matter: a systematic procedure would soon reveal the existing dependences. However, elements usually hang together by groups, some can be varied only along with others: each element is usually influenced by several others and in different ways. Thus we have to combine variations, and with an increasing number of elements the number of combinations to be tested by means of experiment grows so rapidly (a simple calculation shows this), that a systematic treatment of the problem becomes increasingly difficult and in the end practically impossible. (Mach 1905[1976]: 202–203, Engl. Transl. p. 149, translation slightly modified)

As far as robustness-as-stability is concerned, its importance for scientific experiment is evident from the fact that a *reproducible* procedure that reveals "the existing dependences" is part of what makes a good experiment, that is, an experiment able to deliver sound conclusions. From Galilei's experiments on the vacuum to experiments in quantum physics—where we are able to establish relations, not between events, but between the frequency distributions of observed events (on robustness in quantum physics, cf. De Raedt et al. 2014) —the experimenter, on the basis of a hypothetical plan of action, intervenes in a certain way on some aspects (or variables) of the experimental apparatus and notes the corresponding variations of certain other of them. In all cases, namely, a (perhaps only statistically significant) reproducibility is an elementary condition of scientific significance.

However, in so doing, can the experimenter proceed without any reference at all to robustness-as-consilience? It seems to me that it does not, for the following reasons.

Mach's method of variation holds only on the understood condition that no disturbing force intervenes, and in order to assume this with a degree of certainty that is sufficient for his/her purposes, the experimenter might find himself obliged to go beyond the limited conception of robustness-as-stability here considered. As Mach noticed, usually the scientist can vary some elements only along with others because they interact with several others and in different ways. For this reason, we have to "combine variations" (Mach 1905[1976], 203, Engl. Transl. 149), that is to say, we have to combine and to compare different experimental interventions on different experimental apparatuses and laboratories in order to distinguish, in a system of correlated variables, causal from non-causal relationships, and therefore causally dependent variables from causally independent ones.

Simply varying conditions in the same experimental set-up seems clearly to serve the purpose of increasing robustness-as-stability, but varying the experimental apparatus and/or the laboratory is clearly relevant to the aim of increasing the degree of robustness-as-consilience, in the hope that facts already experimentally established (and for this reason already relatively independent) are found to be reproducibly consistent with one another. This follows from the fact that natural laws are in principle to be concretely exemplified in the functioning of 'technical machines' or technical apparatuses built and mastered by the scientists, in which that law is present and operates in a controllable form (for instance, a pendulum for the laws of pendulum motion; determinate measuring instruments for the laws of free fall etc.).[8] But if natural laws can only exist as exemplified in concrete technical apparatuses or measuring instruments in which that laws operate in a reproducible form, a change of experimental apparatus and/or laboratory involves a change in the law, and, more precisely, if this leads to success in reproducing the same outcome, it involves an extension of the old law to a new field of phenomena.

We may illustrate the point from a slightly different point of view. As Gooding rightly pointed out,

> [e]ach test of what may be called the same theory in different laboratories will invoke different background knowledge, enabling assumptions, local resources, and competences […] Experimentation is largely about identifying just the assumptions that matter in *the world as engaged* in that particular laboratory. (Gooding 1992, 69).

In order to obtain a well-reproducible and exemplary experiment, according to Gooding, scientists have to eliminate "nature's recalcitrances", which, on the one hand, impede them in the search after generality, but, on the other hand, indicate a discrepancy between theory and practice and help identify those aspects of the world that are peculiar to contingencies about a particular laboratory at a particular time (cf. Gooding 1992, 69).

Now, it should be obvious on a little reflection that, if scientists can reproduce an experimental result in different laboratories—with different background knowledge, enabling assumptions, local resources, and competences-, they, *in principle*, have already overstepped the limits of robustness-as-stability as usually understood.

Admittedly, there is a huge difference between, on the one hand, the generalisation and extension of a law beyond the limiting conditions of one laboratory, and, on the other hand, Newton's physics, which unified, to use Whewell's words, "remote and unconnected quarters" (Whewell [1840] 1847, Vol. 2, 65), such as Kepler's Laws, the fact that the planets would slightly disturb one another's motions, the perturbations of the moon and planets by the sun and by each other, the fact of the precession of the equinoxes, etc. But this difference, however great it may be, is only of degree, and not a conceptual one. The extensions of important experiments to different laboratories and the unification of disparate phenomena differ only in degree and not in kind.

Finally, the same result may be reached by still another consideration. Polanyi pointed out that reproducibility may depend "on the presence of an unknown and uncontrollable factor which comes and goes in periods of months or years and may vary from one place to another." (Polanyi 1946, 79.) On the one hand, this does not prove anything against our preceding argument, according to which intersubjective reproducibility can be regarded as the most valuable clue to the existence of independent empirical contents if it is taken in relation to the interactions of human beings with the surrounding world. From this point of view, Polanyi's remark only means that the ultimate criterion of claims about experience can only be experience, and exactly for this reason there can be no final certainty in experimental knowledge any more than in any other field of human life. We can never attain absolute

[8]On this point, cf. Buzzoni (2008, Chap. 1).

certainty, since the degree of certainty to which we may come always depends upon the extent to which we are able to make our experience intersubjectively reproducible, and more precisely upon the extent that is sufficient for our practical purposes.

On the other hand, however, it is exactly the possibility of conceiving any experimental reproducibility as a mere coincidence which involves the impossibility of confining intersubjective reproducibility to robustness-as-stability. Again, robustness-as-stability is to be supplemented with some degree of robustness-as-consilience.

For this reason, to suggest, as Hudson does, that it is *always* better to attempt to improve the reliability of a single procedure than to examine the convergence on the same result of different procedures (see e.g. Hudson 2014, 7–8), is not only a hasty and sweeping generalisation, but it is also wrong. In fact, contrary to Hudson (cf. Hudson 2014, 145, with reference to the existence of dark matter), one has to recognize that improving the reliability of a single procedure is hardly decisive or sufficient if taken in isolation. In this sense, Soler is right in complaining that Hudson only shows the deficiencies of what she aptly calls "blind robustness" (Soler 2014, 210), that is, a form of robustness that pretends to be the only method of interpreting intersubjective reproducibility as the distinctive trait of that which we believe to be real (or, what comes to the same thing, of that which we, with a greater or less degree of certainty, assert to be true).

Therefore, it is no fatal objection that robustness-as-consilience reasoning cannot be credited with epistemic warrant if one lacks a minimally reliable observational procedure (which is an expression of robustness-as-stability), since what is important in science is not robustness-as-consilience in itself, but only the synergy (or, if you prefer, the robustness of higher order) between robustness-as-consilience and robustness-as-stability, which are two complementary aspects of the same technical-experimental reproducibility in its more general sense. Robustness-as-stability, if understood as an aspect of experimental reproducibility, is not an alternative to robustness-as-consilience, but it is one of the most important requirements that must be met by a piece of evidence before entering into relations of consilience, which remains an essential integration in the search for more reliable knowledge. What Hudson believes to be "a key part" of his critique of robustness (Hudson 2014, 6), is only a reason for maintaining that it is not sufficient to distinguish between two senses of robustness; it is also necessary to connect them with one another.

Thus, robustness as stability of results is only one aspect of a relation, but an aspect that, in the concrete, must be, at least in principle, inseparably bound up with the other aspect, that is, robustness as consilience of different pieces of evidence. For this reason, scientists look for interdependence and mutual growth and development, that is, for synergy, of the two elements upon which the aforementioned senses of robustness are based.

To sum up, robustness-as-stability and robustness-as-consilience are two side of *the same experimental and intersubjective reproducibility, which, in this sense, may be plausibly presented as the core of robustness in science.* In other words, in principle intersubjective-experimental reproducibility is the most general condition of robustness, which applies to both senses of robustness with which we have been dealing.

5 The Independence Condition for Robustness and the Spiral Synergy of Robustness-as-Stability and Robustness-as-Consilience

However, one might object that the different kinds of testing strategies presuppose each other in a vicious circle. Is there a vicious circle here? The answer to this question will throw into proper perspective the objection raised by the independence condition for robustness.

From an operational point of view, it is easy to acknowledge that there are not only vicious circles, but also fruitful ones, in which self-correcting or spiral procedures take place, with qualitatively new results that each procedure alone could not produce. Everyday life offers plenty of examples. Forging a hammer is a well-known example in this context. One must not already have a hammer to make a new one, for tools that are more rudimentary can be used. Even though similar from some points of view (e.g. from the point of view of beating and forging), they are however different from different points of view (e.g. relatively to their impact resistance). So as one does not already have a hammer to forge a hammer, so robustness in the first sense must not be already definitively founded to reinforce robustness in the second sense, and vice versa.

In the famous fable "The Father and His Sons", Aesop gives another beautiful example taken from practical life:

> A father had a family of sons who were perpetually quarrelling among themselves. When he failed to heal their disputes by his exhortations, he […] one day told them to bring him a bundle of sticks. When they had done so, he placed the faggot into the hands of each of them in succession, and ordered them to break it in pieces. They tried with all their strength, and were not able to do it. He next opened the faggot, took the sticks separately, one by one, and again put them into his sons' hands, upon which they broke them easily. (Aesop [1867]).

The fable has an obvious moral and political meaning ("union gives strength", as Aesop himself commented according to a different version (cf. Aesop [1894]), but it can be used to express well the nature of robustness and the role it plays in experiment and in the experimental sciences. The principle on which the fable rests is also well-known both in the field of technical applications—from braided (metal) ropes to parallel computing: on the one hand, a bundle of sticks is not as easily broken as the sticks taken separately; on the other hand, their combined strength also depends upon the fact that any stick possesses its own relatively autonomous strength.

As we have seen, even though similar in their intersubjective testability, the two senses of robustness do realize testability in different contexts. This difference makes it possible that each element supports the other (or others) to obtain a novel effect, in a fruitful spiral. Tests that are robust in one sense of robustness may be based on tests that are robust in the other sense, and *vice versa*, according to a principle of reciprocal and growing integration. Even though they are reciprocally connected, methods that are robust in different senses can work together to validate empirical claims practically and operationally. Thus, there is no vicious circle here.

This last consideration puts us in a position to deal with the second objection we have still to consider: robustness-as-consilience in science (and the "no-miracle" argument for scientific realism) can be credited with epistemic warrant only if the multiple kinds of available and consilient evidence are independent; but, even if it were true that any single piece of evidence was highly reliable, it still would be very difficult, if not impossible, to ascertain when the independence condition for robustness is satisfied.

Obviously, it is impossible to settle once for all that different causal chains, though experimental—operationally well-established, are independent. In this sense, I fully agree with Nickles: "no improvements in robustness can render […] processes or their products invulnerable to failure" (Nickles 2012, 329). However, the first thing that we should do in meeting this objection fairly is to remove the exaggeration in it. That we cannot prove anything beyond doubt does not mean we do not know anything at all: when both kinds of robustness are properly balanced or proportioned, we get rather convincing evidence as the outcome.

If we interpret the independence condition for robustness in this weakened form, and if what we have said until now about intersubjective reproducibility as interpreted from an experimentalist point of view is generally correct, the following two considerations seem to be sufficient to obtain a fairly satisfying answer to the objection with which we are concerned.

First, Whewell's too-simple view that—as in the case of the Newtonian unification of various disparate laws of mechanics—the consilience of "rules springing from remote and unconnected quarters" can only be explained by the truth of the theory that made it possible (Whewell [1840]1847, Vol. 2, 65) is untenable, at least at the present state of the philosophical discussion. However, as already mentioned, when both kinds of robustness are properly balanced and are progressing in a fruitful circle, we get, for the time being, rather convincing technical-experimental evidence. In other words, what Whewell stresses in this passage is sufficient to give us reasons to believe that scientists have acquired some knowledge about nature that they can provisionally accept today and reason from. And this is what scientists must do if they wish to avoid always coming back to the same point from which they started. In this way, the independence of different pieces of evidence or sources of knowledge is treated *as if*, at least for the time being, it were exceptionless, that is, it is used as a hypothetical and heuristic starting point of later efforts in scientific knowledge. While scientists, at any moment of scientific development, may challenge the particular content of any one claim about the independence of different pieces of knowledge that they have taken as their starting point, they nevertheless have to assume this independence as a regulative ideal for the growth of scientific knowledge (unless there is good reason to suspect there is a common cause able to explain it).

A second point to notice is this. That two fields of our experience are independent is a relative belief or claim, and more precisely one which is relative to the development of our knowledge. The (empirical-experimental) proof that some pieces of evidence were mistakenly believed to be independent *necessarily presupposes new 'robust' findings, robust in both senses of the word, though in different proportion in the different cases.* This also is a fallible judgment, and its reliability is relative to *robustly reproducible* experiences. More generally, any process of testing needs some kind of 'base', and in fact, *at any particular time*, thanks to the mentioned synergy (or robustness of higher order), we can have at our disposal a great number of reliable findings we are not willing, although only provisionally, to put in question.

To maintain that different kinds of robustness will have to cooperate, so far as possible, with one another does not mean that there can be no tension between the two aspects of robustness here considered. On the contrary, the fact that the two senses have to cooperate presupposes that they can be at odds or in tension with each other. Thus, what Nickles notices about robustness as stability—that [r] obustness in one dimension can render a system more vulnerable to catastrophic change in another dimension" (Nickles 2012, 329)—holds also for the relationship between robustness-as-stability and robustness-as-consilience. However, in attempting to solve possible tensions and to build synergy (or, if you prefer, robustness of higher order), it is again plain that, if we wish to remain within the boundaries of scientific discourse, we must recur to arguments that are always robust in both senses of the word, though probably in different proportion.

6 Conclusion

The main purpose of this paper was to cope with some difficulties with which robustness as consilience between different and independent pieces of evidence is confronted, difficulties which have indirect consequences for the problem of scientific realism.

For this purpose, two important meanings of "robustness"—robustness regarded as reproducible stability as against perturbations and variations in parameter values (robustness-as-stability) and robustness as consilience of results from different and independent pieces of evidence (robustness-as-consilience)—was examined.

Following Hacking 1983, who implicitly referred to robustness in his criticism of the no miracle argument, I have argued that, from an experimentalist point of view, the epistemological meaning of robustness-as-stability is not foreign to robustness-as-consilience. More precisely, robustness-as-stability and robustness-as-consilience are to be seen as two cases or specifications of the same thing, that is, of intersubjective reproducibility as the most valuable clue to reality. As such, though conceptually distinct, they are not only intimately connected, but need one another and supplement one another.

There is no vicious circle involved here, but a technical-practical synergy–or a robustness of higher order-, which is at the heart of the experimental method, and which can help us out of two main difficulties concerning robustness-as-consilience. On the one hand, it is no objection that robustness-as-consilience reasoning cannot be credited with epistemic warrant if one lacks a minimally reliable observational procedure, since what is important in science is not robustness-as-consilience in itself, but only in its synergy with robustness in the sense of technical-experimental reproducibility. On the other hand, it is true that it is impossible to settle once for all that different causal chains, though experimental-operationally established, are also independent. However, at any particular time, thanks to the mentioned synergy, we have at our disposal a great number of reliable findings that we are not willing, at least provisionally, to put in question. Thus, from a technical-operational point of view, we may say that the truth is something like 'what shows itself most robust' (in both senses of the world) in the widest context and the longest run.

Acknowledgements I presented an earlier (and much briefer) version of this paper at the Congress of the "Académie International de Philosophie des Sciences" at the University of A Coruña, Ferrol Campus, Spain (September 23–26, 2015). Thanks to all those who contributed to the discussion of the paper during and after the conference, among which Thomas Nickles and Alberto Cordero deserve special mention. I am grateful to Mike Stuart, who read a draft of this article and provided extensive and helpful comments. Italian Ministry for Scientific Research (MIUR) provided funds for this research (PRIN 2012).

References

Agazzi E 2014. *Scientific Objectivity and Its Contexts*. Springer, Berlin Heidelberg New York.

Aesop [1867]. *Aesop's Fables*. Literally translated from the Greek by G. F. Townsend. Routledge, London/New York.

Aesop [1894]. *The Fables of Aesop*, by Joseph Jacobs with illustrations by Richard Heighway. Dover, New York.

Boon M 2012. Understanding Scientific Practices: The Role of Robustness Notions. In: Soler L at al (eds) 2012: 289–315.

Box G 1953. Non-normality and Tests on Variances. *Biometrika*, 40: 318–335.

Bridgman PW 1927. *The Logic of Modern Physics*. New York, MacMillan.

Buzzoni M 2008. Thought Experiment in the Natural Sciences. Würzburg, Königshausen + Neumann.

Buzzoni M 2015. Thought Experiments and Computer Simulations. In: Magnani L, Casadio C (Eds.), Model-Based Reasoning in Science and Technology. Springer, Heidelberg-New York, pp. 57–78.

Calcott B 2010. Wimsatt and the Robustness Family: Review of Wimsatt's Re-engineering Philosophy for Limited Beings? *Biology and Philosophy* 26: 281–293.

Clausing DP. 2004. Operating Window: An Engineering Measure for Robustness. Technometrics 46: 25–29.

Crowe JH, Crowe LM 2000. Preservation of mammalian cells-learning nature's tricks. Nat Biotechnol. 18: 145–146.

De Raedt H, Katsnelson MI, and Michielsen K 2014. Quantum Theory as the Most Robust Description of Reproducible Experiments. *Annals of Physics* 347: 45–73.
Dewey J 1938. Logic. The Theory of Inquiry. Holt, New York.
Duhem P [1906]1914. *La Théorie physique: son objet et sa structure*. Rivière, Paris.
Glymour C 1974. Freud, Kepler, and the Clinical Evidence. In R. Wollheim (ed.), *Freud. A Collection of Critical Essays*, 285–304. New York, Anchor.
Glymour C 1980. *Theory and Evidence*. Princeton, Princeton University Press.
Gooding D 1992. Putting Agency Back into Experiment. In: Pickering A (ed.). *Science as Practice and Culture*. Chicago, Univ. of Chicago Press.
Hacking I 1983. *Representing and Intervening: Introductory Topics in the Philosophy of Natural Science*. Cambridge, Cambridge University Press.
Hacking I 1992. The Self-Vindication of the Laboratory Sciences. In: Pickering A (ed) Science as Practice and Culture, pp. 29–64. Chicago, University of Chicago Press.
Hacking I 1999. The Social Construction of What? Cambridge and London, Harvard University Press.
Hudson R 2014. *Seeing Things: The Philosophy of Reliable Observation*. Oxford/New York, Oxford University Press.
Hughes RIG 1999[2010]. The Ising Model, Computer Simulation, and Universal Physics. In: MS Morgan and M Morrison (eds.), Models as Mediators, pp. 97–146. Cambridge, Cambridge University Press. Quotations are from R.I.G. Hughes, The Theoretical Practices of Physics: Philosophical Essays. Oxford, Oxford University Press, 2010: 164–209.
Janich P 1997. Das Maß der Dinge. Suhrkamp, Frankfurt a.M.
Kitano H 2004. Biological Robustness. *Nature Reviews Genetics* 5: 826–837.
Kosso P 1989. Science and Objectivity. *The Journal of Philosophy* 86: 245–257.
Kosso P 1992. *Reading the Book of Nature. An Introduction to the Philosophy of Science*. Cambridge, Cambridge University Press.
Kuhn TS [1962]1970. The Structure of Scientific Revolutions. Chicago, University of Chicago Press (2nd ed. 1970: quotations are from this edition).
Levins R 1966. The Strategy of Model-Building in Population Biology. *American Scientist* 54: 421–431.
Mach E 1905[1976], *Erkenntnis und Irrtum*, Leipzig, Barth, 1905, 5th ed. 1926. English transl. T.J. McCormack, *Knowledge and Error*, Dordrecht and Boston, Reidel 1976.
Morrison M 2009. Models, measurement and computer simulation: The changing face of experimentation. Philosophical Studies, 143: 33–57.
Nickles T 2012. Dynamic Robustness and Design in Nature and Artifact." In Soler et al. (eds) 2012, 227-266.
Norton S, and Suppe F 2001. Why Atmospheric Modeling is Good Science. In: C.A. Miller and P. Edwards (eds.), Changing the Atmosphere: Expert Knowledge and Environmental Governance, pp. 67–106. Cambridge, MA, MIT Press.
Parker WS 2009. Does Matter Really Matter? Computer Simulations, Experiments, and Materiality. Synthese, 169, 483–496.
Perrin J [1916]. Atoms. (D. L. Hammick, Trans.). New York, Van Nostrand.
Poincaré H 1905. *La valeur de la science*. Paris, Flammarion.
Polanyi M 1946. Science, Faith and Society. Oxford University Press, London.
Popper KR 1959[2002]. *The Logic of Scientific Discovery*. London, Hutchinson (quotations are from the 2002 edition, London, Routledge).
Popper KR [1963]1972. *Conjectures & Refutations: The Growth of Scientific Knowledge*, London, Routledge & Kegan Paul (IV ed., 1972: citations are from this edition).
Popper KR 1972. *Objective Knowledge: An Evolutionary Approach*, Oxford, Clarendon Press.
Psillos S 2011. Moving Molecules Above the Scientific Horizon: On Perrin's Case for Realism. Journal for General Philosophy of Science 42: 339–363.
Putnam H 1975. What is Mathematical Truth? In: *Mathematics, Matter and Method: Philosophical Papers*, Vol. I, pp. 60–78. Cambridge (Mass.), Cambridge Univ. Press.

Salmon WC 1984. *Scientific Explanation and the Causal Structure of the World*. Princeton (N.J.), Princeton Univ. Press.
Sellars W 1963. *Science, Perception and Reality*. New York, Humanities Press.
Singer MA, Lindquist S 1998. Thermotolerance in Saccharomyces cerevisiae: the Yin and Yang of trehalose. Rends Biotechnol 16: 460–468.
Smart JJC 1963. *Philosophy and Scientific Realism*. London, Routledge and Kegan Paul.
Soler L 2014. Against Robustness? Strategies to Support the Reliability of Scientific Results. *International Studies in the Philosophy of Science* 28: 203–215.
Soler L, Trizio E, Nickles T, and Wimsatt WC (eds.) 2012. *Characterizing the Robustness of Science*. Dordrecht, Springer.
Stegenga J 2009. Robustness, Discordance, and Relevance. *Philosophy of Science* 76: 650–661.
Stelling J, Sauer U, Szallasi Z, Doyle III FJ, and Doyle J 2004. Robustness of Cellular Functions. *Cell* 118: 675–685.
Strand A, Oftedal G 2009. Functional Stability and Systems Level Causation. Philosophy of Science 76: 809–820.
Whewell W [1840]1847. *The Philosophy of the Inductive Sciences. Founded upon their History*, London, Parker (2nd ed. 1847; quotations are from the second edition).
Wilke CO. 2006. Robustness and Evolvability in Living Systems. *BioScience* 56: 695–696.
Wimsatt WC 1981[2012]. Robustness, Reliability and Multiple-Determination in Science. In: Brewer M and Collins B(eds) *Scientific Inquiry and the Social Sciences: A Volume in Honor of Donald T. Campbell*, 124–163. San Francisco, Jossey-Bass. Repr. [as "Robustness, Reliability, and Overdetermination"] in Soler et al. 2012: 61–88 (quotations are from this edition).
Wimsatt WC 2007. *Re-engineering philosophy for limited beings: piecewise approximations to reality*. Cambridge, Harvard University Press.
Wimsatt WC 2012. Robustness: Material, and Inferential, in the Natural and Human Sciences? In: Soler et al. 2012: 89–104.
Woodward J 2006. Some Varieties of Robustness. *Journal of Economic Methodology* 13: 219–240.

Cognitive Illusions and Nonrealism: Objections and Replies

Thomas Nickles

Abstract I adopt an agnostic position concerning scientific realism, partly for historicist reasons. In work more fully developed elsewhere, I suggest that cognitive illusions (e.g., the flat future or end-of-history illusion, the maturity illusion, the fish-in-water illusion), often involving insensitivity to past and future history, make realism look more plausible than it is. Strong realists, in effect, claim to be able to foretell the future in denying that there will be either revolutionary or long-term evolutionary change in mature theoretical sciences—no matter how far into the future we go. Most of this chapter is devoted to answering selected objections to the agnostic position. All research tools available to realists are also available to pragmatists, and, contrary to common perception, pragmatists can be tougher minded than strong realists.

1 Are Strong Scientific Realists Tempted by Cognitive Illusions?

When it comes to deep, postulatory theories I am a nonrealist. I deny that we have sufficient evidence and argument to conclude with confidence that even our most mature theories are true, or very nearly, that at best minor tweaking will be necessary.[1] But I am not a die-hard anti-realist. For all I know, some theory might be true. However, we can never be sufficiently sure to say that we *know* it is true. Moreover, whether or not it is true simply *does not matter* for human action, either to further research or to applications. Or so I claim. Thus I remain agnostic, although I place the burden on the side of the strong realists to make their case,

[1]In the future I shall assume the qualifying phrase without repeating it.

T. Nickles (✉)
University of Nevada, Reno, USA
e-mail: nickles@unr.edu

E. Agazzi (ed.), *Varieties of Scientific Realism*,
DOI 10.1007/978-3-319-51608-0_8

including the case for why it matters.[2] Theirs is, after all, the stronger position and thus the one that most needs defending. So, if forced to say so, I do lean toward the anti-realism side of nonrealism. Prominent versions of allegedly strong realism are not even genuine realisms comparable to good-old-fashioned scientific realism.

As I am using the term, strong realism divides into two somewhat distinguishable components, strong epistemological or "epistemic" realism and strong ontological or "ontic" realism. A strong epistemic realist holds that we can, in some cases, *know* that a theoretical claim is true. Meanwhile, a strong ontic realist claims that we can know that the postulated entities and/or processes exist *and that we understand what they are*. That is, the entities are intelligible to the relevant experts, as a scientific replacement for the metaphysics of old, not merely a vague or ambiguous reference to whatever entities or processes may satisfy abstract equations (shades of Frank Ramsey). There are also cases in which we now have good observational or experimental access and control of entities once postulated. In these cases, the theoretical realism is now justified, but it is shallow because it is no longer deeply theoretical. But even in some of these cases ontic realism fails. Do we really even understand what electrons are, let alone quarks? Without entering into further discussion of it, I note that structural realism gives up the "metaphysical" pretensions that strong realists attempt to keep.

There are several good reasons, in my opinion, not to be a strong realist. In this section I focus on one of those reasons, or lines of reasoning, but only briefly. Details can be found in a companion paper (Nickles 2016) where I present a dozen or so cognitive illusions that appear to make strong realism more plausible than it is. We only seem to see scientific truth circling around us everywhere.

Many of the illusions derive from a failure to be sensitive to our place in a longer history, including future history. What I call *the flat future* or *end-of-history illusion* leads us to believe that the future will be relatively flat, uneventful, in relevant respects (cf. Quoidbach et al. 2013). This is understandable because we are concretely aware of past dynamic changes in our personal lives and in world historical development; but we do not have, and cannot have, such knowledge of the future. Our tendency is to project our present views and preferences forward, as if the transformative changes are now over. Applied to science, the temptation is to see the future of a currently successful field as similarly flat—as if the dynamical, innovative history of the field is now ended. On this view, future science will likely produce more applications and connections to other work, but the science a

---

[2]But I am definitely *not* agnostic in the sense that Laudan (1984, 93) has in mind when he criticizes relativists.[T]he relativist insists on remaining agnostic about the respective merits of different

> methodological and evaluative strategies for testing claims about the world… [and] denies in principle that there can be any way of showing one doxastic or belief-forming policy to be superior to another, or one set of methods to be objectively preferable to another.

The proper pragmatist response is to say that there are often clear ways to showing that one process or product works better for specified purposes than another. "It works!" is down-to-earth working in context, not some high-flown theory of rationality that decides the issue.

thousand years in the future will be basically the same as today. Nothing transformative will occur either by rapid revolution or by slow evolution. In short, such a field is over—sterile—as a research frontier.

The flat future is an illusion owing to our particular historical perspective.[3] It helps to remain nonrealist here by remembering that just about every attempt to predict long-term futures has been a ludicrous failure, including scientific and technological futures and even when 'long term' is only 100 years. Ditto for attempts to identify an "end of history." Think of Hegel or Marx or Fukuyama.

The maturity illusion has a similar basis. Why should we think that today's best science is true when past scientists believed the same of their science—which we reject today as badly wrong? The answer some strong realists give is that today's science is now mature, whereas theirs was not. After all, we now know of flaws in their theories, their instrumentation, their experimental design, their goals and standards, etc. But wait! What is to keep our distant successors from saying the same about us? Just because today's most successful theoretical claims seem practically flawless to us does not mean that they really are. Realists who make the maturity move are violating what Mary Hesse once called "the principle of no [historical] privilege" (Hesse 1976, 264). They are treating maturity as an absolute, ahistorical threshold that we have crossed. Any claims that people make about maturity are historically relative and should be historically indexed, for it is only maturity as they characterize it at their particular historical moment.

We are like fish in water in that key parts of our intellectual environment are invisible to us. In some cases, they are probably not yet articulable by us. Therefore, we are unable fully to locate ourselves in historical development, obviously in future development but also in relation to the past—in the sense that future historians and culture commentators (who are also fallible, of course) will make insightful observations about us in relation to our past that we have failed to see—and that perhaps will only become articulable in a still undeveloped future. In that sense we cannot escape from history, even though we can identify progress in overcoming previous historical blindness. Or, to put the point more provocatively, we can overcome history incompletely and only to the extent that we become strong historicists, able to take into account deep and subtle historical perspectives.

What I call "the Copernican illusion" is the idea that, via an extended "Copernican" program, we can gradually identify every human perspectival element and eliminate it, so that, finally, our science will be completely objective, without an ounce of anthropomorphism (Nickles 2017). Our unavoidable historical perspectives already doom such a program. We are not able to stand outside history like gods. The same conclusion follows from the models-all-the-way-down point stressed by Paul Teller (e.g., 2001) and others. Step-by-step removal of approximations and simplifications until the perfect representation of the target system is realized is an increasingly unlikely scenario. The claim that our models will, quite

[3] For this and other points about perspective, see Nickles (2016).

generally in mature science, become as realist as you please is a very strong claim that has not received the degree of justification that it requires.

In my view strong realism reduces historical sensitivity in both directions—future as well as past. Most of the discussion has concerned the past, but insensitivity to the future leads to a tame conception of scientific frontiers, one that tends to replace genuine uncertainty (where even the probabilities and utilities are unknown and often undefined) with mere risks (as 'uncertainty' and 'risk' are defined in standard rational decision theory).

I hope that these examples convey the flavor of the many cognitive illusions presented in Nickles (2017).

2 Objections to the Nonrealist—and Replies

Scientific realism has been the subject of hundreds of papers and dozens of books. I have space to consider only a small selection of objections. The reader might imagine the following as the text of an extended interview.

a. **Realism, antirealism, nonrealism**

   i. *You reject scientific realism, so that makes you an antirealist. Why, then, do you insist on calling your position nonrealist?*

Because I *am* a nonrealist, not an antirealist, although admittedly on the "left," anti-realist side of nonrealism. A person who rejects theism is not automatically an atheist, for agnosticism is a respectable alternative. The same holds for scientific realism. Nonrealism is agnostic concerning the debate between realists and nonrealists (Fine 1986; van Fraassen 2002). Thus I do not say that current theory and practice are false regarding the existence of entities and processes and truth claims about them. I do not deny that black holes and quarks and mirror neurons exist. But neither do I accept that current claims about them are the final, representational truth about these matters, or nearly so. I don't see the point of doing either.

I do not object to "intentional realism," whereby scientists (and science commentators) personally believe that they are seeking the truth about the world and interpret current work realistically, although it is not the best way to describe the aims of science, in my opinion. I do object to what I have termed "strong realism."

   ii. *Several authors agree that global scientific realism is problematic, since there are fields such as quantum mechanics where realism is difficult to defend. These authors therefore retreat to local realism. But you go overboard in the opposite direction, ending up as a global nonrealist and therefore a global skeptic about science. Isn't this a "retro" move that ignores the differences that local realists have noted?*

Local realism is admittedly a difficult issue for me and (I suspect) for many others. I do not think there is any neat line of demarcation that can be drawn.

Instead, we must proceed on a case-by-case basis. In some instances, where we have good experimental control as well as good understanding of what is being claimed about the entities and processes involved, I can be a realist—at least on Mondays, Wednesdays, and Fridays. But, as indicated above, I regard this as shallow realism, precisely because we have such good access. It is not (or no longer) theoretically deep.

Although a nonrealist my position is very far from global skepticism. On the contrary, it is based on comparative reliability claims as well as forward-looking heuristic appraisals. To say that we have good empirical evidence that item A works better than item B for purpose P or that option O is currently more promising than option R does not express a skeptical position. I have no problem with pragmatic claims that we can genuinely justify, but I am indeed skeptical of strong realist claims that no future research will seriously affect present "mature" results.

My realist critic would like to say that I am making assumptions about the future in order to bias the case against realism. However, I am happy to leave the question of the future completely open. My point turns the tables on the objector, for it is the realists, not I, who are making far stronger claims about the future than they seem to realize. Treating all of future history as unknown and unknowable is quite understandable, but to make strong claims about it is quite another matter.

b. **Truth, reference, and pragmatic instrumentalism**

i. *Abandoning truth as a goal throws everything into chaos. We admire science as that human endeavor (alongside the law) that strives for objectivity and honesty and attempts to minimize error, to detect and correct it when it occurs. Denying that we can know scientific truth plays into the hands of the anti-science crowd, the purveyors of scientific uncertainty.*

The objector confuses truth with truthfulness.[4] While I am often agnostic about theoretical truth claims, I fully support truthfulness in the sense of honesty, clarity, and openness to criticism. As I see it, it is more empirically objective and hence more honest to speak in down-to-earth, pragmatic terms of what works better or best among current options than to couch accomplishments in terms of absolute, abstract, ahistorical truth-claims. Besides, the current realist orthodoxy does nothing to stop the naysayers. You may think of my position as treating the products of science as closer to human artifacts such as computers, medicines, accounting methods, and message delivery services.

ii. *You seem to agree with Richard Rorty* (1991) *that truth is only an empty compliment that we pay ourselves when our scientific claims satisfy the usual standards of consistency, high confirmation, and the like.*

[4]See Williams (2002) and Brandom (2009, Chap. 6), "Why Truth Is Not Important in Philosophy".

Yes, I believe that Rorty is basically correct on this point.[5] In deep theoretical matters we have no direct access to the truth. Hence it cannot play a directly guiding role in research. When people are tempted to speak of truth in theoretical contexts it is because the methodological indicators that are available to us (predictive success, problem-solving effectiveness, fertility, etc., plus many lower-level checks in scientific practice) point in favorable directions. In my view (which may also have been Rorty's), truth-talk is a *façon de parler* for reliability and for historically-indexed claims that X is the best theory or model or explanation (or whatever) yet available. Talk of truth is a sort of heuristic short cut!

My own pragmatic orientation again comes into play here. One reason why the Copernican dream of complete objectivity can never be realized (Nickles 2016) is that the dream is a moving target. Over time our scientific interests change, our goals and standards change. The natural world shapes our interests, to be sure, but it is we humans who decide where our research priorities lie. In that respect, we remain in control. A second reason is that frontier research is full of satisficing moves, as that great pragmatist, Herbert Simon, argued (e.g., Simon 1983). Scientific progress is more a matter of moving rapidly through a fertile research terrain, pausing only long enough to get answers good enough to take the next step. Foundational investigations purely for foundational epistemological reasons are mostly a waste of time. Had early foundational epistemology fully shaped methodology, modern science would never have gotten off the ground.

iii. *But surely you believe in well-established theoretical entities? For example, in the decades around 1900 there was a serious debate about whether atoms and molecules exist. French physicist Jean Perrin was able to show that they do exist and are responsible for the random jostling of visible particles in a liquid. Moreover, Perrin was able to establish the value of Avogadro's number—the number of molecules in one mole of a substance—in many different ways. This robust result convinced even noted skeptics that atoms really exist. Surely you can't deny that Perrin produced a strong case for the reality of atoms and molecules!*

Perrin's work was very impressive. It provided the best warrant then available for taking seriously the postulation of atoms and molecules and the statistical apparatus developed by Boltzmann and others, including Einstein, for dealing with large populations of such particles. However, Perrin's work did not really establish what atoms are, and so the depth of the warranted realist conclusion was quite limited (cf. van Fraassen 2009). Under some conditions the "particles" posited by the best physicists are not even particles, and some such particles do not obey classical Maxwell-Boltzmann statistics. In these and many other cases of impressive theoretical and experimental work we lack basic understanding of "what is really going on"—a phrase that I find quite useful in the realism discussion. And, yes, over the rest of the

[5]Although not on some others (see e-i below). See Rorty (1991, 6, 24) and Rorty and Engel (2007).

century scientists learned a great deal about atoms, molecules, and chemical bonds, and so on. However, even today, there is much that we don't know about the exact nature of chemical bonds, even when we can manipulate the chemicals quite reliably. And do we today have a truly intelligible conception of atoms, of the subatomic particles that constitute them, and how all these things interact? I would submit not. We can make amazingly precise predictions in many cases but words fail us when we attempt to provide a literal description about what is going on—literal for the relevant expert scientists, who today still disagree profoundly about the "metaphysical" interpretation of the equations (Brockman 2015). Prediction and control and even "explanation" do not equate to the kind of understanding that I am attributing to strong realism.

I am not arguing that what we now claim to know is wrong. Rather, I am making the two-fold point, first, that such knowledge claims remain a far cry from "good old fashioned realism" (another useful phrase) that claimed that we were discovering—and understanding—what is really going on; and, second, that as science advances it gets stranger and stranger, i.e., harder than ever to understand. In this regard, scientific progress has *not* been in the direction of genuine realism. Core physics is where realists often take their stand on "mature" science. It is not a wise choice, for, insofar as they are correct that quantum theory (quantum field theory, the standard model, etc.) will last, they undermine their own strong realist position.

Do we really know what a molecule is? Molecules are made of atoms, which are made of electrons, protons, etc., which are made of quarks. Do we really know what an electron is in the sense of good old fashioned theoretical realism? The deeper and more "mature" our best science becomes, the weirder and less intelligible are the items it postulates in the old-fashioned sense of realism.

c. History, historicity, and progress

i. *You say that you are a "complete historicist" who gives equal attention to future history and to past history. Letting pass that 'future history' sounds oxymoronic, there are several difficulties with your position. First, future history does not yet exist, so there is little we can say about it that bears on realism. (More below.)*

Admittedly, historians themselves are not in the business of forecasting the future and are professionally hesitant to talk about it. My point, again, is that strong realists themselves make much stronger claims about the future of science than some of them seem to realize. For if they truly believe that a deep theoretical claim is true, or so near the truth that only a bit of tinkering would make it true, then that theoretical domain has reached its terminus, the end of its dynamic history. In other words, strong realists predict (at least tacitly) that there will be (or need be) no significant changes in that domain in the entire future of scientific research. Scientists thousands of years in the future will still recognize these results as correct.

But such a claim is not really a scientific prediction, I would claim, but a forecast or even a prophesy. And, as noted above, forecasts of future knowledge and technical capabilities tend to be ludicrous, even fifty to a hundred years out. Yes, there is little that we can now say about the distant knowledge future, but the strong realist does say things that bear quite strongly on future history. Strong realists are acting as prophets, not warranted scientists.

ii. *Second, you exacerbate the first difficulty by asking us to think about science a thousand years or even tens of thousands of years into the future. That is surely a silly exercise.*

Why? History is (or will be) history, and realists are making claims about the far future, as we have just seen. It is their problem, not mine. I hereby confess to being an extreme historicist, a *complete* historicist.

iii. *Third, appeals to possible future historical change do not raise specific objections to scientific claims today and therefore should not cast doubt on present science. Only actual difficulties need be taken into account.*

The grain of truth here is that working scientists can address only actual (or concretely imaginable) difficulties. But strong realism is not working science in this sense. Strong realists cannot simply write off the future as irrelevant. Thus work such as Kyle Stanford's on unconceived and underconceived alternatives (including those that will never happen to be conceived in the future) has considerable bite (Stanford 2006). The future remains highly contingent. It is rife with uncertainty, not merely risk (in the decision-theory sense). No one has carte blanche to ignore the future. Heightening sensitivity to future history—historicizing the future—is, indeed, a problem for all of us, for this very reason. We have so little information that is concrete enough to make the future "come alive" in the way that our past is (or was) alive; but we must do the best we can.

In Bayesian approaches to science the difficulty of establishing the truth or high probability of a hypothesis, given that only a few serious hypotheses have yet been formulated in a given domain, is known as the problem of the "catchall" hypothesis—the collection of all alternatives to the one being tested. Here is Wesley Salmon's comment on the problem (Salmon 1991, 329):

> What is the likelihood of any given piece of evidence with respect to the catchall? This question strikes me as utterly intractable: to answer it we would have to predict the future course of the history of science. No one is ever in a position to do that with any reliability.

iv. *Fourth, despite your claimed neutrality about the future, your own position is, to some degree, based on your anticipations of what the future will bring. In particular, you seem committed to Thomas Kuhn's claim in* The Structure of Scientific Revolutions (1962) *about occasional scientific revolutions in perpetuity.*

That is a possibility that cannot be ignored, but I am not committed to it. It is also possible that the sciences will evolve slowly; and it is a basic lesson from evolutionary biology that very slow evolution can be as transformative as you please, given enough time. What we may call *the evolutionary illusion* undercuts both the revolutionary view and the strong realists' static view. Evolutionary change can be so slow that it is hard to recognize within a single, productive, human lifetime. The evolutionary scale here is obviously much faster than much biological evolution, yet it is still very different from the scale of ordinary human experience. Note that the realist response to Kuhnian revolution claims—that there was, in fact, a good deal of continuity between the predecessor and successor theory or paradigm—does nothing to address the long-term evolutionary point. You can have all the continuity you want between temporally adjacent work, but over a long enough time the changes can be radical.

Strong realists (those who claim that we know that we already have the truth, or something very close) are committed to denying that the conditions for evolution (any longer) apply to mature fields. Yet all it takes for evolution to occur are variations, selections from among those, some of which are retained as a basis for future variation. Darwin himself pointed out that when these conditions occur (with the right sort of linkage), evolution is not improbable; on the contrary, it is virtually unstoppable.

v. *How can nonrealists account for scientific progress? Hilary Putnam (*1975*, 73) once claimed that "Realism is the only philosophy that does not make the success of science a miracle." Even if the "miracle" language is too strong, nonrealists and antirealists surely have a hard time explaining scientific success, whereas realists do not.*

Here I must be extremely brief. I first note, on the negative side, that Putnam (1988) later abandoned this strong scientific realism for what he called "internal realism." Second, I agree with those commentators who hold that the whole problem of explaining success is a pseudo-problem insofar as it seeks a monolithic answer in terms of Truth or a specially truth conducive Method unique to science. Scientists engage in all sorts of imaginative moves and checks on them. I repeat that it is strong realists who carry the burden of explanation if success means approximation to the truth, since nonrealists do not accept this strong view of progress. Third, I agree with Arthur Fine (1986), Bas van Fraassen (2002), and others that the miracle argument is fatally defective, or at best too weak to support the strong claims made by realists. As Fine noted early on, nonrealists do not regard apparent explanatory success as sufficient reason to conclude truth, owing to underdetermination, etc., yet the inference to truth as the best explanation for the success of science makes a similar move at the metalevel. Fourth, it is also gratuitous for those trying to understand how science works, since appeal to truth provides no new research tool.

Here an ambiguity in "explaining the success of science" emerges. Are we talking about the success of a particular theory or model? Then the realist faces the challenge of explaining the success of Newtonian theory, given how wrong we now think it is. Are we talking about the research activities of scientists, activities such as improved experimental design, better instrumentation, integration that makes possible more cross-checking and hence more robust results, more rigorous reviewing of publications, etc.? Appeal to truth is no help here. Or are we asking the transcendental question: What are the necessary presuppositions for a given science to have gotten progressively better over time? Strong realists appeal to truth in this latter sense, which (if I and other critics are right) has nothing to do with how science works in the previous sense. In saying that approach to the truth is the only, or most important, explanation of progress, realists are transcending working science.

To state the difference between the two positions in dramatic form, nonrealists want to know "the particular go" of science[6] whereas strong realists want to believe that they have seen the Forms before they die. The trouble with this second desire, based on the idea that science has been moving toward the truth over a succession of long-term developments, is that the basic ontologies of these developments have changed radically (even among contemporary experts) and do not resemble successive approximation to a final, stable truth.

On the positive side, I suggest that we think of scientific progress in terms of ongoing research fertility and pragmatic progress in craft traditions and in translational research rather than in terms of progress toward a universal, final truth. We have ways of recognizing improvement, innovation, when it comes to opening new research domains and design of new products and services and their modes of production and distribution, without needing to bring in truth with a capital T. Why can't we evaluate scientific products and processes in the same way? In point of fact, I suggest that we already do, often without realizing it. (Realists will reply that technical progress also calls for a truth explanation, and so we fail to engage again, across the two sides of the above ambiguity.) Pragmatic success must indeed engage the world,[7] but success is also conditioned by human goals and standards as conditioned by global, local, and personal histories.

George Reisch (2016, 15) quotes from a letter Thomas Kuhn wrote to his mentor, James Bryant Conant, who urged Kuhn to drop the paradigm language from a draft of what would become *The Structure of Scientific Revolutions*: "Would you say that home industry was *merely* a less effective way of doing what the factory system later did?" Perhaps not, but my view remains closer to Kuhn's than to Conant's.

d. **Alleged practical virtues of realism**

    i. *Realists want to know the truth about reality. For them research into what really exists is an inspiring quest that energizes them. By contrast, nonrealism dampens curiosity about our universe and thus diminishes motivation for research into the unknown. This was one of Popper's objections to instrumentalism* Popper (1963, Chap. 3).

First, I am not rejecting intentional realism, which is the form of realism that Popper had in mind. Popper was a nonrealist in my sense, given his denial that we can ever know the theoretical truth. If anything, he was a stronger agnostic than I

---

[6]Cf. James (1907, Lecture VI) on young James Clerk Maxwell.

[7]This view leads me to accept a vague sort of pragmatic or instrumental realism.

am. Second, scientists and science analysts who are nonrealists are still attempting to find the best answers available to questions that they (we) find important; and we will continue to do so as curious people can conduct free inquiry. Progressives are always trying to improve on current capabilities. In fact, I believe that the position I defend encourages talented investigators to be bolder, to treat the future as still open to significant change, even in currently mature fields. It encourages people, from experts to the general public, to regard the sciences as an ongoing process of inquiry rather than as a body of established truth, thus in a more prospective and a less retrospective manner.

ii. *Realism helps with what you call heuristic appraisal and is not opposed to it, as you seem to think. Crucial decisions at the research frontier involve which claims are stable enough to use as a basis for other work rather than being the focus of additional testing. Thus realism provides a good guide for what is hot and what is not. Moreover, finding what is real helps to constrain the other parts of the research. In short, realism enables us to distribute research resources more efficiently than does non-realism.*[8]

We get the same results with ordinary experimental and theoretical justification. Calling the result real does not add any epistemic warrant that was not already there. It is just a verbal label that we attach to mark an already made decision to change focus to something now considered more fruitful. Besides adding nothing to the research process, realism poses a danger of changing the focus too soon or too much, thereby discouraging fundamental research and thereby possibly distorting other work.

e. **Postmodernism and policy concerns**

i. *You describe yourself as a pragmatist. Rorty dismisses the debate between realists and nonrealists ("constructivists") on pragmatic grounds, as a difference that does not make a difference* (Rorty 1991, 2; 2007, 34).

While I have some sympathy for Rorty's position (see a–ii above), I think he goes too far here.[9] There is a difference between strong realism and nonrealism (and also antirealism) that can make a difference, both to science policy and to scientific practice in the research choices that scientists make. For example, funding agencies are unlikely to pour massive funding into areas declared to be certifiably true, thus at the end of their innovative history and hence sterile in terms of fundamental research (Nickles 2009, 2017). These also tend to be the most expensive specialties

[8]Thanks here to NorettaKoertge.

[9]Although Rorty is here talking about both realism and antirealism as representational positions. I have no problem speaking of models and such as representations in the sense of our human "take" on the world (the better ones being easy to use and heuristically fruitful). This is different than the realist sense of a representation as an attempt to find a symbolic isomorphism or analogue to a real-world structure, something that "mirrors" nature. Ditto for Rorty's point about objectivity as intersubjectivity ("solidarity") versus objectivity as accurate representation of the world, absent all human perspective.

to support. And even accomplished scientists shift to new problem areas that they judge both more challenging and more fertile, once their old specialty seems to have reached its current goals.

ii. *Realism provides a simplified model of science for public understanding of science and against the anti-science naysayers on evolution, climate change, and much else. The public is fascinated by scientific discovery because they believe it really is deep* discovery *of the wonders of our universe. To reject realism threatens to reduce public support for science, including government support.*

There is a grain of truth to the point about public interest in science, but even among nonrealists there is great interest in learning the latest ideas about what our universe might be as well as in finding out what our universe is not. This last, a sort of "negative realism," is something that I can accept. It is basically Popperian realism: denial that we are likely to *know* the positive theoretical truth in deep domains while being confident that some deep theoretical claims have been refuted. For example, useful as it is, I readily agree that our universe is not Newtonian, that it is shockingly weird compared to the classical universe.

iii. *Opponents of scientific realism are often thought to be intellectually "soft," having fallen into radical (as opposed to moderate) postmodern social constructivism and relativism.*

No doubt some are. But it seems to me that the situation is largely just the reverse—that it is strong and deep realists who cannot fully resist the temptations of the cognitive-historical illusions and emotional psychological factors such as the satisfaction of predictive success and of neat problem solution, factors that go beyond the hard-nosed respect for empirical information and clear logical and mathematical thinking that the sciences, above all, are supposed to honor.

Aknowledgment I would like to thank Marlo Alai, Marco Buzzoni, Wenceslao González, and Tomasz Placek for helpful conversation.

References

Brandom, R (2009) Reason in Philosophy: Animating Ideas. Belknap Press, Harvard: Cambridge, MA
Brockman, J (2015) This Idea Must Die: Scientific Theories That Are Blocking Progress. Harper, New York
Fine, A (1986) Unnatural Attitudes: Realist and Instrumentalist Attachments to Science. Mind 95: 149–179
Hesse, M (1976) Truth and the Growth of Scientific Knowledge. PSA 1976, vol. 2.261–280
James, W (1907) Pragmatism. Longmans, Green, New York
Laudan, L (1984) Explaining the success of science. In Cushing, J, Gutting, G, Delaney, C et al. (eds), Science and Reality, pp. 83–105. University of Notre Dame Press, Notre Dame, IN
Nickles, T (2009) Life at the Frontier: The Relevance of Heuristic Appraisal to Policy. Axiomathes 19: 441–464

Nickles, T (2016) Perspectivism versus a Completed Copernican Revolution. Axiomathes 26(4): 367–382
Nickles, T (2017) Strong Realism as Scientism: Are We at the End of History? In: Boudry, M, Pigliucci, M (eds) Science Unlimited? The Challenges of Scientism. University of Chicago Press, Chicago
Popper, K (1963) Conjectures and Refutations. Basic Books, New York
Putnam, H (1975) Mathematics, Matter and Method: Philosophical Papers, vol. 1. Cambridge University Press, Cambridge
Putnam, H (1988) Reality and Representation. MIT Press, Cambridge, MA
Quoidbach, J, Gilbert, D, Wilson, T (2013) "The End of History Illusion." *Science* 339 (4 January) 2013: 96–98
Reisch, G (2016) Aristotle in the Cold War: On the Origins of Thomas Kuhn's *The Structure of Scientific Revolutions.* In Richards, R, Daston, L (eds), Kuhn's *Structure of Scientific Revolutions* at Fifty. Univ. of Chicago Press, Chicago
Rorty, R (1991) Objectivity, Relativism, and Truth: Philosophical Papers, vol. 1. Cambridge University Press, Cambridge
Rorty, R, Engel, P (2007) What's the Use of TRUTH? Columbia University Press, New York
Salmon, W. (1991) The Appraisal of Theories: Kuhn Meets Bayes. PSA 1990, vol. 2, pp. 325–332
Simon, H (1983) Reason in Human Affairs. Stanford University Press, Stanford
Stanford, PK (2006) Beyond Our Grasp: Science, History, and the Problem of Unconceived Alternatives. Oxford University Press, New York
Teller, P (2001) Twilight of the Perfect Model Model. Erkenntnis 55(3): 393–415
van Fraassen, B (2002) The Empirical Stance. Yale University Press, New Haven
van Fraassen, B (2009) The Perils of Perrin, in the Hands of Philosophers. Philosophical Studies 143: 5–24
Williams, B (2002) Truth and Truthfulness. Princeton University Press, Princeton

Why Do Theories Fail? The Best Argument for Realism

Gerhard Vollmer

Abstract There are arguments for and against realism. None of them is conclusive, but some are better than others. Which one is the best argument for realism? According to some the best argument is the *success* of scientific theories. This is a mistake. The best argument is not the success, but the *failure* of realistic theories. First of all there are much more false theories than true ones. But it is not a question of quantities, but rather a question of logic: Since theories cannot be proven, it may happen that theories are confirmed although they are false. But if they are disproved they must be false. Thus the realist can *explain* why a theory fails (because the world is different from what the theory says). But the antirealist cannot explain the failure. Hence realism has more *explanatory power* than antirealism. That's why we should keep to realism.

1 Introduction

In 1976 the philosopher Hilary Putnam wrote: „Idealism makes the success of science a miracle" (Putnam 1976). And in 1980 the philosopher Bas van Fraassen talked about the success of science as the „Ultimate Argument for realism" (van Fraassen 1980). But there are doubts: Even a false theory *might* be successful. It might be worthwhile then to look for a better argument.

In this paper we try to give a better argument. The best arguments for realism is not the *success* of scientific theories, but the *failure* of so many more theories. In order to understand this we should say what we mean by „success" and „failure".

A scientific theory is *successful* if it helps us to reach our goals. We distinguish cognitive and practical goals. Cognitive goals are:

- useful descriptions,
- testable hypotheses,

G. Vollmer (✉)
Technische Universität Braunschweig, Braunschweig, Germany
e-mail: gerhardvollmer@gmx.de

E. Agazzi (ed.), *Varieties of Scientific Realism*,
DOI 10.1007/978-3-319-51608-0_9

- satisfying explanations,
- correct predictions,
- simplifying theories,
- unifying theories.

Practical goals are:

- simplifying life,
- mastering nature,
- avoiding and evading dangers, mishappenings, disasters,
- diagnosing and healing illnesses.

These goals are not always reached. The relevant theories then are not successful: they *fail*. Normally we prefer successful theories. But failing theories have at least one advantage: They teach us that theories can and do fail and they provoke the question *why* theories fail. This is the great moment for the realist: He has an answer to this question: Whereas most other positions—constructivism, idealism, phenomenalism, positivism, solipsism, transcendentalism—have no answer, the realist does have an answer, and a simple one at that: Theories fail, because they are wrong! Or, in the realist's language, because they are simply not true!

2 Facets of Realism

It is usual to distinguish at least three types of realism: ontological, epistemological, and methodological realism. What do they say?

Ontological (or *metaphysical*) *realism*: There is a real world „outside". It is there even if I don't look and even if nobody looks, and it was there before there were living or conscious beings. This world is the result of a long evolution. It is structured in time and space, by matter and energy. Its existence and structure does not depend on conscious or knowing systems, in particular not depend on human beings.

An alternative position would be *idealism*: The world is (nothing but) a creation of my mind. My mind could exist even if there were no matter. Such positions were invented or even defended by philosophers like George Berkeley, Immanuel Kant, Friedrich Wilhelm Schelling, Georg Wilhelm Friedrich Hegel, Arthur Schopenhauer.

Epistemological realism: This world is *knowable*, at least in parts and approximately. It is reconstructible from its interactions with cognitive systems.

An alternative is *positivism* in its different forms as operationalism, logical empiricism, constructivism, especially radical constructivism, conventionalism. Positivists were Ernst Mach, Ludwig Wittgenstein in part, Percy Williams Bridgman and several members of the Vienna Circle; a conventionalist – at least with respect to the geometrical description of the world – was Henri Poincaré.

Methodological realism: Scientific theories describe (i.e. reconstruct, predict, retrodict, explain) structures of the real world. The termini „true“ and „false“ are to be understood in the sense of correspondence theory: A statement is factually true if and only if the real world has indeed the properties ascribed to it by the statement. (In the long run we search not only for true, but for *minimal* descriptions of the world.)

An alternative is *instrumentalism:* Theories are (nothing but) economic condensations of past experiences and useful instruments for the prediction (or avoidance) of future events. Instrumentalists, at least in some sense, were Andreas Osiander (1498-1552), in his strive to present Copernicus' system as a useful model of the world without truth pretension), John Dewey (1859-1952), thinking and cognition being *instruments* for coping with and adapting to the world).

These three kinds of realism—ontological, epistemological, methodological—depend on their respective predecessors: We cannot perceive or know something if it doesn't exist, and we cannot recognize a factual statement as *true* if the objects talked about don't exist. It is true: even the hardest realist cannot *prove* that the three principles presupposed by him are true. But that any of them is false cannot be proved either. Therefore the realist should present his principles not as certain, but as useful presuppositions of his worldview.

We might be tempted to compare the realist's position with that of Gorgias of Leontinoi (ca. 480–380b.c.), a well-known sophist who declared: „There is nothing. If there were something, it could not be recognized. And if it could be recognized, it could not be communicated.“ With some tolerance we could relate the realist's three principles to Gorgias' three negations. (The third element does not correspond exactly.) We may conclude that the problems concerning realism and its critiques did engage philosophers since long. And we might wonder that the relevant problems are still not solved. Let's try at least to collect some *arguments* in favour of realism.

3 Arguments for Realism

If someone claims to be a realist he might be asked *why* he is one. This question may have two meanings. It might ask for an *explanation*: How come? What happened to you that you became a realist? Are there *causes* which made you a realist? Was it an episode, an insight, a special discovery, an accident? Did something change your opinion or did you just realize that you had been a realist before and became conscious of your own opinion only now? Without striving for completeness we shall collect some possible answers to these *causal* questions. Because the answers *explain* a realist's position, we mark them by an E for „Explanation“.

E1 *Psychological evidence*

There is psychological evidence suggesting, even forcing us to take things we see or hear or feel for real. Although we know that we *might* be wrong, this kind of psychological evidence is very strong, and it is very difficult to overcome it even if there are rational reasons against it.

E2 *Realism is rooted in our language*

More often than not things with a name are taken as real. This can be seen clearly in languages with a definite article „the"—Greek, German, English, and even Romance languages (but not Latin!)—where things with a definite article seem to be concrete, even tangible, hence real.

Since the realistic weight of our senses and of language are so strong we normally do not doubt what they teach us. Thus under normal conditions the *burden of proof* is not with the natural realistic position, but with the sceptical side. For most people, especially for children, it is much easier to trust than to doubt.

E3 *Realism is rooted in our perceiving and handling the world*

The reason is simple: In evolution it paid to be a realist. Doubting and investigating takes time, and if you prefer to check whether something looking like a tiger is *really* a tiger, you might not survive and not bring your sceptical genes into the next generation. In short: It is better to loose some calories by hiding or running than to loose your life. Evolution made us realists. This *explains* why we are realists, but there is no ultimate proof that we are right.

Up to now we asked for the *causes* of realistic thinking. But the question „Why are you a realist?" might also go in another direction: It may not ask for historical events or for evolutionary explanations, but for *arguments*: Are there good *reasons* to be a realist? Are you logically *entitled* to be a realist? Could your reasons be good enough not to *persuade* people but to *convince* them? In what follows we shall try to answer these questions. And we shall number our answers by A for „Argument". But let it be clear from the start that there is *no final argument* proving that realism is right. That makes the case more difficult, but also more interesting.

A1 *Simplicity*

As we saw, the realistic position is simpler than any other competing position because we are born as realists. This fact alone however does not mean that realism must be right. But it is easier to handle.

This leads to another advantage:

A2 *Simple theories, if wrong, are easier exposed as wrong*

If the realistic position is wrong then it should be easier to be *recognized* as wrong than positions which are more complicated and can be adjusted by playing with open parameters. According to Karl Popper (1902–1994) we should prefer simple theories against more complicated ones because with simpler candidates we have better chances to find out what is wrong with them. And knowing this we will have more motivation to search for better theories.

A3 *Which kind of realism?*

There are many positions between pure solipsism (I am the only thing existing) and naive realism (the world is just as it seems to be). If we had to choose we would start with extremes. One extreme is solipsism. Although not strictly refutable, it is no serious choice. The other extreme—naive realism—is easily refuted. In order to make a tenable choice, we suggest to choose a position as realistic as possible, that is compatible with our factual knowledge. One example would be *hypothetical realism*, another *scientific realism.*

Why so cautious? Quantum mechanics is a very successful physical theory without a serious competitor, However, up to now not all problems with its interpretation are solved. It seems that on the microscopic level the concept of realism must be refined. Meanwhile the strategy for a useful concept of realism will be: „let's be as realistic as possible".

A4 *Convergent functions of sense organs in organismic systems*

Simple example: There is a drinking-cup. We can see it *and* feel it (and sometimes —alas—even hear it). This convergence of our perceptions is simply *explained* by the hypothesis that there is just *one real object*: a cup.

A5 *Constancy achievements, especially of perception (in German „Konstanzleistungen")*

An example: Eye and brain inform my consciousness that a moving object moves. If the object is *not* moved, but my head is, then the brain tells this to my consciousness, although the signal reaching my retina moves! Such constancy achievements work in all higher organs concerning colors, forms, depth, direction, objects. They make sense only if there are indeed real objects.

A6 *Invariants in science*

Many things, if not all things, change. But there are some properties which do not change. We call them *constants of nature* or even *laws of nature*. There are conservation laws saying that energy, momentum, angular momentum, electric charge, and others, are constant. This suggests that there must be something outside us and not depending on our existence, our thoughts, our wishes, ideas, views, points of view, preferences, or even our actions. There are things or relations we may find and measure, but not change. Something real, even something objective.

A7 *Convergence of different measuring methods*

Many things in science may be measured. Sometimes the results depend on the methods we use for measuring. But very often it is possible to use *different methods* of measuring, but to find the same values. Example: The Loschmidt constant (Loschmidt's number) is the number of particles (atoms or molecules) in one mole of an arbitrary gas. (One mole of a gas is its molecular weight in gramme.) This number can be measured in different ways, but always with the same result $L = 6 \times 10^{23}$. Such convergence strongly suggests that there is something real behind.

A8 *Convergence of measuring results*

If scientists measure a special quantity, for instance the speed of light, they should lay open the precision of their results and try to improve them. In fact, if we follow the history of scientific research, we see that measurements got better and better and that the limits of possible mistakes get narrower. In this case we may even say that the values *converge* against an asymptote in the mathematical sense—even if the possible error is never zero. (There are no infinitely exact measurements.) This kind of convergence also suggests that there is a true value, a *real* value of a *real* property.

A9 *Convergence of theories*

Sometimes two or more theories compete with each other. It may happen that they start from different premises, but still deliver the same testable results. They might even *contradict* each other in their premises, but still lead to the same testable predictions. Thus we might fear that to every accepted theory there are competitors. We could not tell then which theory is true because all of them would (not be proven, but) *confirmed* by our empirical tests. Hence we could not decide how the hidden reality is structured. This would be an unpleasant situation for scientists, for teachers, for all curious people.

But the normal situation is different. As Albert Einstein remarked 1918 in his Planck lecture *Principles of research*: „The development of physics has shown that of all conceivable constructions always a single one has proved itself unconditionally superior to all others.“ If only one theory survives, then it is legitimate and usual to suppose that we have uncovered part of reality.

True, there are cases where there is not such a final decision between true and false. The best example is quantum physics where there are two theories accepted: Heisenberg's discrete matrix approach of 1925 and Schrödinger's continuous differential equation of 1926. But in this case it turned out (and was proven by Schrödinger himself) that both approaches are empirically *equivalent*. They can be regarded as the same theory in different clothes. There is a difference in the presentation, but no contradiction.

A10 *Increasing coherence of theories*

In addition, scientists sometimes succeed in formulating *more general* theories, to combine two or more theories to one overarching theory. This is also a hint to the existence and uniqueness of reality. Isaac Newton (1643–1727) formulated his law of gravitation for the astronomical objects known at his time. Later it turned out that they apply also to meteorites, moons, stars, double stars, star clusters, „nebulae“, galaxies, clusters of galaxies, even to the universe. And James Clerk Maxwell

(1831–1879) discovered that electricity, magnetism and light can be described by *one* theory, formulated in the so-called Maxwell equations.

Scientists try to formulate even more general descriptions of the world. Whether there is a *World Equation* or a *Theory of Everything* (ToE) describing the fundamental principles of the whole world, and whether it can be found by us, is not known. It is clear however that, even if it exists and can be found, it will be very abstract. But it would still yield a further argument for the existence and uniqueness of a real world.

A11 *Growing objectivity of scientific worldview*

For hundreds and thousands of years, philosophers have discussed, which parts of our knowledge are objective and which are contributed by the knowing subject, hence subjective. We don't claim to have a final answer to this question, but we are convinced that we have made progress. As an example we take light: Colours are subjective, wavelengths are objective. For Johann Wolfgang von Goethe this was by no means clear. His polemics were directed against Newton; but he did not realize that he was talking about the *psychology* of colours whereas Newton had studied mainly the *physics* of colours.

Science has not only discovered new things and new facts, but has also made progress in questions of objectivity. With these results science has also brought more objectivity in our worldview.

A12 *Success of realistic theories*:

In the very beginning of our paper we cite Putnam with his *bon mot* „Idealism makes the success of science a miracle", and also van Fraassen talking about the success of science as the „Ultimate Argument for realism". In both cases the *success* of science is claimed to deliver the best argument. We don't criticize that success is used as an argument. In fact, it is just normal that we prefer to rely on things which we know to have worked already. What we *criticize* is the fact that they took the success of realistic theories as the *best* argument for realism.

True, Putnam himself later changed his mind and invented „internal" realism in 1982; but in 1994 he preferred to propagate „direct" realism giving its due to immediate experience or even to naive realism. So if you refer to Putnam in an argument you should tell which Putnam you are referring to. For the moment, we refer to Putnam the realist of 1976.

4 Finally: The Best Argument for Realism

A13 *The failure of scientific theories may be explained by realism*

The realist says: Theories fail because they are wrong. For the average realist this sounds nearly trivial. He would even think that true theories must be successful whereas false theories must fail. But alas! this is not the case. It might happen that a theory is wrong, but doesn't fail. But if it fails it cannot be true. It is simple logic which is used here: If the premises are true and all deductive steps are correct, then the conclusion is true as well. We may therefore say: In correct deductions truth is *hereditary*. That's why logic is so important for science: It gives a conditional *warranty*: *If* premises are true (and all deductions are correct), *then* conclusions are true as well. Unfortunately this does not apply in the opposite direction: If conclusions are true, this does not imply that all premises must be true. Or in the language of the scientist: If the predictions derived from my hypotheses turn out to be correct, this does not mean that these hypotheses are true. That's a pity, but we cannot help.

Fortunately we have another warranty: If *conclusions* are wrong (although all deductions are correct), then at least one of the *premises* must be wrong. That's the simple reason why we can never prove a theory, but may manage to falsify it.

We may deplore that we cannot prove our theories, but we should be happy that we can at least falsify them. Thus we can learn which ideas, concepts, speculations, hypotheses, models, theories don't work. Falsifications improve our knowledge. Even if we don't know which theory is true, we at least know of some theories that they are *not* true. We learn from our mistakes. This led to the saying: „I learned so much from my mistakes. I think I should make even more mistakes." A colleague of mine, Odo Marquard, put it in an even shorter formula: „We err upwords."

Why was this simple insight so long neglected? Why did Putnam, van Fraassen and most other realists put up with the *success* of realistic hypotheses, theories and so on instead of referring to their *failure*? Again the answer is simple enough: They would say that they are interested in *truth*, not in *falsehood*. That‘s honorable. But the argument for realism would be valid even if up to now *no* theory had been successful. For the realist this would mean that up to now none of his theories was right (or true) and that he should try again with new hypotheses. He might despair and abandon every hope. But he could still say that all his approaches were wrong, false, untrue, and that this was the reason why they failed.

Thus the realist has at least an explanation why his approaches failed. Whereupon the antirealist has *no answer* to the question why his approaches failed. He might invent new words for describing his failure: He might say that it went awkward or haywire; he might just confess that he was unsuccessful. But all these formulations are descriptive, *not explanatory*. I had this experience several times when I asked constructivists *why* some of their approaches failed. They could not answer this causal question, but only *translate* their description into other words. But a simple translation is of course no explanation.

Again this does not mean that realism is proved by the failure of realistic theories. That realism *explains* the failure of theories is a good argument, but not a proof. And it is a better argument than the *success* of science because, as we said, even false theories might be successful. The advantage of realism over antirealism is not provability, but *explanatory power*.

In normal teaching, we are not told that the history of science is full of false theories. For good reasons we are taught only theories which are accepted as true. The reason is simple: There is no time to teach and to learn—beyond the theories we accept—also some of the theories we reject. There is one advantage: We don't waste time in learning things we don't really need. But there are also several disadvantages of not learning about the many false theories:

- We don't learn how difficult it was to find the truth.
- We don't learn how easy it is to make mistakes.
- We don't learn how many false theories were created before the good ones were found.
- We don't learn that even the most admired heroes of science made a lot of mistakes and to use this as a consolation and as an encouragement to try again.
- We erroneously conclude that our predecessors were naive or even stupid. What then will our followers think about us?
- We don't learn that we should not be ashamed of having made mistakes and of still making more of them. (But we should take into account that we may make mistakes and try to find and to eliminate them. The search for mistakes is one way of making progress.)
- We tend to distrust realism instead of taking the failures of our theories as *confirmations* of realism.

Regarding all this we see that thinking about realism and considering arguments for and against realism has not only theoretical consequences, but is also of practical value in learning and teaching. To put it in an aphorism: The more theories are exposed as wrong, the better confirmed is realism!

5 Extinction as an Argument for the Effectiveness of Natural Selection

Natural selection is differential reproduction due to varying fitness. According to evolutionary epistemology, cognitive abilities raise fitness; therefore selection works for better cognition, at least in cases where such improvements are available, useful, and not too expensive. As far as our cognitive ability is reliable we may *explain* this reliability asan effect of natural selection.

The fact that man survived evolution under competition makes plausible the reverse conjecture: Our cognition cannot be too bad. This inverted argument is not altogether compelling. But by this argumentative step we may *justify* our (limited) trust in our cognitive apparatus.

What testifies the effectiveness of natural selection? Usually the multiplicity of species counts as the best argument. Wasn't it the different finches on the Galapagos Islands which aroused in Darwin the idea of natural selection? And if we are told that there exist on earth at least five million, possibly even twenty million different kinds of organisms (not to count bacteria or viruses), all occupying their own ecological niches, then we are even more easily convinced of the effectiveness of natural selection.

But again there is an objection: Could there not be several, even many, ways to put up with the same environmental conditions? Could not totally different species occupy the same ecological niche? Is it then a question not of natural selection, but of mere *chance* which species are formed and populate the earth?

There are in fact arguments supporting this interpretation. We have concrete cases where similar ecological niches are occupied by completely different species: The niche of the great pasture animals is occupied in the savannas of Africa by hoofed animals, in Australia by kangaroos. According to the neutral theory of evolution, developed since 1968 by Motoo Kimura, many genetic changes follow pure chance processes. From this slow and uniform „genetic clock", we can even determine the age of a species, that is the time elapsed since it branched off from its next relatives. Is organic evolution a mere chance process with natural selection playing a minor role or none at all?

Again, there is a *better* argument for the effectiveness of natural selection: the *extinction of species*. For that, we must only recall how many species did already die out. Evolutionary biologists take the number of extinct species to be at least one hundred times that of the existing ones. Why did so many species die out?

As with individuals it happens occasionally that species get extinct more or less accidentally, by a flood, by the outbreak of a volcano, or by the impact of a meteorite. As with individuals, we might talk here of *situational death*. It would be absurd, however, to book all extinctions under situational death. In contradistinction to individual aging and dying, there is, as far as we know, no preprogrammed species extinction. Thus, we must look for external causes in most cases. Hence we might also ask: What makes organisms, populations, species fail?

For selectionists the answer is simple: Populations and higher taxonomic units die *either* because they cannot put up (any more) with environmental conditions, first of all if these conditions change relatively fast, *or* because they are displaced by fitter organisms, possibly by superior members of the same species. Both cases instantiate mechanisms of natural selection.

And how do antiselectionists, e.g. neutralists, explain species extinction? Not at all. The reason is not that they couldn't cope with the term 'extinction'. That species become extinct, even antiselectionists can state and find in need of explanation. However, they cannot offer a plausible explanation. The theory of natural selection has higher *explanatory power* than any antiselectionist theory, say the neutral theory.

Selection theories not only explain the success, but also the *failure* of species. Again, there is a pronounced *asymmetry*: For success several explanations are thinkable, but not for failure. The failure of species is therefore a much better argument for the theory of selection, presumably the best one.

It should be evident by now why we made this side-step to biology, to evolution, to the extinction of species, and to the explanatory power of natural selection: The arguments have the same structure, they are *isomorphic*. In neutral terms the argument reads: The best argument for realism/for natural selection is not the *success* of scientific theories/of organic species, but rather the *failure* of so many others. And if we accept this analogy, we may even admit that the argument for realism and the argument for natural selection *support* each other. That's why we made this side-step to biology. And if we accept this mutual support, we may even take it as a further argument A14 for realism.

References

Putnam, Hilary: *What is „Realism"?* Proceedings of the Aristotelian Society 76 (1976) 177–194, p. 177.
Van Fraassen, Bas: *The Scientific Image.* Oxford: Clarendon Press 1980, pp. 37–39.

The Epistemological Problem of the Objectivity of Knowledge

Fabio Minazzi

Abstract The problem of the objectivity of knowledge arises in scientific research, but it is not exclusive of scientific disciplines. The paper analyzes therefore the problem of human knowledge from the point of view of its intrinsic objectivity. From this point of view reflection on human knowledge allows to redefine the map of objective knowledge criticizing all forms of scientism. Knowledge in each subject area is characterized by rigor and public intersubjectivity. Thus, following Husserl, we can illustrate the value of "regional ontologies", while following Bachelard we grasp the precise role of *ontogenèse* in the various disciplines. The comparison with Agazzi's recent analyses of objectivity and its epistemic contexts allows to reconsider also the tradition of criticism, outlining an original theoretical explanation of the values and limits of the kinds of knowledge we build within different disciplines. The logical neo-realism that inspired this reflection derives from a complex tradition of epistemological thought that feeds on different conceptual and classical traditions.

1 Object and Objective

José Ferrater Mora has written:

> 'Objeto' deriva da *obiectum*, que es el participio pasado del verbo *objicio* (infinitivo, *objicere*), el cual significa «echar hacia adelante», «ofrecerse», «exponerse a algo», «presentarse a los ojos». En sentido figurado *objicio* significa «proponer», «causar», «inspirar» (un pensamiento o un sentimiento), «oponer» (algo en defensa propia), «interponer» (come cuando Lucrecio escribe *objicere orbem radiis* [interponer su disco entre los rayos del sol]. Se puede decir que 'objecto' (*ob-jectum*) significa, en general, «lo contrapuesto» (análogamente al vocablo alemán *Gegenstand*, que se traduce comúnmente por 'objeto'). Los sentidos originario de *objecio* y, por derivación, de *objectum* son útiles para entender

F. Minazzi (✉)
University of Insubria, Varese, Italy
e-mail: fabio.minazzi@uninsubria.it

E. Agazzi (ed.), *Varieties of Scientific Realism*,
DOI 10.1007/978-3-319-51608-0_10

> algunas de las significaciones que se han dado al término 'objeto' (y a los correspondientes términos en varios lenguajes) y a los términos 'objetivo', 'objetivamente', etc. (y a los correspondientes términos en varios lenguas). En la historia de la filosofía occidental estas significaciones pueden dividirse en dos grupos: el que puede llamarse «tradicional», especialmente entre los escolásticos, y el que puede llamarse «moderno», particularmente desde Kant y Baugmarten (Ferrater Mora 2001: III, 2603, italic in the text).

Hence the term *Objective* (Italian *Oggettivo*, French *Objectif*, German *Objecktiv*) refers, in the first instance, to *what exists as an object* or to *what possesses an object*, or, again, *to what belongs to an object*. The semantic field of the adjective "objective" therefore appears much broader, more highly articulated, and more dilated that what the corresponding noun refers to, because the adjective, in addition to the meanings of the noun, has been expanded in order to indicate both everything which appears to be valid for everyone, and what appears to be independent of the subject, as well as everything which is "external" with respect to consciousness or thought and, last but not least, everything which is found to comply with certain rules or methods (Abbagnano 1971: 631–632). However, in the light of this same broad semantic spectrum, if we look at the tradition of Western thought we can identify at least three major and different conceptual senses of the term, resulting in three different traditions of thought:

1. objective understood as what exists as an object in its own right;
2. objective as what possesses an object;
3. objective as what appears to be valid for all

The first meaning refers to objectivity as something which exists as the limit or term of a given operation (whether active or passive). In the tradition of late Scholasticism—for instance, in the reflection of an author like the Doctoris subtilii John Duns Scotus, or in that of Durand de Saint Pourçain, or again that of Francesco Majrone—'objective' constantly refers to what exists as an object of intellect, insofar as it is thought or imagined, without that existence implying, in itself, the existence in the real world, or outside the intellect. In this sense, the notion of universality possesses a certain objectivity only and exclusively in the intellect, because the intellect can understand the lion in its universality without referring to this or that specific and particular living lion. Therefore, in this acceptation of late scholasticism, the concept of objective existence coincides with the possibility itself of elaborating a representation or an idea, which are all objects of thought or perception.

This acceptation of what constitutes what is objective has been widely accepted also by several modern philosophers such as Descartes, who reformulates it and recognises it explicitly in his first *Meditationes de prima philosophia* (in particular in the third meditation), and by Spinoza in his *Ethica ordine geometrico demonstrata* (I, 30; II 8 corollary.), and again by a singular thinker like Berkeley, who illustrates it in his *Siris. A Chain of Philosophical Reflexions and Inquiries Concerning the Virtues of Tar Water* (Section 292). In particular Descartes writes about the objectivity of ideas:

> And in order that I may have an opportunity of inquiring into this in an orderly way [without interrupting the order of meditation which I have proposed to myself, and which is little by little to pass from the notions which I find first of all in my mind to those which I shall later on discover in it] it is requisite that I should here divide my thoughts into certain kinds, and that I should consider in which of these kinds there is, properly speaking, truth or error to be found. Of my thoughts some are, so to speak, images of the things, and to these alone is the title "idea" properly applied; examples are my thought of a man or of a chimera, of heaven, of an angel, or [even] of God. But other thoughts possess other forms as well. For example in willing, fearing, approving, denying, though I always perceive something as the subject of the action of my mind, 15 yet by this action I always add something else to the idea16 which I have of that thing; and of the thoughts of this kind some are called volitions or affections, and others judgments. Now as to what concerns ideas, if we consider them only in themselves and do not relate them to anything else beyond themselves, they cannot properly speaking be false; for whether I imagine a goat or a chimera, it is not less true that I imagine the one that the other. We must not fear likewise that falsity can enter into will and into affections, for although I may desire evil things, or even things that never existed, it is not the less true that I desire them. Thus there remains no more than the judgments which we make, in which I must take the greatest care not to deceive myself. But the principal error and the commonest which we may meet with in them, consists in my judging that the ideas which are in me are similar or conformable to the things which are outside me; for without doubt if I considered the ideas only as certain modes of my thoughts, without trying to relate them to anything beyond, they could scarcely give me material for error (Descartes 1964–1974, vol. VII, pp. 36–37).

> Nunc autem ordo videtur exigere, ut prius omnes meas cogitatione in certa genera distribuam, & in quibusnam ex illis veritas aut falsitas proprie consistat, inquiram. Quaendam ex his tanquam rerum imagines sunt, quibus solis proprie convenit idea nomen: ut cum hominem, vel Chimaeram, vel Coelum, vel Angelus, vel Deum cogito. Aliae vero alias quasdam praeterea formas habent: ut, cum volo, cum timeo, cum affirmo, cum nego, sempre quidam aliquam rem ut subjectum meae cogitationis apprehendo, sed aliquid etiam amplius quam istius rei similitudinem cogitatione complector; & ex his aliae voluntates, sive affectus, aliae autem judicia appellantur.

> Jam quod ad ideas attinte, si solae in se spectentur, nec ad aliud quis illas referam, false proprie esse non possunt; nam sive capram, sive chimaeram imaginer, non minus verum est me unam imaginari quam alteram. Nulla etima in ipsa voluntate, vel affectibus, falsitas est timenda; nam, quamvis prava, quamvis etiam ea quae nusquam sunt, passim optare, non tamen ideo non verum est illa me optare. Ac proinde sola supersunt judicia, in quibus mihi cavendum est ne fallar. Praecipuus autem error & frequentissimus qui possit in illis reperiri, consistit in eo quod ideas, quae in me sunt, judicem rebus quibusdam extra me positis similes esse sive conformes; nam profecto, si tantum ideas ipsas ut cogitationis meae quondam modos considerarem, nec ad quidquam aliud referrem, vix mihi ullam errandi materia dare possent (Descartes 1964-1974, vol. VII, pp. 36-37).

Thus for Descartes ideas, if considered in themselves, can never be false, because they are always images; nor, in his view, should we fear lies in relation with volitions and desires, because even if, for example, we wish bad things, nonetheless, we actually desire and want them. For Descartes we must instead pay close attention to our judgments in relation to which we must be careful not to deceive ourselves. According to Descartes the main mistake in this area is precisely to consider our ideas, which are in us, as authentic copies of the things that are outside us. In short, in his view, only the illusion of dominating this conceptually problematic relationship between the ideas that are in us and what exists outside us, generates the most valid premises (and the deepest ones from a metaphysical point of view) producing a real error. Thus the Cartesian dualism—between *res cogitans* and *res extensa*—is rooted in the image of knowledge itself, giving rise to the classical gnoseological dualism whichhas variously marked the history of modern epistemology.

On the same, overall metaphysical horizon is located a thinker like Spinoza, who, in the corollary previously indicated of his *Ethica*, recognises that:

> Hence it follows that as long as individual things do not exist except insofar as they are comprehended in the attributes of God, their being as objects of thought- that is, their ideas —do not exist except insofar as the infinite idea of God exists; and when individual things are said to exist not only insofar as they are comprehended in the attributes of God but also insofar as they are said to have duration, their ideas also will involve the existence through which they are said to have duration (Spinoza, *Complete works*, 2002, p. 248).

> Hinc sequitur, quod, quamdiu res singulares non existunt, nisi quatenus in Dei attribuitis comrehenduntur, earum esse objectivum, sive ideae non existunt, nisi quatenus infinita Dei idea existit; & ubi res singulares dicuntur esistere, non tantum quatenus in Dei attributis comprehenduntur, sed quatenus etiam durare dicuntur, earum ideae etiam existentiam, per quam durare dicuntur, involvent (Spinoza 2010: II, 8, corollary, 1232).

For Spinoza, too, objectivity therefore coincides with ideas which exist only because they are included in the attributes of divinity. In any case, for all these very different authors of the modern era, "objective" always stands only for what constitutes an intellectual object, and thus does not have anything to do, in the first instance, either with what is real, or with what appears to be unreal. Which, of course, does not preclude the possibility that, on second and fuller consideration, what is an object of the intellect may also be either real or unreal, as the case may be.

Immanuel Kant moved, by contrast, explicitly *against* this traditional view of the objective. He held that objective always refers to something that does not exist only subjectively in the intellect, but which relates to something real and objective. Kant insists in fact in emphasising that the object of knowledge is, and cannot be anything but a "real" object, empirically determinable, as part of a real experience. Thus for Kant "objective" refers to that which has as its object its own particular and specific reality, empirically established, experimentally given and circumscribed. Which, of course, is directly connected with the problem of the "limit" which characterises, more generally, the whole new Kantian conception of knowledge, since for Kant knowledge can be established solely and exclusively within a well-defined and precise "boundary" that we must be able to trace, *normatively*, within the scope of possible experiences. Kant in fact writes, in the first section of the first book of Transcendental Dialectic in the *Critique of Pure Reason*,

> The genus is representation in general (*repraesentatio*). Subordinate to it stands representation with consciousness (*perceptio*). A perception which relates solely to the subject as the modification of its state is sensation (*sensatio*), an objective perception is knowledge (cognitid). This is either intuition or concept *(intuitus vel conceptus*). The former relates immediately to the object and is single, the latter refers to it mediately by means of a feature which several things may have in common. The concept is either an empirical or a pure

> concept. The pure concept, in so far as it has its origin in the understanding alone (not in the pure image of sensibility), is called a notion (*notio*)? A concept formed from notions and transcending the possibility of experience is an *idea* or concept of reason (Kant, english translation by Kemp Smith, 1929, p. 314).

> Eine P e r z e p t i o n, die sich lediglich auf das Subjekt, als die Modifikation seines Zustandes bezieht, ist E m p f i n d u n g (*sensatio*), eine objektive Perzeption ist E r k e n n t n i s (*cognitio*). Diese ist entweder A n s c h a u u n g oder B e g r i f f (*intuitis vel conceptus*). Jene bezieht sich unmittelbar auf der Gegenstand und ist einzeln; dieser mittelbar, vermittelst eines Merkmals, was mehreren Dingen gemein sein kann. Der Begriff ist entweder ein e m p i i s c h e r oder r e i n e r be g r if f, und der reine Begriff, so fern er lediglich im Verstande seinen Ursprung hat (nicht in reinen Bilde Sinnlichkeit) heisst *Notio*. Ein Begriff aus Notionen, der die Möglichkeit der Errfahrung übersteigt, ist die I d e e, oder der Vernunftbegriff (Kant 1787: A 320-B 377).

For Kant, therefore, an objective perception constitutes knowledge only if there is a conceptual mediation by which a concept can refer to several realities unifying them. If, in fact, by perception we refer to a single reality it is only by judgment that we come to an objective understanding, thanks to which the multiplicity of intuitions is combined in a concept capable of pointing out the characteristic traits common to several realities. In this perspective, the idea, by presuming to go beyond the limits of possible experience, inevitably places itself beyond the plane of objective knowledge, because the latter is always a synthesis by which human rationality is configured as a function of critical unification of what is experienced by the senses. In any case, for Kant the conceptual mediation which unifies experience is the ground on which one builds an objectivity which always takes into consideration the equally fundamental dimension of possible and actual experiences.

In this way Kant introduced a very important epistemological breakthrough, because thanks to his innovative critical perspective, objective validity coincides with the reality itself empirically known and tested, where the latter is therefore no longer conceived metaphysically as reality itself or as the classical noumenon (*Ding an sich*), but, precisely as a given empirical reality actually known, through conceptual mediation, always in an objective, experimentally verifiable and controllable way. In other words, for Kant objectivity always coincides with empirical reality (cfr. *infra*).

After Kant's arguments—but, in part, also thanks to them–a third, different, sense of objectivity was delineated, by which it coincided with what appears to be "valid for all" within a specified area of investigation, in which a limited number of rules and also some methods shared by the vast majority of the experts in a particular area of research play a unique heuristic role. Henrì Poincaré was one of the first epistemologists who, in *La valeur de la science* (1905), has best expressed this concept:

> Cette harmonie que l'intelligence humaine croit découvrir dans la nature, existe-t-elle en dehors de cette intelligence? Non, sans doute, une réalité complètement indépendante de l'esprit qui la conçoit, la voit ou la sent, c'est une impossibilité. Un monde si extérieur que cela, si même il existait, nous serait à jamais inaccessible. Mais ce que nous appelons la réalité objective, c'est, en dernière analyse, ce qui est commun à plusieurs êtres pensants, et pourrait être commun à touts: cette partie commune, nous le verrons, ce ne peut être que l'harmonie exprimée par des lois mathématiques. C'est donc cette harmonie qui est la seule

> réalité objective, la seule vérité que nous puissions atteindre; et si j'ajoute que l'harmonie universelle de monde est la source de toute beauté, on comprendra quel prix nous devons attacher aux lentes et pénibles progrès qui nous la font peu à peu mieux connaître (Poincaré 1905; 11–12).

If Poincaré, in his epistemological reflection, had the mathematical sciences as a reference point, very soon his concept of objectivity as the sharing of certain rules within a given field of research was, however, widely accepted, endorsed and claimed also by many other scholars. For example also by a scholar of the social sciences such as Max Weber, who claimed the validity of scientific objectivity in the context of the methodology of the social sciences and even within the social policy itself, and he advanced this claim by appealing to the observation made by Poincare that scientific truth coincides substantially with everything which proves to be valid and shared by all those who are investigating a particular field of study.

This kind of objectivity is configured, therefore, as a form of *intersubjectivity*, because in this acceptation "valid for all" comes to coincide exclusively with what appears to be "intersubjectively valid", and this appears to be because it is configured as something that turns out to be "in conformity with a specific and precise method of investigation". In this particular acceptation, objectivity as intersubjectivity shared by a certain community of scholars, then ends up also by absorbing the concepts traditionally associated with objectivity, that is both its "independence of the subject," and its characteristic (of Cartesian origin) of being "outside the consciousness" of the investigator. The inter-subjective value of knowledge imposes itself on all researchers regardless of their subjectivity, and also of their preferences and of their often very different and conflicting assessments. Moreover, this acceptation of objectivity as mere intersubjectivity comes to present an obvious problem, at least insofar as it seems to have formed as a kind of authentic epistemological weakening of the second sense of objectivity, namely of the realability of human knowledge to grasp effectively objective aspects—and not only inter-subjective ones—of the reality which is studied and investigated. This should then lead us to take into serious and attentive epistemological consideration precisely the notion of objectivity in its second acceptation, in order to determine the true nature of scientific knowledge, which turns out to be true because it is capable of understanding some effective aspects of the world studied by human beings.

2 The Problem of Knowledge in the Western Tradition

Humanity in the course of the very complex and articulated history of knowledge, at least as it has unfolded in the context of the Western tradition, has generally fluctuated between two antithetic metaphysical polarities, like reversed and mutually exclusive mirror images. In one of these polarities human beings have claimed to be able to actually know reality absolutely and definitively, as can be deduced already by the observations with which Galileo introduces his well-known distinction between *sapere intensive* (intensive knowledge) and *sapere extensive*

(extensive knowledge), as explained at the conclusion of the first day of his less famous *Dialogue Concerning the Two Chief World Systems* of 1632 (Galileo 1968, VII, 128–131), where we read specifically the following:

> [...] the human understanding can be taken in two modes, the intensive or the extensive. Extensively, that is, with regard to the multitude of intelligibles, which are infinite, the human understanding is as nothing even if it understands a thousand propositions; for a thousand in relation to infinity is zero. But taking man's understanding intensively, in so far as this term denotes understanding some proposition perfectly, I say that the human intellect does understand some of them perfectly, and thus in these it has as much absolute certainty as Nature itself has. Of such are the mathematical sciences alone; that is, geometry and arithmetic, in which the Divine intellect indeed knows infinitely more propositions, since it knows all. But with regard to those few which the human intellect does understand, I believe that its knowledge equals the Divine in objective certainty, for here it succeeds in understanding necessity, beyond which there can be no greater sureness (english translation by Stilmann Drake).

> [...] l'intendere si può pigliare in due modi, cioè *intensive* o vero *extensive*: e che *extensive*, cioè quanto alla moltitudine degli intelligibili, che sono infiniti, l'intender umano è come nullo, quando bene egli intendesse mille proposizioni, perché mille rispetto all'infinità è come zero; ma pigliando l'intendere *intensive*, in quanto cotal termine importa intensivamente, cioè perfettamente, alcuna proposizione, dico che l'intelletto umano ne intende alcune così perfettamente, e ne ha così assoluta certezza, quanto se n'habbia l'istessa natura; e tali sono le scienze matematiche pure, cioè la geometria e l'aritmetica, delle quali l'intelletto divino ne sa bene infinite proposizioni di più, perché le sa tutte, ma di quelle poche intese dall'intelletto umano credo che la cognizione agguagli la divina nella certezza obiettiva, poiché arriva a comprenderne la necessità, sopra la quale non par che possa essere sicurezza maggiore (Galilei, 1968: VII, 128-129).

This epistemological approach also explains why Galileo himself had then developed a substantially "cumulativistic" conception of the history of science itself, because he believed any knowledge could only be "absolute", and could constitute an outcome on which new achievements and new knowledge, just as absolute and unchangeable, would be superimposed. Which also explains why, in the last phase of his life, after being condemned by the Catholic Church to life imprisonment, the Pisan scientist still held to the certainty that he had not wasted his life. If, in fact, his research programme certainly suffered a dramatic setback with the trial, his abjuration and his condemnation by the Catholic Inquisition, yet psychologically Galileo could still appeal to the satisfactory realisation that during his life he had nonetheless achieved "half a dozen true things", and that these results would remain a definitive achievement for all humanity. For this reason Galileo could rightly conclude his argument:

> I conclude from this that our understanding, as well in the manner as in the number of things understood, is infinitely surpassed by the Divine; but I do not thereby abase it so much as to consider it absolutely null. No, when I consider what marvelous things and how many of them men have understood, inquired into, and contrived, I recognize and understand only too clearly that the human mind is a work of God's, and one of the most excellent (english translation by Stilmann Drake).

> Concludo per tanto, l'intender nostro, e quanto al modo e quanto alla moltitudine delle cose intese, esser d'infinito intervallo superato dal divino; ma non però l'avvilisco tanto, ch'io lo reputi assolutamente nullo; anzi, quando io vo considerando quante e quanto maravigliose

cose hanno intese investigare ed operate gli uomini, pur troppo chiaramente conosco ed intendo, esser la mente umana opera di Dio, e delle più eccellenti (Galilei 1968, VII, 130).

This conception of the claimed absoluteness of human knowledge (which, from the theological point of view, is indeed true blasphemy, because it still puts man on the same level as divinity through science), had therefore already emerged with the genesis of modern science in the seventeenth century. But it has not disappeared since then, as it again coincides with the positivist dream itself (and also with the neo-positivist one), respectively of the nineteenth and twentieth centuries, on the basis of which the interplay of knowledge appears to these movements as a sort of puzzle, albeit extremely complex, in relation to which, however, humanity, in the course of time and the centuries, will finally succeed in gaining all the pieces. Then, by putting them together, humanity will finally obtain an entirely conclusive and absolute picture of the knowledge of the world.

In opposition to this tradition of metaphysical thought, decidedly optimistic about the overall result of human knowledge, there emerged a very different and conflicting metaphysical movement of thought which has constantly insisted rather on the intrinsic limits to our ability to know the world. Ever since the Greek origins of Western thought, the genesis of the tradition of scepticism has also constantly shattered precisely that metaphysical dream of absolute knowledge with undoubted acuteness and original criticism, showing all its many difficulties, contradictions, and undue assumptions, often completely unjustified (Dal Pra 1989). But in delineating, with undoubted acumen and absolute critical originality, this meritorious *pars denstruens*, scepticism has ended up most often by falling out of the frying pan of absolute metaphysical knowledge into the fire of a pragmatic, hermeneutic and equally absolute and metaphysical relativism which denies, aprioristically and prejudicially, the very possibility of human knowledge about the world. Thus knowledge, as already stated by sophists in the fifth century BC and by supporters of hermeneutics in the twentieth century, is ready to reduce knowledge to mere opinion, to a senseless pun, reducing everything to an endless semiosis which would always be devoid of any objective referent, leading to the weakening of human rationality by presenting it as a feeble and inadequate tool for investigating the world objectively.

But between these two antithetical, symmetrical and mutually mirrored polarities there is a third theoretical possibility, one inaugurated, since its Greek origins, by the Socratic research, which seeks rather to pursue a different and much more complex representation of objective knowledge, which is not metaphysically absolute, even though it vindicates the opportunity to delineate some form of objective knowledge conceived as coinciding with the search for truth itself: a search, therefore, that, by its very nature, *can never end*. Within this complex and labyrinthine interplay of knowledge, we must therefore change our traditional image of metaphysical knowledge, *à la* Socrates, by seeking to see the profound and even disturbing connections which always relate our knowledge to our ignorance. In fact, however much we may have undoubtedly increased, in the course of our history of human beings, our technical skills and knowledge, at the same time, as indeed was duly noted by Socrates, we have inevitably increased the knowledge of our own ignorance. Therefore knowledge and ignorance are not on oppositional and confrontational metaphysical horizons, precisely because they coexist and live together, always intertwined, on the same problematic horizon of our limping desire to know the world into which we have been cast. To better understand the critical subtlety of the sophisticated epistemological perspective inaugurated by Socrates, we can consider a sphere and imagine that its content matches overall our technical knowledge and skills, while its surface indicates the border area, i.e. that space, always moving and dynamic, which separates our knowledge from the scope of our own ignorance. Well, the increase in the overall volume of the sphere will mean, inevitably, the increase of our own ignorance, because knowledge and ignorance are two sides of the same coin, i.e. they are two components within which the very possibility of any increase of our though vague, but objective knowledge of the world is developed. So we can no longer think that knowledge and ignorance are two separate and clearly antithetical worlds, since the interplay in which we are involved is much more complex, because knowledge and ignorance mingle and intertwine constantly realising a composite reality of knowledge and technical skills within which knowledge turns up among error, but error too turns up among knowledge. Therefore we must be constantly vigilant to gather this complex interweaving critically, without ever attributing the status of absolute knowledge to this or that metaphysical alternative in a dilemma. Precisely because, as Socrates taught, the more we know, the more we become aware of our own ignorance…

3 Objective Knowledge According to Kantian Criticism

The "Copernican revolution" of Kant programmatically stems from an innovative critical reflection on the nature, the limits, the value, the articulation and meaning of the knowledge which science has made available to man since the birth of modern science. Because in his opinion

> When Galileo caused balls, the weights of which he had himself previously determined, to roll down an inclined plane; when Torricelli made the air carry a weight which he had calculated beforehand to be equal to that of a definite volume of water; or in more recent times, […] a light broke upon all students of nature. They learned that reason has insight only into that which it produces after a plan of its own, and that it must not allow itself to be kept, as it were, in nature's leading-strings, but must itself show the way with principles of judgment based upon fixed laws, constraining nature to give answer to questions of reason's own determining (Kant 1787, B XIII).

> als G a l i l e i seine Kugeln die schiefe Fläche mit einer von ihm selbst gewählten Schwere herabrollen, oder T o r r i c e l l i die Luft ein Gewicht, was er sich zum voraus dem einer ihm bekannten Wassersäule gleich gedacht hatte, […] so ging allen Naturforschern ein Licht auf. Sie begriffen, dass die Vernunft nur das einsieht was sie selbst nach ihrem, dass die Vernunft nur das einsieht, was sie selbst nach ihren Entwurfe hervorbringt, dass sie mit Prinzipien ihrer Urteile nach beständigen Gesetzen vorangehen und die Natur nötigen

müsse, auf ihre Fragen zu antworten, nicht aber sich von ihr allein gleichsam am Leitbande gängeln lassen müsse (Kant 1787, B XIII).

Kant has indeed clearly understood the *synthetic* nature of scientific knowledge, which can never proceed from empirical observation conducted randomly and haphazardly, because in fact it always originates from a precise *conceptual mediation* through which it is possible to read, in an original and innovative way, the empirical reality which is always "read", "reconstructed" and "normed" within a specific theoretical paradigm. Besides, Kant is also well aware that, within this perspective framework in which we must always *be able* to see the world in the light of a determined and circumscribed theory, also the empirical experimental dimension plays an equally important and indispensable role, because only experimental laboratory tests are capable of answering—positively or negatively—our questions, so providing an equally decisive and irreplaceable contribution:

Reason, holding in one hand its principles, according to which alone concordant appearances can be admitted as equivalent to laws, and in the other hand the experiment which it has devised in conformity with these principles, must approach nature in order to be taught by it. It must not, however, do so in the character of a pupil who listens to everything that the teacher chooses to say, but of an appointed judge who compels the witnesses to answer questions which he has himself formulated (Kant 1787, B XIII).

denn sonst hängen zufällige, nach keinem vorher entworfenen Plane gemacht Beobachtungen gar nicht in einem notwendigen Gesetze zusammen, welches doch die Vernunft sucht und bedarf. Der Vernunft muss mit ihren Prinzipien, nach denen allein übereinkommende Erscheinungen für Gesetze gelten können, in einer Hand, und mit dem Experiment, das sie nach jenen ausdachte, in der anderen, an die Natur gehen, zwar um von ihr belehrt zu werden, aber nicht in der Qualität eines Schülers, der sich alles vorsagen lässt, was der Lehrer will, sodern eines bestallten Richters, der die Zeugen nötigt, auf die Fragen zu antworten, die er ihnen vorlegt (Kant 1787 B XIII).

For Kant scientific knowledge proceeds therefore "swayinglike a sailor", precisely because it rests, alternately, on the one hand on the "necessary demonstrations" (i.e. all mathematical and deductive inferences), and on the other hand on "certain experiences" (i.e. the experimental dimension which is realised in a scientific laboratory), as by the way Galileo had clearly argued and illustrated in *The Assayer* (1623). Thus scientific knowledge is established exactly at the problematic critical junction between these two opposite polarities of human rationality and the dimension of experiences: precisely the peculiar nature of this crucial critical "pattern" created by plastic rationality and experimentation determines the specific and detailed configuration of each different and independent area of scientific knowledge.

Even physics, therefore, owes the beneficent revolution in its point of view entirely to the happy thought, that while reason must seek in nature, not fictitiously ascribe to it, whatever as not being knowable through reason's own resources has to be learnt, if learnt at all, only from nature, it must adopt as its guide, in so seeking, that which it has itself put into nature. It is thus that the study of nature has entered on the secure path of a science, after having for so many centuries been nothing but a process of merely random groping (Kant 1787, B XIV).

> Und so hat sogar Physik die so vorteilhafte Revolution ihrer Denkart lediglich dem Einfalle zu verdanken, demjenigen, was die Vernunft selbst in die Natur hineinlegt, gemäss, dasjenige in ihr zu suchen (nicht ihr anzudichten), was sie von dieser lernen muss, und wovon sie für sich selbst nichts wissen würde. Hierdurch ist die Naturwissenschaft allererst in den sicheren Gang einer Wissenschaft gebracht worden, da sie so viel Jahrhunderte durch nichts weiter als ein blosses Herumtappen gewesen war (Kant 1787, B XIII-XIV).

This its ability to investigate nature in accordance to what reason itself invests it with, coincides exactly with the discovery of the new heuristic plane of *transcendentality*, by which Kant constructs the overall theoretical framework of his epistemological meta-critic reflection, deeply innovating not only the whole concept of knowledge, but also the style and modes of human rationality. The onset of this complex tradition of rationalist thought can in fact be identified, quite correctly, in that innovative Socratic criticism by which reason turns into a privileged critical-dialogical investigative tool, where, in fact, rationality is expressed in the ability to establish a critical argued comparison among different and even conflicting positions—according to the well-known saying of Heraclitus that "*polèmos* is the father of all things and king of all" (Diel-Kranz 2006: Heraclitus B 53, 353). Compared to this tradition of critical rationalism Kant, however, introduces an important innovation, identifying the level of transcendence as coinciding with that which operates a priori in every possible experience. In this way, transcendentality, with its normative contribution, makes experiences actual, just because by its intrinsic nature transcendentality is apriori constitutive of any possible and effective knowledge. The transcendental dimension does not concern the object of knowledge in itself, but the modalities with which knowledge is conceived by human beings in their cognitive relationship with the world.

In his reflection Kant therefore highlights precisely the *conceptual dimension* of science with full critical evidence, since in his view science is what it is precisely because—whatever may be Heidegger's opinion—it is always capable of thinking its object by constructing it through a plastic critical interplay of continuous comparison with the experimental dimension. If for Heidegger "die Wissenschaft denkt nicht", for Kant, on the contrary, science is always able to think, because without scientific thought there would be no objective knowledge of our world. However from a critical perspective we should avoid any possible transcendental amphiboly, confusing for example the empirical use of the intellect with its transcendental use, because for Kant only and exclusively the transcendental use of the intellect enables us to underline the indispensable conceptual component of all objective knowledge.

From this precise and innovative critical-transcendentalist perspective, the object (*Gegenstand/Objekt*) for Kant coincides with that in whose concept multiple aspects of an intuition are unified (Kant 1787: 137 B). Of course the object is still only offered through the receptivity of the sense impressions, but we must also add, it is always exclusively *thought* through the spontaneity of concepts (B 74). Therefore, for Kant objects constitute representations determined by the concepts of space and time according to the laws of the unity of experience (B 522). The object of knowledge then coincides with a *phenomenon* (that which appears, *Erscheinung*, or that which is always the result of an interaction); the latter is always conceived by Kant, in strict harmony with his transcendentalist approach, as a *reality of relationship*, i.e. the specific and heuristically valuable normative and borderline reality, within which the objectivity of knowledge is constructed. For this reason the phenomenon is nothing in itself, precisely because it must always be conceived as a set of relational representations of apprehension (B 236), while the empirical object cannot but coincide, rigorously, with a phenomenon (B 299). In the reciprocal relationship which is established between our knowledge and objects, therefore, at least two different cases are possible: first that in which the object makes the representation possible, in which case the representation will be purely empirical, precisely because it is aprioristically not possible. Or, second, representation makes objects possible, determining the latter *normatively*, in accordance with the dictates of a prescriptive epistemology. Naturally, in this second case the representation does not produce the existence of objects as such, but instead makes their aprioristic knowledge possible, which is always realised under the constraint of two conditions: the presence of *the intuition of the senses* by which the object of knowledge as a phenomenon is offered, and the *concept*, by which an object is *thought* corresponding to the intuition of the senses (B 124–125).

In this way the dual structure of the Kantian conception of objective knowledge emerges again, because the concepts of objects in general, are always an aprioristic foundation of any possible, eventual empirical knowledge (B 126). For Kant the real object is therefore created when the concept appears to be in connection with perception and, through the latter, is determined and regulated conceptually through the intellect (B 286), while the necessary object is determined by means of a connection among perceptions implemented according to the categorical structures specific to concepts. On the contrary the transcendental object is configured, however, necessarily, as a merely intelligible cause, a sort of unknown x, about which we do not know, nor can we know, anything. At most it can only be configured as a correlation of the unit of apperception in relation to the unity of multiplicity, perceived through the intuition of the senses, by which the function of critical integration performed by *Verstand,* in fact, unifies the multiplicity of sensible intuitions into the concept of an object. Therefore, using Kantian terms, the two words *Gegenstand* and *Objekt* seem to be interchangeable.

Exactly on this epistemological basis Kant, already in the first edition of the *Critique of Pure Reason*, explicitly declares to undoubtedly support a distinctive form of *transcendental critical idealism* which delineates a specific *empirical realism*:

> The transcendental idealist, on the other hand, may be an empirical realist or, as he is called, a dualist] that is, he may admit the existence of matter without going outside his mere self-consciousness, or assuming anything more than the certainty of his representations, that is, the cogito, ergo sum. For he considers this matter and even its inner possibility to be appearance merely; and appearance, if separated from our sensibility, is nothing. Matter is with him, therefore, only a species of representations (intuition), which are called external, not as standing in relation to objects in themselves external, but because they relate perceptions to the space in which all things are external to one another, while yet the space itself is in us (Kant 1787, A 370).

> Der transzendentale Idealist kann hingegen ein empirischer Realist, mithin, wie man ihn nennt, ein D u a l i s t sein, d. i. die Existenz der Materie einräumen, ohne aus dem blossen Selbstbewusstsein hinauszugehen, und etwas mehr, als die Gewissheit der Vorstellungen in mir, mithin dal *cogito ero sum*, anzunehmen. Denn weil er diese Materie und sogar deren innere Möglichkeit bloss vor Erscheinung gelten lässt, die, von unserer Sinnlichkeit abgetrennt, nichts ist: so ist sie bei ihm nur eine Art Vorstellungen (Anschauung), welche äusserlich heissen, nicht, als ob sie sich auf a n s i ch selbst ä u s s e r e Gegenstände bezögen, sondern weil sie Wahrnehmungen auf den Raum beziehen, in welchem alles ausser einander, er selbst der Raum aber in uns ist (Kant 1787: A 370).

This delineates one of the most important, innovative and even controversial points of the critical perspective, which, not surprisingly, has often been misinterpreted by many commentators. The empirical realism about which Kant writes seems to be an ambiguous perspective, and as such it is not able to satisfy either the tradition of classical metaphysical realism, or even less the tradition of modern empiricism (also metaphysical, as it was delineated by Hume until the logical positivists of the *Wiener Kreis*). Nor is it enough: because it is this demand for a realistic empirical perspective, intrinsic to his critical-transcendental idealism, which similarly enabled representatives of the traditional classical metaphysics to criticise Kant for having remained entangled in a form, however complex, of the metaphysical Cartesian dualism. In this regard, all Kant's stands setting a critical distance between himself and those very different traditions of thought were not sufficient to free him from many critical comments, which variously reduce his perspective to that of idealism or to that, though decidedly antithetical, of metaphysical realism. Despite all these misinterpretations of the criticism, Kant in fact delineated a new and very fruitful epistemological horizon which enables us, even today, to have a better understanding of the intrinsically critical nature of human knowledge, liberating the very notion of objectivity from any metaphysical assumption. Kant stated clearly that

> The transcendental idealist is, therefore, an empirical realist, and allows to matter, as appearance, a reality which does not permit of being inferred, but is immediately perceived. Transcendental realism, on the other hand, inevitably falls into difficulties, and finds itself obliged to give way to empirical idealism, in that it regards the objects of outer sense as something distinct from the senses themselves, treating mere appearances as self-subsistent beings, existing outside us. On such a view as this, however clearly we may be conscious 1 of our representation of these things, it is still far from certain that, if the representation exists, there exists also the object corresponding to it. In our system, on the other hand, these external things, namely matter, are in all their configurations and alterations nothing but mere appearances, that is, representations in us, of the reality of which we are immediately conscious (Kant 1787, A 371–372).

> Also ist der transzendentale Idealist ein empirischer Realist und gesteht der Materie, als Erscheinung, eine Wirklichkeit zu, die nicht geschlossen werden darf, sondern unmittelbar wahrgenommen wird. Dagegen kommt der transzendentale Realismus notwendig in Verlegenheit, und sieht sich genötigt, dem empirischen Idealismus Platz einzuräumen, weil er die Gegenstände äusserer Sinne vor etwas von den Sinnen selbst Unterschiedenes, und blosse Erscheinungen vor selbständige Wesen ansieht, die sich ausser uns befinden; da denn freilich, bei unserem besten Bewusstsein unserer Vorstellung von diesen Dingen, noch lange nicht gewiss ist, dass, wenn die Vorstellung existiert, ach der ihr korrespondierende Gegenstand existiere; dahingegen in unserem System diese äussere Dinge,

> die Materie nämlich, in allen ihren Gestalten und Veränderungen, nichts als blosse Erscheinungen, d. i. Vorstellungen in uns sind, deren Wirklichkeit wir uns unmittelbar bewusst werden (Kant 1787: A 371–372).

Here we can grasp all the revolutionary character of Kant's epistemological stance, which delineated a new image of objective knowledge by liberating it from all the traditional metaphysical assumptions which actually lead to making knowledge absolute. Kant's objective is certainly not easy, because he seeks to maintain the objective cognitive scope of science, liberating it, however, from undue—and traditional—metaphysical absolutisation. Which is not without consequences for the complex relationship with the tradition of scepticism itself, which is criticised by Kant because it claims to aprioristically deny the very possibility of knowledge, but which he nevertheless appreciated to the extent that it helps us to liberate ourselves from all the "metaphysical cramps" of our reason. As Jules Vouillemin rightly observes.

> Avant Kant, la philosophie classique essaie, une fois ébranlés les systèmes théologiques du Moyen Age, de découvrir un absolu susceptible de fonder la vérité. Par exemple, les concepts de substance, de cause, de force, de nécessité reçoivent ce rôle de substituts de Dieu. L'acte révolutionnaire de Kant dans l'histoire de la pensée, sa «révolution copernicienne», a consisté, en reprenant l'analyse de ces différentes notions par rapport à la fonction qu'elles exercent dans la connaissance objective, à montrer que, loin de monnayer l'absolu, elles ne conservaient de signification que dans les limites de l'expérience possible, c'est-à-dire si on les coupait de leur contexte théologique. A cet égard, la théorie kantienne de la connaissance est la première théorie conséquentet vraiment philosophique d'une connaissance sans Dieu. [...] la génie critique a consisté à refuser de replacer le problème de la vérité par celui de la convention ou de la commodité, à maintenir donc la question de la différence entre le réel et l'apparent, entre le nécessaire et le contingent, à l'intérieur d'une philosophie qui s'interdit de parler des choses en soi et qui fonde toute sa physique sur la relativité du mouvement (Vuillemin 1955: 3658–359, ma cfr. anche Holzhey 1970, *passim*).

To this decisive point, Kant himself returned on several occasions, especially in the writings devoted to the criticism composed in the latter part of his life. For example, in his invaluable notes, which he did not publish but were later published by Rink in 1840, he prepared to answer the famous question asked by the Royal Academy of Sciences of Berlin in the last years of the eighteenth century, about *Welche sind die wirklichen Fortschritte, die die Metaphysick seit Leibnizens und Wolf's Zeiten in Deutschland gemacht hat*? In this text Kant intended to show how criticism led metaphysics to take a decisive step forward, enabling it to move from a "critique of metaphysics" to the definition of a "critical metaphysics", because in his view "real metaphysics" cannot but recognise the limits of human reason, delineating the possibility of a new critical ontology, no longer metaphysical. This ontology consists precisely in developing a systematic, critical meta-reflection on the different theories developed in various disciplines in order to finally identify the constitutive transcendental structures of the various disciplines. Kant also claims that a positive result of his investigation lies precisely in having determined that the theoretical knowledge of pure reason can never go beyond the objects of the senses and, in this perspective, he also repeatedly underlines that there is always a close correlation between empirical intuitions and intellectual categories, for it is through the intuition which conforms to a concept, that the object is actually *given*, while if the intuition of the senses is missing the object is thought as empty (because it is just thought). The objectivity of knowledge is built precisely within this critical interplay between pure concepts and the intuitions of the senses, while Kant claims that his criticism makes it possible to avoid both the despotism of empiricism and also the anarchic excesses of a limitless philodoxy. Moreover, Kant realised that his critical perspective allows us to understand the precise heuristic role that "critical metaphysics" always plays within scientific theories, enabling us to construct scientific disciplines which, in order to know the world, have to introduce regulatory epistemological concepts through which we are able to read and interpret the world cognitively, at least to the extent that these same concepts are intertwined with the results of different experimental verifications. Thus Kantian criticism, which also has the limitation of having never investigated the role of technology within the dynamic growth of knowledge made available by scientific research, however, had the merit of underlining that the key problem of scientific knowledge is rooted precisely in its own objectivity. Therefore Kantian criticism gives the subsequent debate—and also the contemporary one—the valuable suggestion to rethink the objectivity of scientific knowledge, liberating it from the traditional metaphysical reductionism of empiricism, and also, conversely, from any undue absolutisation typical of the composite tradition of positivism. Certainly in the Kantian reflection there is no critical awareness that the transcendental structures are not to be conceived as the "fixed stars" of thought, because they too are historical, relative and conventional. However, this our different epistemological awareness can only be based on Kant's discovery of a "Copernican revolution" which constantly reminds us of how the objectivity conquered by scientific knowledge can never be confused with absolute knowledge. But then, how can we conceive the objectivity of knowledge, taking into account its historical relativity and the conceptual changes which characterise the history of science?

4 Can Objectivity Exist Without Objects?

In any case the Kantian critical stance, and its revolutionary epistemology, in which his critical-transcendental idealism is combined—as we have seen—with an original form of empirical realism, was deeply misinterpreted, at least in accordance with two different interpretations. A first interpretation—developed by Friedrich Heinrich Jacobi—attributes in fact, mistakenly, to Kantian criticism the assumption of the existence of objects which would be found *outside* any possible experience, with the known result, paradoxically, that "without the thing itself no one can enter Kantian criticism, but with the thing itself no one can remain within it." Thus, with the notion of the *Ding an sich*, we would be facing an eminently antinomical outcome, because on the one hand the concept of phenomenon can only refer to something that would be situated *behind* the known phenomenal object; on the other hand this noumenal object is defined as something which, in principle, goes beyond all possible experience. So Jacobi concludes that Kant's philosophy necessarily leads to the noumenon, whose assertion is a total denial of the whole criticism, since the notion of a phenomenon always involves a reference to the noumenon, which is, however, out of the reach of any possible knowledge. Kantianism would lead to a sophisticated form of scepticism which denies men any actual knowledge of reality. On the other hand in Kant's criticism other commentators—especially those with a metaphysical orientation—have found the perpetuation of that antithetical realism of the metaphysical "dualist gnoseology" introduced by Descartes with his *cogito* arising from the metaphysical opposition between *res cogitans* and *res extensa*. In this way Kant would have done nothing but perpetuate a metaphysical form of the traditional dualistic realism, becoming embroiled, therefore, in a problem which appears to be unsolvable by its very conceptual approach in principle. According to this interpretation the Kantian *phenomenon* is thus reduced to mere appearance, ending up even coinciding with the secondary qualities of which Galileo had already spoken, contrasting them to definitely *measurable primary qualities* on which the necessary, universal scientific knowledge was indeed based.

In contrast with these two typical and classic misinterpretations of Kantian criticism it is clear that Kant precisely tried to safeguard the possibility of forming a new conception of the objectivity of knowledge which does not deprive it at all of the ability of referring *cognitively* to real objects on which different scientific disciplines are focusing in their investigations and studies. In this sense, the stance of Kantian criticism is in full agreement with the position held by a scientist like Galileo who, though denying that science could grasp the underlying metaphysical essence of reality, was equally sure, however, that scientific knowledge was indubitably able to speak to us of the *passions* of the physical world, i.e. its real features, actual and intrinsic (though not substantial, in the sense of the traditional metaphysical ontology) (see Minazzi 1992: *passim* and Agazzi 1994). But this interesting and fruitful realistic harmony between Kant and Galileo was instead systematically denied by those who misinterpreted Kantian criticism, believing that, for the philosopher from Königsberg, "objective" merely implies a reference to a universal notion, necessary and independent from individual subjects. However, from this point of view, the misinterpretation of Kantian criticism appears to be in complete harmony with the overall evolution of scientific thought, which during the eighteenth and nineteenth century, not to mention the early twentieth century, for reasons deeply connected with the development of different scientific theories, progressively abandoned any strong pretension to "realism", thus undoubtedly relinquishing the notion of objectivity for the weaker notion of mere intersubjectivity. Two different meanings of objectivity have emerged: a strong (or substantial) conception which has gradually and historically been confronted by a weak (or formal) conception. Thus the formal characteristics of knowledge (universality, necessity and independence from the subject) have ended by engulfing the essential characteristics, precisely those which involved a precise reference to the object which thus was comprehended cognitively.

The paradoxical notion of *objectivity without objects* was thus established, progressively, mainly in the development of the physical sciences—especially in the crucial phase at the turn of the nineteenth and twentieth century, during the transition from Newtonian physics to relativistic physics and, even more, to quantum physics—also favouring a misinterpretation of Kantian criticism that intertwined with the Romantic idealist deformation of Kantianism which, in the name of the *Ich Denke* requirements, ended with absorbing Kant's empirical realism into a decidedly idealistic and metaphysically absolutist perspective. The general misinterpretation of Kantianism put down its roots precisely within this convergence between the development of scientific thought and the development of philosophical thought which has finally created the paradoxical image of objectivity without objects.

5 Agazzi: Scientific Objectivity and Its Contexts

Evandro Agazzi in his works (Agazzi 1969, 2012, 2014), while adhering to the traditional interpretation of Kantian metaphysics (comprehended as the fruit of the Cartesian "dualistic gnoseology"), however, on the epistemological and philosophical level has always defended a substantial (strong)conception of the objectivity of scientific knowledge, which has since led him to elaborate a coherent form of scientific realism (Agazzi 1989 and 2014). Taking into account some observations developed in many of his previous works (see Minazzi 2015 Id. 2007), in his most recent, systematic and comprehensive monograph devoted to *Scientific Objectivity and Its Contexts* of 2014, Agazzi draws attention to the importance of the distinction between the *nature* of the physical object and its *structure*:

> We must remember that, according to our view, operations determine the *nature* of the scientific object – or its ontological status as we shall call it later – (as they 'clip it out' of reality, and determine the basic attributes that constitute it), while logical and mathematical construction determine its *structure* (that is, the structure of the set of operational and non-operational attributes involved) (Agazzi 2014, p. 109).

Therefore, the delineation of the mathematical model of a particular aspect of reality gives us only a *structure*, but does still not provide us with a defined range of objects to which we can attribute this same structure. On the other hand if we consider only operating criteria, we can certainly obtain a specific collection of empirical data, thus constituting a material whose *nature*is determined, defining also its belonging to a particular scientific discipline, although its specific structure is still unknown. Therefore the *nature* and *structure* of a particular scientific object may legitimately be distinguished, since we can also have mathematical structures which adapt to different empirical fields, while, on the other hand, a specific constellation of empirical data can be compatible with different mathematical models. This interesting epistemological conception is by the way reconnected by Agazzi to a traditional and classical philosophical approach, which had already matured within mediaeval scholastic philosophy:

> To use a traditional distinction, the concretely existing things, which are immediately present to us in an *intentio prima* (knowledge by acquaintance), cannot be investigated without the elaboration of a conceptual picture of them which can be intellectual scrutinised and is universal and abstract (*intentio secunda*). However, the results of our scrutiny do not concern the conceptual picture, but the concrete referents of the *intentio prima*. In the case of modern science, the *intentio prima* does not properly consist in *perpeptual acts*, but in *operational procedures*, starting from which we elaborate a conceptual model which we then proceed to study (*intentio secunda*). As a result of our study we attribute to certain referents those properties which are compatible with the operational procedures constituting the real tools of our *intentio prima*, and which do not necessarily meet the usual requirements of the *perceptual* (typically, visual) structure of this *intentio* (Agazzi 2014: 113–114, italics in the text).

For Agazzi it is therefore imperative to clearly separate the notion of "thing" from that of "object", maintaining the epistemological awareness that the second term is obtained, through certain *operational procedures*, which enable us, in fact, to "cut out", from among the things of the world, a series of "objects" belonging to different scientific disciplines. In this sense, the "object" is different from the "thing", because some specific properties of "objects" are not "inherent" to "things". Or rather, to bring out an "object", science must be able to identify a specific set of properties by which a particular cognitive objectification of reality is realised. Therefore, in Agazzi's opinion, the scope of reality turns out to be much broader than the horizon of objectivity, because everything pertaining to an "object" is always real, while not everything which is "real" is also an "object". Indeed, the challenge of knowledge consists precisely and exactly in the ability to identify new features of the world of objects within the real world.

This epistemological approach, in addition to its intrinsic merit, can also overcome some common misinterpretations which are often supported by many epistemologists. For example, it is often claimed that classical mechanics might be *falsified* by quantum mechanics, not realising that the two mechanics in question do not refer to the same object at all. Only in this case there would, in fact, be a direct antonymic conflict between the two theories, while, in reality, classical mechanics and quantum mechanics deal with different objects and, therefore, quantum mechanics produces no falsification of classical mechanics, since the two theories allow us to investigate different features of different objects which refer to the world of real things, while configuring different objectifications of the world.

And if we refer to the history of philosophical thought, paying specific attention to the issues of realism that originated in the field of epistemological dualism of Cartesian origin, it is then easy to realise how Agazzi eludes the classical, much debated, dualistic (and metaphysical) problem of modern epistemology, regaining a perspective already discussed by medieval scholasticism, which on the one hand dates back to Aristotle's teaching, while on the other it was recovered and developed in the course of the philosophical debate of the last century, mainly due to the genesis of the research programme on phenomenology delineated by Husserl's early work. Because if, according to the epistemology of Cartesian orientation, things are only known through their representation elaborated by our minds, Agazzi refers instead to the classical approach (which, albeit in different forms, is present both in Greek reflection and in medieval philosophy), according to which knowledge arises from the fact that things are present to our minds. This presence of things in our minds is brought about precisely through an *intentional identity* of thought and reality. Agazzi observes

> In a perception, or in own intellectual intuition, our cognitive capacities 'identify' themselves with objects, thought remaining ontologically distinct from them. This ontological distinction furnishes the correct meaning of "the 'external' world", which, otherwise, would mean everything 'outside my skin'. The *representation* of modern epistemology, from this 'classical' point of view, is simply a thing's 'way of being present' to our cognitive capacities, and is ontologically depend on both, though not produced by either. Modern epistemology, having lost the notion of the intentional identity, gives to representations the status of being direct objects of knowledge that we encounter in our mind (Agazzi 2014: 246, italics in the text).

From Agazzi's perspective the success of the idealistic tradition, especially during the last centuries of modernity, derives from the gnoseological orientation of dualism which undoubtedly favoured the reduction of reality to thought, while, on the contrary, even the classical tradition of the intentional identity of thought and reality can still be blamed for not being able to better define the specific nature of intentional identity. This situation has however changed both with the rediscovery of the classical notion of intentionality made by Husserl's phenomenology, and also with many researches resulting in the philosophy of the mind and also in the field of all cognitive sciences

> aiming at understanding in what this marvellous process (i.e. knowledge) consist, a process throughout which certain beings are able to 'interiorise' the external world without destroying it in order to 'assimilate' it (Agazzi 2014: 246–247).

Certainly, precisely on the basis of these considerations, Agazzi, as already mentioned, however, inclined to consider Kantian reflection as a specific result of the tradition of epistemological dualism, because, in his opinion, the *phenomena* mentioned by Kant would be equivalent to the *secondary qualities* mentioned by Galileo, contrasting them with the measurable *primary qualities* (see Agazzi 2014: 249). Thus the constitutive *relationality* of the *phenomenon* mentioned by Kant is ignored, but this relationality, as we have seen, is indeed essential, because it refers not only to the specific *normativity* of the objects of scientific knowledge, but also enables us to understand the reasons for which Kant always insisted on the *universal* and *necessary* nature of objective knowledge which science is actually able to achieve (as we have explained in the previous Sect. 3). From this point of view Kant created a significant epistemological (rather than simply philosophical) resistance to empiricist deviation, according to which the universality and necessity of scientific knowledge were instead replaced by a sort of *generality* conceived in accordance with the classical traditional and the empiricist conception of induction.

In any case, this highly original philosophical and epistemological anti-Cartesian approach allows Agazzi to assume an epistemological stance according to which scientific realism is based on a different understanding of the role and the heuristic function of scientific theories and which can be well summarised by the following assumptions:

> However, these sentences do not express the *Gestalt* simply as result of *logical* connections. Thus: (a) the aim of theories is far from that of telling a 'literally true story' concerning the world, but is rather of giving the most faithful depiction of a certain (partial) vision of the world under a specific point of view, usually in order to *explain* – often by indicating causal relations between the constituents of the picture – certain empirically accessible features of the world; (b) theories are therefore neither true nor false, but only more or less 'adequate' or 'tenable': (c) nevertheless, *certain singles sentences* of a theory ma bed true or false, and this implies [...] that the objects referred to in *these* sentences exist and have the properties ascribed to them (if sentence is true), or do not exist, or do not possess these properties (if the sentence is false). Clearly, we can agree that theories do not tell a 'literally true story? About the constitution of the world, but this does not commit us to rejecting the several sentences *in* theories are true or false, nor that this has consequences for our appreciation of the real constitution of the world (Agazzi n 2014: 256-257, italics in the text).

From this perspective it is possible to propose a different and innovative image of scientific theories because

> Theories are proposed as hypothetical constructs *intentionally* directed towards the world (i.e. a domain of *referents*); and if we have good reason for accepting a theory, for the same good reason we must accept that their referents exist (Agazzi 2014: 257).

In other words it can be argued that a theory can never be conceived, in a Cartesian way, as a mere representation in itself, because this theory, if anything, can only be a representation thanks to an intentionality which refers directly to the *sense* (what a logician as Gottlob Frege termed Sinn) which is always connected, in turn, with a precise *Bedeutung* (see Frege 1892). Using Husserl's terminology, we could say that it is only thanks to *noemata* that the theory may refer, according to a precise intentionality, to the hyletic world we want to learn and study. In fact these theoretical constructs, intentionally oriented towards the world, can then be more or less "filled" by a hyletic-material component, subsumed within a particular function, that is, within a peculiar *morfé*. Therefore all these different theories turn out to be always oriented towards a particular field of *referents*: "their objects have a kind of intentional or *noematic* reality, and may at best be approximated by concrete objects which sufficiently accurately instantiate the properties these abstracts objects encode" (Agazzi 2014: 259).

Thus, again in this case, Agazzi shows he is the intelligent heir to a classical philosophical tradition which, also through Husserl's phenomenology and the previous reflection in logic by Gottlob Frege, draws directly from Aristotle's *Organon* (1955), because in this way Agazzi fully recovers the fundamental Aristotelian distinction between *semantic logos* and *apophantic logos*. The first limits itself in fact merely to "meaning", while the second "states", i.e. affirms or denies. The *semantic logos* is therefore limited to affirming meanings, without ever posing the problem of the truth or falsity of its utterances, while the *apophantic logos* necessarily and always implies the affirmation or denial of the truth or falsity of a given statement (Agazzi 2012: 109–130). So if the *semantic logos* has to do solely with the *meaning* of linguistic expressions and therefore investigates the precise *understanding* of sentences, on the contrary the *apophantic logos* mainly studies the reference associated with these expressions and then there arises a question relating to their truth or falsity. In Frege's reflection this Aristotelian distinction returns to play a precise role, especially to the extent that for each linguistic expression or sign [*Zeichen*], the German logician distinguishes, in fact, the *meaning* [*Sinn*] from the *referent* [*Bedeutung*]. Frege intended to study above all the *objective contents of thought* [*Gedanken*], so that his semantics highlighted the objective scope of the *meaning* which referred to objective conceptual contents, through which the referents are addressed, according to a certain conceptual mode. Or rather: for Frege, *referents* can only and exclusively be grasped through the fundamental heuristic mediation of *meaning*. But this fundamental level of conceptual mediation was gradually lost sight of by extensional semantics for formal systems which, from Russell to Tarski, have come to reduce the meaning of linguistic signs to their referents or denotational meanings, ignoring the fundamental function of conceptual mediation exerted by *Sinn*. In this way the three-layer semantics of Frege was gradually reduced to a two-level semantics which neglected the fundamental conceptual component of knowledge. And this has occurred precisely because, while the crucial semantically fundamental problem concerns meaning as such, on the contrary the problem regarding reference is not reduced to the semantic dimension (although, of course, it is linked to semantics), because it implies precisely the capacity to grasp the referent, a capacity which takes place outside the scope of semantics, as it involves access to an operational and pragmatic dimension through which theories "grasp" their referents. So Agazzi's epistemology stems also from the need to understand the relationship established between meaning and reference, bearing in mind that semantics is not really connected with reference, because it is concerned first of all with *meaning* (Minazzi 2012). Therefore it is necessary to study the correlation of these three different levels (sign, meaning and reference), as well as understanding the relative autonomy which characterises both the moment of semantic logos and that of apophantic logos. Nor is this all: because for Agazzi semantic analysis then has to be integrated with epistemological analysis, which finally expands to ontological analysis (see Agazzi 2012: 243–264), because he holds that "the thesis of the referentiality of scientific language is the expression of the thesis of scientific realism when one moves from the epistemological level to that of the philosophy of language" (Agazzi 2014: 270). Thus we can justifiably criticise the excessive claims of contemporary epistemological contextualism, and at the same time equally dissociate ourselves from the so-called "linguistic breakthrough" in epistemological analysis, which has typically sought to unilaterally reduce scientific theories to the sole semantic logos. On the contrary, Agazzi writes,

> we recognised then that any science necessarily studies abstract objects, but with the intention of knowing an extra mental reality to which it 'refers', and in which it intends to find 'concrete objects' that are 'referents' exemplifying its abstract objects (Agazzi 2014: 279).

Consequently, in line with this epistemological approach, in each scientific discipline objects cannot but coincide with a set—more or less structured, depending on the degree of accuracy of this same discipline—of attributes which are recognised operationally within a given reality, precisely because they are operationally related to the objects themselves. So these *attributes* are attributed to these *objects* through an *operational* mediation (not based on a mere operation of thought). Which, however, does not preclude a particular referent from possessing also other and different properties, which can be studied by other sciences, or which might be the subject of other possible discourses.

> This means that the *referent* one is reaching thought being 'encountered' by means of certain operational procedures, is much richer than the bundle of operationally defined characteristics or attributes that those procedures are able to demonstrate and 'sum up' in the objects. This does not mean, however, that this same referent cannot be further investigated by means of other criteria of referentiality and become in such a way the subject-matter (the object) of other objectification procedures. Our position could be expressed by saying that there is a *distinction* (but not a *separation* 9 between the realm of objectivity and that of reality in this precise sense: the domain of objectivity is always much more *restricted* than the domain of reality (do not forget that, according to our definition, reality coincides with existence, and therefore encompasses the total domain of being), and it can never be brought coincide with it. Indeed any objectification depends on a point of view *within another point of view* (that is, the broader point of view in which 'things' are given, which is in itself 'contingent' upon a certain historical situation and never encompasses 'the whole' of reality). This must not be understood however, as if there were secluded parts of reality perpetually immune to any objectification. On the contrary, there is no part of reality which may be thought of as not being able *in principle* to undergo objectification (such a claim would be a concealed form of *epistemological dualism*) (Agazzi 2014: 282, italics in the text).

In other words, we fall back into epistemological dualism if we think that behind an electron—identified by its properties—there is a mythical substance which we are never able to know because we can only know its specific properties. Thus we can no longer conceive an electron as a thing to which some properties are attributed, because it is necessary instead to understand an electron as an object which is being built thanks to and through these properties.

> An object is to be considered as the 'structured' totality of the objectively affirmable properties and not as a mysterious substratum of these properties. This might sound as a Humean positivism, but it is not, since we do not maintain that such properties are exclusively our perceptions: they are ontological aspects of reality, and may even be perceptually unattainable (Agazzi 2014: 283).

6 Logical Neo-realism and the Problem of Objectivity, Between Husserl and Socrates

In the light of the considerations treated in the previous paragraphs, the inherently problematic complexity of the objectivity of human knowledge emerges. And if we especially consider the explosive evolution of knowledge since the birth of modern science intertwined, *ab origine*, with increasingly pressing and sophisticated technological devices, we can better understand how the problem of the objectivity of knowledge coincides with the understanding of the dynamics of the critical-conceptual growth of the technical assets of knowledge (for an in-depth analysis of this subject see the precise observations of Geymonat 1977 and 1970–76, on whose work see also Minazzi 2001, 2004b, 2010).

A good critical understanding of the specific nature of objective knowledge made available by science is the decisive aspect of our problem: *hic Rhodus, hic salta* (to quote Marx). Incidentally to deal with this problem we can still refer to the interesting and precise indications which Galileo gave in *The Assayer* (1623), in which, from a purely methodological point of view, he consistently insisted in reaffirming that scientific knowledge always proceeds from a critical and problematic intertwining between "necessary demonstrations" and "meaningful experiences". Mathematics and the experimental dimension are the opposite poles, and yet mutually integrated, of a new style of research, the very one that was inaugurated by the scientific approach, in whose name these two opposing polarities, referring on the one hand to the autonomous and creative force of mathematical thinking and, on the other hand to the constraints established by experimental verification, are capable of being integrated with each other so happily and fruitfully that we can better understand the world and realise how things really are. The interesting aspect of Galileo's approach is rooted precisely in his programmatic refusal to unravel, with an algorithmic formula, the very nature of this relationship which can be established between strictly deductive mathematical inferences and the multiple practices of technological and scientific laboratory experiments. In other words, Galileo did not want to take that (metaphysical) step which René Descartes delineated in his famous *Discours de la méthode* (1637). For Descartes, science can and must certainly be reduced to his method. In this way he helped to spread an authentic "Cartesian syndrome" (see Pera 1992), by virtue of which, from Descartes up to Popper, almost every epistemologist, for three centuries, debated—and often quarrelled about—what the "real" and "authentic" method of science can be. On the contrary, Galileo escaped from this misleading methodological approach, and preferred to delineate only the opposing polarities within which the most advanced and original scientific discourse is continuously organised in ever-renewed original forms. Did he do it because he lacked an adequate critical awareness of the methodological problems of science, or, on the contrary, because he had conducted different scientific investigations and, therefore, *in corpore vili* of the activity of the militant scientist, he gradually developed a more articulated sophisticated and critical awareness of the complexity of the methods which every scientific discipline must always put in place and construct in order to achieve an objective knowledge of the world which it seeks to investigate?

If we opt for the second answer, Galileo's methodological indications appear to be in perfect harmony with the more mature considerations of another great Western physicist, Albert Einstein's. In fact Einstein, reflecting on different epistemological perspectives legitimately suggested by his many fundamental scientific contributions, realised how the activities of a militant scientist may seem, at least in the eyes of systematic epistemologists, to be the result of the attitude of an "unscrupulous opportunist" (see Einstein 1949). An "unscrupulous opportunist", because militant scientists can be *realistic*, because they seek to describe a world that exists independently of the acts of perception, or can be *idealistic*, because they consider theories as the result of the free invention of the human imagination, or else they can be *positivist*, because they believe concepts are justifiable only to the extent that they provide a logically rigorous representation of the relations which can be established among sensory experiences, or can even be *Platonic* (or *Pythagorean*), at least insofar as they consider the criterion of logical simplicity as the preferred means of scientific research. In the militant action of a scientist all these very different and contrasting epistemological positions are indeed possible, because scientists are little interested in establishing themselves as systematic and consistent epistemologists: their problem is different, namely to learn about the relevant and objective aspects of the world. Hence scientists strive as far as possible to adhere to their objects of study, the way a limpet does to its rock. Or, if you prefer another comparison, the militant scientist is perhaps like a lover (of knowledge!), ready to make any move, even the most unscrupulous one (be it realistic, idealistic, positivistic, Platonic, Pythagorean, etc.), in order to capture "the object of their love", that is the effective objective increase of our technical-cognitive patrimony.

Ergo, we cannot do innovative scientific research if we stick to an epistemological belief to which we swear eternal fidelity, because militant scientists who significantly innovate the scientific tradition, must always be able to build and intertwine, case by case, discipline by discipline, that particular relationship which could possibly be structured between the polarity of creative thinking and the rigid constraints of experimental verification. Exactly as Galileo argued in 1623, since during his own scientific work he had evidently experienced the truth of this inherent flexibility of scientific practice. Moreover, unsurprisingly, during his lifetime Galileo worked on very different scientific disciplines, passing from astronomy to the dynamics of rigid bodies, from some biological observations to the discussion of the problem of the flotation of solids on liquids, from the consideration of problems of mathematical analysis to the study of the resistance of materials, etc. etc. The extent and complexity of his investigations must have led Galileo to develop a full critical awareness that the claim to arbitrarily and unilaterally reduce science to this or that specific method (inductive, deductive, conventionalist, abductive, verificationist, falsificationist, idoneist, etc.) was an approach which appeared to be quite inadequate to explain the actual and intrinsic complexity of scientific research.

Reaching this more sophisticated methodological awareness, Galileo, *contra* Descartes (and also *contra* the endless array of later epistemologists, who on the contrary usually shared the "Cartesian syndrome"), Galileo thus opened up the prospects for a new and very different epistemological, methodological and philosophical assessment of scientific knowledge. With wise methodological caution, Galileo in fact reminds us, *negatively*, that the objective knowledge we can really achieve in different areas of investigation can never be coerced into this or that abstract method, into this or that rigid methodological rule, which we should then limit ourselves to applying systematically as a kind of template. *Positively*, it reminds us that knowledge is always the result of a free, creative and oppositional interplay (with the natural world, which by its nature is always "deaf and inexorable"), within which we must be able to use all our intelligence, all our creative imagination, all our technical skills and also all our tenacity in order to finally achieve—if we're lucky—an objective increase in our knowledge of the world.

Moreover, Galileo himself, at least in relation to this perspective opening up of the horizon, took some steps backward. He did so, for example, as already mentioned, when he believed he could still attribute to human knowledge (understood as *sapere intensive*) an absolute and unchangeable scope, and as a result elaborated a coherent cumulativistic picture of scientific knowledge. But this is now an absolutistic conception of scientific knowledge, which can no longer be accepted for several reasons. If anything, always on this terrain of contemporary scientific knowledge, we will feel a greater philosophical and critical harmony with Einstein's different positions. He openly recognises that the objectivity of knowledge can only be established in the effective and complex intertwining realised by associating different accentuations of the various components which always exist between the dimension of the *Lebenswelt* (common sense especially woven together and built by the impressions of the senses and by the way of living of all people), and that of pure ideas, identified thanks to an effort of creative imagination, by which we can then precisely and *deductively* develop theories whose consequences, through the fundamental practical mediation of technology and experimental trials, enable us to return again to that same *Lebenswelt* in which we live, and perhaps to amend it in the light of new knowledge and new tools. But, again, this very complex and detailed pattern of knowledge delineates a particular interplay between the world of ideas and that of the senses, pushing us back to that critical perspective from which Kant started in order to delineate his "Copernican revolution". If anything, Einstein observed, our distance from the Kantian architectonic system can best be measured in our awareness that the ideas and categories of thought are not unchangeable constellations of fixed stars, because they are, on the contrary, free creations and, as such, they delineate an aprioristic *historical* and *relativistic* structure, i.e. one that is conventional.

The critical understanding of this very complex universe of scientific discourse did not fail to be present in some writers and philosophers of the twentieth century belonging to the current of thought of European rational criticism. While appreciating the rigour of the neo-positivistic lesson, they also saw its limitations rooted in its (metaphysical) inability to understand the heuristic value of the notion of Kantian transcendence, thus remaining a prey to a radically and rigidly empiricist approach. A radical empiricist vision, in whose name they claimed, invariably, that we can actually reduce the abstract level of ideas and theories to empirical facts alone. As is well known, precisely in an effort to implement this impossible and utopian programme of epistemological reductionistic research, neo-positivism went through different *phases* and *forms*, through which it gradually and constantly weakened precisely the radical empiricist reductionism of its origin, expressed perhaps in the most abrasive terms in the first schematic formulation of the Viennese verification principle, the legitimate offspring of the equally metaphysical approach in Ludwig Wittgenstein's *Tractatus Logico-Philosophicus* (1921).

Hencelogical empiricism, in the course of its history, finally realised, especially in its American phase, that between the abstract level of theories and that of experimentation there is always a relationship of reciprocal relative autonomy. But this correct epistemological awareness naturally and inevitably coincided with the theoretical dissolution of the neo-positivist research programme itself, as it emerges from the contributions of an author like Carl Gustav Hempel.

Those who were never uncritically ensnared by the metaphysical approach of the neo-positivists found themselves in the best position to appreciate the fruitfulness of their research, at the same time devising a critical, sophisticated and aware reactivation, of the tradition of transcendentalism. This complex and fruitful operation emerges for example, with undoubted clarity, in the reflections of the philosopher Giulio Preti, who unsurprisingly developed a very interesting and fruitful form of historical objective transcendentalism in the light of which he proposed an interesting and fruitful research programme of neo-realistic logic, which makes it possible to reconsider the problem of the objectivity of knowledge by tying up the loose ends of all our previous considerations. To define his starting point Preti wrote as follows:

> It is rather a historical-objective transcendentalism, which surveys the constructive forms of the various universes of discourse through a historical-critical analysis of rules of method that have been imposed historically and still apply in knowledge, etc. In short, it is a transcendental Ontology (or rather transcendental ontologies) which does not claim to understand the forms and structures of a Being in itself, but seeks to determine the way (or ways) in which the category of being is enacted in the historically mobile and logically conventional (arbitrary) construction of the ontological regions by scientific knowledge (in particular) and culture (in general) (Preti 2011: 297).

This interesting programme was born within a deep, free, original and fruitful critical reassessment, based on various constructive hybridisations, drawn from different and even opposing traditions of thought, which range from the tradition of neo-Kantian transcendentalism to Husserl's early phenomenology, from the tradition of Dewey's and Marx's pragmatism to logical empiricism and analytical philosophy. But Preti did not limit himself to dialoguing with major contemporary traditions of thought, because he was also able to establish his research programme —which at one point he named with the emblematic expression of transcendental logical neo-realism—with some of the major conceptual traditions of classical and medieval thought (for an in-depth study of his thought see Minazzi 1984, 1994, 2004a, 2011, and Minazzi [ed.] 1987, 2009, 2015). But what did Preti mean by neo-realism? Here is his answer:

> "neo-realism" is a relatively new name for a very old doctrine. It is a position that in fact the author of these essays has derived from meditation on and discussion of the more strictly philosophical (theoretical) problems of the contemporary analytic philosophy, logic and epistemology (of Moore, Russell, Carnap, Ayer, etc. etc.) in the light of the doctrines of the early Husserl (of the *Logische Untersuchungen* and the *Ideen*). But it goes all the way back to fourteen-century scholasticism (Preti 2011: 37).

According to this perspective, the object-of-knowledge coincides, therefore, with knowledge itself, namely technical and scientific competence in a given historical society. The issue is, therefore, no longer to investigate an "external" or "internal" structure of knowledge, because the way knowledge is conceived has itself changed. Knowledge, then, can only start from itself, because the justification of scientific knowledge cannot be located either in the notion of an alleged superior divinity (which would be the guarantor of knowledge), or, even less, in referring to experiences which, as a factual and experimental basis, would justify a certain knowledge in themselves. On the contrary, the epistemological perspective delineated by Preti reminds us that the foundation and most authentic and reliable justification of science—in its objective cognitive scope—is rooted in its own autonomous plan of transcendentality, within which a relatively autonomous knowledge is established, which may be ultimately based only on itself (both in relation to the theoretical factors and in relation to those inherent in the world of praxis and its more or less effective technological working principles). In this perspective, then, the objectivity of knowledge is rooted in objective unsaturated paradigms, which though they have no substantial and metaphysical unity, yet enable us to gather a few threads of truth in relation to the real world that we wish to study and investigate. Offering an example, inspired by the famous medieval dispute over universals and their significance, Preti then underlines how the innovative solution proposed by neo-realist mediaeval logicians consisted in rejecting both the solution of traditional metaphysical Platonic realism (which reduces a dog to its unchangeable, eternal and substantial *eidos*) and the solution of radical nominalism (which on the contrary reduces a dog to a mere *flatus vocis*, a comfortable summation, obtained from simple induction, of every possible experience we have of flesh and blood dogs). But on the contrary.

> For the neo-realist there instead returns the idea of the objective paradigm (and this is why we also call them "realists"), but not as a substantial unity "in itself". The *significatio* (or concept) of "dog" is to dogs as, say, the project of a building designed by an architect is to building (or even the potentially unlimited class of buildings) which is actually built to that project. So signification and denotation are not hard direction of reference to different metaphysical realities: the ultimate reference is always to dogs (to buildings) in the flesh (in stones and mortar). This is expressed in the distinction (to which that between *significatio* and *suppositio* seems to be reduced) between *suppositio pro significato non ultimato* and *suppositio pro significato ultimato*. The *ultimatio* is the complete intuitive fulfilment of that "project" that was the concept of meaning in the name (in the categorematic term), and when the term stands for the content of this type, it denotes. The meaning differs from the denotation not by genus but by species: it is an incomplete a denotation, one not completely fulfilled, and therefore in a certain sense, vague (it contains notes that remain indeterminate, and therefore variables) (Preti 2011: 42).

Thus the semantic incompleteness always connected to a given term allows us, when it is properly transposed onto the plane of linguistic objectivity, to comprehend the epistemic problem of the objective constitution of things. There emerges again the problem—drawn from Cartesian philosophy—of the relationship between *esse obiectivum* and *esse formale* on which the issue of the constitution of the objects of knowledge might be based. But in order to grasp the object of knowledge, we then have to follow the approach suggested by Kant with his "Copernican revolution", and understand that our natural orientation and its ingenuous realism has to be suspended to adopt the perspective of an analytical level of meta-reflection on different disciplines, to understand all the different elements which structure the object-of-knowledge. The hyphens used in this expression are intended to show how this object-of-knowledge cannot even be understood if we are not able to grasp the precise "regional ontology" (to use one of Husserl's categories again) within which the knowledge proper to each discipline is constructed. The consequences of this transcendentalist approach are dual: concepts are in fact intended as *unifying functions* (or as the result of different *unifying functions*), while the object-of-knowledge is configured as a continuous task and an open project with the aim of learning to start from knowledge itself, that is from its technical and scientific competence and its various conceptual frameworks. The history of modern science since the seventeenth century to the present is in line with this; it testifies that the justification of scientific knowledge and of its technological practices can never be supplied *from outside* science itself. In this way, as science is justified by its history, so our objective knowledge of the world is likewise rooted in language universes, conceptual categories, problems, verification and falsification methods, which were devised by a certain technical competence and a type of knowledge at some stage of its cognitive and pragmatic development. In this perspective, the truth of objective knowledge can then no longer be thought of as being "commensurate" with, or as being a "correspondence" with, an object metaphysically given beyond our own possible experiences (as Kant taught us); because, if anything, the objectivity of our knowledge appears as an infinite and always open research programme where knowledge is "configured as 'a commensuration of the actuation of knowledge to its own intentional direction'" (Minazzi 2011, p. 161). Which brings us back, first, to the most authentic Socratic perspective, according to which the *search never ends*, because *truth coincides with the search for truth*. But now this Socratic awareness about knowledge and its relative autonomy is firmly held together by the lesson of the early Husserl, who perceived the plurality of the different levels of transcendentality within which different *morfé* are established, within which *hyletic data* are systematically subsumed, creating the objective world in which we live (Minazzi 1996 and 2004c).

References

Abbagnano, Nicola (1971), *Dizionario di filosofia*, Second revised and expanded edition, Turin: Utet.

Agazzi, Evandro (1969), *Temi e problemi di filosofia della fisica*, Milan: Manfredi (reprinted in 1974, Roma: Abete).

Agazzi, Evandro (1989), *Filosofia, scienza e verità* (with Fabio Minazzi and Ludovico Geymonat), Milan: Rusconi.

Agazzi, Evandro (1994), Was Galileo a realist?, *Physis*, 31 (1), 273–296.

Agazzi, Evandro (2012), *Ragioni e limiti del formalismo. Saggi di filosofia della logica e della matematica*, edited with a *Prefazione* by Fabio Minazzi, Milan: Franco Angeli.

Agazzi, Evandro (2014), *Scientific Objectivity and Its Contexts*, Cham-Heidelberg-New York-Dordrechte-London: Springer.
Alai, Mario – Buzzoni, Marco – Tarozzi, Gino (2015), *Science Between Truth and Ethical Responsability. Evandro Agazzi in the Contemporary Scientific and Philosophical Debate*, Cham Heidelberg New York Dordrecht -London: Springer.
Aristotele (1955), *Organon*, introduction, translation and notes by Giorgio Colli, Turin: Einaudi.
Buzzoni, Marco (2015) see Alai (2015).
Dal Pra, Mario (1989), *Lo scetticismo greco*, Rome-Bari: Laterza.
Descartes, René (1964–1974), *Oeuvres de Descartes* publiéespar Charles Adam & Paul Tannery, Paris: Librairie Philosophique J. Vrin, 11 vol. (english translation: *Descartes' Philosophical Writings*, edited by Norman Kemp Smith, 1958 ed.).
Diel, Hermann – Kranz, Walter (2006), *I Presocratici. Prima traduzione integrale con testi originali a fronte delle testimonianze e dei frammenti nella raccolta di H. Diels e W. Kranz*, edited by Giovanni Reale, Milan: Bompiani.
Einstein, Albert (1949), *Albert Einstein: Philosopher-Scientist*, edited by Paul Arthur Schlipp, Illinois: Evanston (*The Library of Living Philosophers*, VII).
Ferrater Mora, José (2001), *Diccionario de filosofía,* Nueva edición revisada, aumentada y actualizada por el professor Josep-Maria Terricabras, supervisión de la profesora Priscilla Cohn Ferrater Mora, Barcellona: Editorial Ariel S. A., 4 voll.
Frege, Gottlob (1892), Über Sinn und Bedeutung, *Zeitschrift für Philosophie und philosophische Kritik,* NF 100, S. 25–50.
Galilei, Galileo (1968), *Le opere di Galileo Galilei*, national edition directed by Antonio Favaro, Firenwe: G. Barbèra Editore, 20 vol. (english translation by Stillman Drake, *Dialogue Concerning Two Chief World System*, Published by University of California Press, Cambridge 1967).
Geymonat, Ludovico (1970–76), *Storia del pensiero filosofico e scientifico*, Milan: Garzanti, 7 vol.
Geymonat, Ludovico (1977), *Scienza e realismo*, Milan: Feltrinelli.
Geymonat, Ludovico (1989) see Agazzi (1989).
Holzhey, Helmut (1970), *Kant Erfahrungdsbegriff. Quellengeschichtliche und bedeutungsanalytische Untersuchungen*, Basel-Stuttgart: Schwabe – Verlag.
Kant, Immuel (1787), *Kritik der reinen Vernunft.* Zweite hin und wieder verbesserte Auflage, Riga: Johan Friedric Hartknoch (english transaltion: *Immanuel Kant's Critique of pure reason*, translated by Norman Kemp Smith, MacMillan and Co., Limited, London 1929).
Minazzi, Fabio (1984), *Giulio Preti: bibliografia*, Milan: Franco Angeli.
Minazzi, Fabio (1987), ed., *Il pensiero di G. Preti nella cultura filosofica del Novecento*, Milan: Franco Angeli.
Minazzi, Fabio (1989) see Agazzi (1989).
Minazzi, Fabio (1992), *Galileo «filosofo geometra»*, Milan: Rusconi.
Minazzi, Fabio (1994), *L'onesto mestiere del filosofare*, Milan: Franco Angeli.
Minazzi, Fabio (1996), *L'oggettività della conoscenza scientifica*, edited by F. Minazzi, Milan: Franco Angeli
Minazzi, Fabio (2001), *La passione della ragione. Studi sul pensiero di L. Geymonat*, Milan-Mendrisio: Thèlema-Università della Svizzera italiana.
Minazzi, Fabio (2004a), *Il cacodémone neoilluminista. L'inquietudine pascaliana di G. Preti*, Milan: Franco Angeli.
Minazzi, Fabio (2004b), *Contestare e creare. La lezione epistemologico-civile di L. Geymonat*, Naples: La Città del Sole.
Minazzi, Fabio (2004c), *Le saette dei tartari. Il problema epistemologico dell'oggettività*, Milan: Franco Angeli.
Minazzi, Fabio (2006), *L'épistémologie comme herméneutique de la raison*, Naples-Paris: La Città del Sole- J. Vrin.
Minazzi, Fabio (2007), ed., *Filosofia, scienza e bioetica nel dibattito contemporaneo. Studi internazionali inb onore di Evandro Agazzi*, Rome: Presidency of the Council of Ministers, Department of Information and Publishing, Istituto Poligrafico e Zecca dello Stato.

Minazzi, Fabio (2009), sous la direction de M. Brondino et F. Minazzi, *Le mektoub tunisien de G. Preti. La vie et l'oeuvre d'un philosophe italien rationaliste*, Paris: Édition Publisud.

Minazzi, Fabio (2010), *Geymonat epistemologo. Con documenti inediti e rari (un inedito del 1936, il carteggio con Moritz Schlick, lettere con Antonio Banfi e Mario Dal Pra)*, Milan-Udine: Mimesis.

Minazzi, Fabio (2011), *Suppositio pro significato non ultimato. Giulio Preti neorealista logico*, Milan-Udine: Mimesis.

Minazzi, Fabio (2012), *Prefazione* ad Agazzi (2012), pp. 11–34.

Minazzi, Fabio (2015), ed., *Sul bios theoretikós di Giulio Preti*, Milan-Udine: Mimesis 2 voll.

Minazzi, Fabio (2015), *Evandro Agazzi Philosopher. An Overview of His Thought* in Alai-Buzzoni-Tarozzi (2015), pp. 1–17.

Pera, Marcello (1992), *Scienza e retorica*, Rome-Bari: Laterza (English translation, Chicago: Chicago University Press).

Poincaré, Henri (1905), *La valeur de la science,* Paris: Ernest Flammarion Éditeur (ed. 1948).

Preti, Giulio (2011), *Philosophical Essays. Critic Rationalism Historical-objective Transcendentalism*, edited by Fabio Minazzi, translation from Italian by Richard Sadleir, Bruxelles – Bern – Berlin – Frankfurt am Main – New York – Oxford – Wien: P.I.E. Peter Lang.

Spinoza, Baruch (2010), *Tutte le opere. Testi originali a fronte,* introductory essay, presentations, notes and apparatus by Andrea Sangiacomo, translations by Various Authors, Milan: Bompiani (english translations: Spinoza, *Complete works*, with translations by Samuel Shirley, edited, with Introduction and Notes, by Michael L. Morgan, Hackett Publishing Company Inc., Indianapolis, Cambridge 2002).

Tarozzi, Gino (2015) see Alai (2015).

Vuillemen, Jean (1955), *Physique et métaphysique kantiennes*, Paris: Presses Universitaires de France.

Part II
Recent Conceptions of Scientific Realism

Scientific Realism and the Mind-Independence of the World

Stathis Psillos

> The characterization of some philosophical view concerning a particular topic as realistic has substance only by contrast with an opposing view that has been maintained or is at least imaginable.
>
> Michael Dummett

Abstract In this paper, I analyse the metaphysical component of scientific realism. I develop and defend the view that the realist claim of mind-independence is captured by what I call 'the possibility of divergence', viz., the possibility of a gap between what there is in the world and what is issued (or licensed) as existing by a suitable (even ideal) set of epistemic practices and conditions. I break up the realist commitment to mind-independence into two components: irreducible existence and objective existence. I then show how various versions of anti-realism (in particular, idealism and verificationism) compromise one or both of these conditions. I also show that a verificationist version of scientific realism, though honouring the condition of irreducible existence, compromises the condition of objective existence; hence it is in conflict with the realist demand for mind-independence.

1 Introduction

I have argued in my (1999) that scientific realism consists of three theses.

The Metaphysical Thesis: The world has a definite and mind-independent structure.

The Semantic Thesis: Scientific theories are truth-conditioned descriptions of their intended domain. Hence, they are capable of being true or false. The

S. Psillos (✉)
Department of Philosophy and History of Science,
University of Athens, Athens, Greece
e-mail: psillos@phs.uoa.gr

E. Agazzi (ed.), *Varieties of Scientific Realism*,
DOI 10.1007/978-3-319-51608-0_11

theoretical terms featuring in theories have putative factual reference. So if scientific theories are true, the unobservable entities they posit populate the world.

The Epistemic Thesis: Mature and predictively successful scientific theories are well-confirmed and approximately true. So entities posited by them, or, at any rate entities very similar to those posited, inhabit the world.

Briefly put, the rationale for these three theses is the following. Scientific realism is a tri-partite philosophical view. It is a view about the world; a view about scientific theories; and a view about the cognitive achievements of science. Being a view about the world, scientific realism is a species of *realism*; hence, it must imply some commitment to a realist understanding of the world and the relation of science to it. More specifically, it must honour the realist commitment to the mind-independence of the world. Being a view about scientific theories, scientific realism should be committed to realist semantics for scientific theories, viz., a commitment to an irreducible, truth-conditional account of the meaning of theoretical terms. Being a view about the cognitive achievements of science, scientific realism should be committed to some kind of epistemic optimism, viz., the claim that science has been on the right track; it has succeeded in offering us knowledge of the world and in particular knowledge of the parts of the world which are not detectable by the naked senses.

Though conceptually distinct, these three theses form a tight network. For instance, what's distinctive about scientific realism is the view that the world of appearances (the world as it is revealed to us through our senses) does not exhaust the content of the world—this content being, by and large, unobservable, though no less real. Moreover, this world is by and large independent of us and the various ways scientists use to describe it and it is this independent world which renders theories true, insofar as they are true. Taking science as a cognitive endeavour which succeeds in describing an independent world is not a priori true—nor historically constitutive of science. It is a philosophical stance towards science which has its basis on the fact that scientists developed theories which posited unobservable entities to explain the behaviour of observables; but it is also grounded on a certain semantic attitude towards theories—viz., taking them in an irreducible (face-value) way. This implies that theories—*qua* attempts to describe the independent world—have excess content over and above whatever can be expressed in an observational language, insofar as we can make sense of a theory-free observational language.

In my own work over the years I have mostly dealt with the epistemic thesis and to a lesser extent with the semantic thesis. Insofar as I have dealt with the metaphysical thesis, I have taken it to be a minimalist one. It amounts to a declaration of independence: the world is mind-independent. In this paper, I want to say a bit more about the metaphysical component of scientific realism and in particular to develop and defend the view that the realist claim of mind-independence is captured by what I have called 'the possibility of divergence', viz., the possibility of a gap between what there is in the world and what is issued (or licensed) as existing by a suitable (even ideal) set of epistemic practices and conditions.

To fix our ideas, I will break up the realist commitment to mind-independence into two components: irreducible existence and objective existence.

Irreducible existence means existence in its own right; that is, not dependent on the existence of something else. To say that a kind of entities X exist irreducibly is to say their existence is not grounded, in whatever way, in the existence of another kind of entities Y; or it does not depend on the existence of another kind of entities Y. Hence, to say that entity K exists reducibly is *not* to say that K is not real but that its reality depends on (or is reducible to) the reality of some other entity (or entities). As we will see in more detail later, for traditional idealism material objects are not irreducibly real, while for traditional materialism, they are.

Objective existence means existence independently of epistemic or cognitive conditions that require the verification, recognition or knowledge of existence. To say that a kind of entities X exist objectively is to say that their existence is not in any way constitutively connected with epistemic procedures Φ that allow or enable cognizers to decide, or otherwise certify, their existence. As we shall see in more detail later on, verificationist anti-realism denies the condition of objective existence, without thereby denying the condition of irreducible existence. (It is also consistent to deny irreducible existence without denying objective existence—an example would be the mind-body identity thesis.)

In light of Michael Dummett's quotation in the beginning of the paper, I want to contrast scientific realism with two opposing views: idealism and verificationism. I will be relatively brief with idealism and I will focus my attention on verificationism. My chief aim will be to show that a verificationist version of scientific realism, though honouring the condition of irreducible existence, compromises the condition of objective existence and hence it is in conflict with the realist demand for mind-independence.

2 Berkeley's Idealism

One way to construe the contrary claim that the world is mind-dependent is to take seriously traditional idealism. On this view, what exists is mental, and hence mind-dependent: it cannot exist without the mind. As is well-known, Berkeley was an immaterialist. He denied the existence of matter *qua* an unthinking substance: an inert, senseless substance which supports (bears) extension, figure and motion (Principles §9).[1] As he put it: "extension, figure, and motion are only ideas existing in the mind, and that an idea can be like nothing but another idea, and that consequently neither they nor their archetypes can exist in an unperceiving substance".

[1]The references are from Berkeley's *A Treatise Concerning the Principles of Human Knowledge*, second edition 1734, reprinted in Berkeley (2008).

Now, 'idea' is a technical term in Berkeleyan philosophy (Principles §§38–39). Calling something a *thing* has a connotation of existing without the mind. In this sense, ideas are *not* things. They cannot exist unperceived or without the mind. The sensible qualities of 'things' are ideas—hence the 'things' are collections of sensible qualities which are called ideas. Significantly, ideas are "the objects of sense"—to sense is to perceive; hence what one perceives is an idea. This technical use of the term 'idea' removes the paradox that is usually associated with idealism: after all, we are fed with food and not with *ideas*. As Berkeley put it: "we are fed and clothed with those things which we perceive immediately by our senses. The hardness or softness, the colour, taste, warmth, figure, and such like qualities, which combined together constitute the several sorts of victuals and apparel, have been shown to exist only in the mind that perceives them". Given that there are only two kinds of Being, "spirits and ideas", everything is mental. Spirits "are active, indivisible substances"; ideas "are inert, fleeting, dependent beings, which subsist not by themselves but are supported by, or exist in, minds or spiritual substances". (Principles §90)

Berkeley, notably, intended to make his idealism consistent with corpusculari-anism—which posited *invisible* corpuscles.[2] But corpuscles, if they exist at all, must be ideas (collections of ideas, to be sure); hence they require minds or spiritual substances "to exist in". Yet, they are not ideas (or collections thereof) that are imprinted on our own senses. They can only exist in the spirit that perceives them, and this can only be God.[3] So, plausibly, corpuscles 'exist in' God; they are dependent on him. Hence corpuscles are OK-Berkeleyan entities since they cannot exist without the mind. For Berkeley it is enough for something to exist to be perceived by "the eternal mind of the Creator".

But in this sense, corpuscles can be deemed "'external' with regard to their origin" (§90) in that the relevant ideas are "not generated from within the mind itself" (meaning: the human mind); they can also be deemed to exist "without the mind" in the sense that they [the corpuscles] "exist in some other mind", viz., God's mind.

Berkeley does not deny that unthinking things are in some sense *real*. But he's careful to note that, according to him, "the unthinking beings perceived by sense,

[2]This might sound controversial, but it isn't. Berkeley's chief aim was not to argue against corpuscularianism but, instead, to show that there is no argument for matter stemming from mechanical philosophy. That is, that mechanical philosophy—or the "clockwork of nature", as he put it—is consistent with immaterialism. This "clockwork of nature" is such that "great part whereof is so wonderfully fine and subtle as scarce to be discerned by the best microscope" (Principles §60). His chief target then was to show that accepting that there is no matter does not entail that one has to abandon the view that there are mechanisms in nature. On the contrary, he wanted to show that if mechanism is seen as a material mechanism "without the mind" it won't be able to produce anything anyway. For some similar considerations, see Garber (1982). For a somewhat differing view see Downing (2005).

[3]For Berkeley, the 'clockwork of nature' is the way God has chosen to produce effects in nature in a regular and orderly way. In particular, God acts in nature via laws of nature and these are 'implemented' by mechanisms, which embody the regular behaviour assigned to them by laws.

have no existence distinct from being perceived, and cannot therefore exist in any other substance, than those unextended, indivisible substances, or spirits, which act, and think, and perceive them" (Principles §91). The ideas that are excited in minds "from the outside" have law-like connections. They are impressed "according to certain rules or laws of nature, speak themselves the effects of a mind more powerful and wise than human spirits". These ideas, then, are not "fictions of the mind". They are not produced "at pleasure". In fact, the ideas that are produced "from the outside", that is sensations, "are said to have more reality in them than the former [ideas excited by the mind "from the inside"]. Hence, the sun is real—that it, when we perceive the sun, we perceive the "real sun" (Principles §36). He couldn't be more upfront:

> In the sense here given of 'reality', it is evident that every vegetable, star, mineral, and in general each part of the mundane system, is as much a real being by our principles as by any other. Whether others mean any thing by the term 'reality' different from what I do, I entreat them to look into their own thoughts and see (Principles §36).

The key point here is that for idealism reality (including reality which is not directly sensed by the human senses, like invisible corpuscles) is mind-dependent because it is mental: it is either mental substances (spirits) or mental attributes (which cannot exist without the mental substances). Hence, mind-dependence should be understood as a claim about what exists, that is about *what kinds of stuff make up reality*.

In light of the distinction drawn in the Introduction, Berkeley's reality is mind-dependent because Berkeley denies the irreducible existence of *things* and posits that insofar as they are taken to be (in some sense) real, they are so because they are reducible to collections of another type of (mental) entity, viz. ideas. Interestingly, Berkeley's position vis-à-vis the objectivity condition is a bit more complex. Insofar as the existence of things (*qua* reducible to ideas) and in particular, the existence of invisible mechanisms, is not tied to them being sensed by human minds, that is, insofar as he denies that they are collections of ideas possessed by *humans*, Berkeley allows that what exists is not tied to a *humanly* realisable epistemic condition, to wit perceivability. But from this it does not follow that they exist objectively since for something to exist it must be perceivable by God even if it is not perceived by humans. Hence, Berkeley comes to compromise the second sense in which reality is mind-independent. For Berkeley there is simply no way in which there might be a divergence between what God (being a supreme mind) knows and what there is.

Though there are important arguments against idealism—going back to G E Moore among other philosophers—my main point here is not that idealism is wrong.[4] My main point is that even though it is consistent with idealism that there are unperceived-by-humans entities and even though an idealist, pretty much like a scientific realist, could believe in the existence of electrons and their ilk, scientific realism is anti-idealism and, by the same token, idealism is inconsistent with

[4]A modern classic refutation of idealism is given in Musgrave (1999, Chap. 9).

scientific realism. The chief reason for this is that idealism compromises the realist commitment to mind-independence of the world, both when it comes to the irreducibility condition and the objectivity condition. The scientific realist claim of the mind-independence of the world is meant to be a declaration that there is irreducibly non-mental stuff in the world and, in particular, that the entities posited by scientific theories are *material*. In this sense, scientific realism is a species of materialism—not in claiming that, necessarily, all there is is material; but in asserting that the material world exists independently of anyone being in a position to perceive it. Hence, the material world is causally and temporally prior to any kinds of minds that there may be part of reality and is irreducible to mental stuff.

3 Verificationist Anti-realism

Going for idealism is not the only way to go for a kind of mind-dependence of the world. The other way is weaker; it does not centre on the irreducibility condition—since it is perfectly OK with it—but on the objectivity condition. Hence it is not a view about what *types* of entity exist (whether they are material or mental or what have you); rather, it ties what exists to what can be known to exist.

There is a long anti-realist philosophical tradition according to which it does not make sense to assert the existence (or reality) of some entities unless this assertion is understood to be connected to…, where the ellipsis is filled with the requirement of fulfilment of a suitable *epistemic/conceptual* condition. Putnam's (in his middle period) favourite filling would be based on the condition of rational acceptability; Dummett's would relate to warranted assertibility; and Kant's own line was related to the possibility of something being encountered in experience.

In the sequel I will deal mostly with Dummett and his disciples. But a quick note on Kant is worthwhile.

3.1 A Note on Kant on Scientific Realism

Kant's Transcendental Idealism is captured by his dictum: "everything intuited in space or in time, hence all objects of an experience possible for us, are nothing but appearances, i.e., mere representations, which, as they are represented, as extended beings or series of alterations, have outside our thoughts no existence grounded in itself." Opposing this view, Transcendental Realism "makes these modifications of our sensibility into things subsisting in themselves, and hence makes *mere representations* into things in themselves" (Critique A 490–1; B 518–9).[5] Kant's conception of representation is quite technical. It *does* imply that the represented—*qua*

[5]All references to Kant's *Critique of Pure Reason* are from Kant (1998).

represented in space and time or as falling under the schematised categories of pure understanding—is in an important sense mind-dependent. And yet, Kant distances himself from Berkeleyan idealism: material objects are not just (collections of) ideas and contents of thought. Even if objects end up being mind dependent (given that space and time are a priori forms of pure intuition) it does not follow that the represented is *constructed* out of acts of representation. Kant insists that "representation in itself (…) does not produce its object as far as its existence is concerned" (A92/B125). Rather, (for Kant) it is only through the representation that it is possible to know anything *as an object*. More specifically, Kant distinguished between "empirically external objects from those that might be called "external" in the transcendental sense" by directly calling the former "things that are to be encountered in space". (A373).

Hence, critical or formal idealism, as Kant called his position, is not a reductive position; it does not concern whether material things exist and whether they are reducible to mental entities. It concerns the conditions under which it is proper to say that objects exist—and these conditions include the properties that have to be predicated of them by virtue of which objects can be known (cf. Kant 2004, §§289–91).

Kant, notably, had no problem at all in accepting the existence of invisible entities posited by the best theories of his time.[6] In fact, they are no less part of appearances than macroscopic entities since they are in space. And they are no less real than them. Would it be then a foregone conclusion that Kant was a scientific realist?

The answer is complicated by the fact that Kant famously allowed the existence of noumena—of things as they are in themselves. This, to be sure, is a limiting concept, but it does allow for an in-principle divergence between how things appear to be and how they are in themselves. (This, in fact, was a main criticism that Hegel levelled against Kant.) But in many ways, the Kantian noumena are an idle wheel in his philosophy of nature (cf. A253/B309). Not because they are in principle unknowable; but because they play no role, except in name, in grounding what is knowable of the world. As Kant put it: "The transcendental object that grounds both outer appearances and inner intuition is neither matter nor a thinking being in itself, but rather an unknown ground of those appearances that supply us with our empirical concepts of the former as well as the latter" (A380).

What is, in principle, knowable of the world cannot possibly, for Kant, be different from what is licensed as knowable by the categories of pure understanding and the a priori forms of pure intuition. The famous Copernican turn suggests as much:

> Up to now it has been assumed that all our cognition must conform to the objects; but all attempts to find out something about them a priori through concepts that would extend our cognition have, on this presupposition, come to nothing. Hence let us once try whether we

---

[6]For more on this see Rae Langton (1998, 6 & 143ff). Langton treats Kant as an outright scientific realist but mainly because he accepts the reality of invisible entities. She does not deal with the Objectivity Condition at all.

> do not get farther with the problems of metaphysics by assuming that the objects must conform to our cognition, which would agree better with the requested possibility of an a priori cognition of them, which is to establish something about objects before they are given to us (Bxvi).

But since there can *in principle* be more content in the world than what is *in principle* knowable by sensible intuition (the noumena as opposed to the phenomena), Kant seems to honour not just the irreducibility condition but something akin to the objectivity condition. Still, the verdict is not straightforward. For the phenomena—which for Kant include everything that exists *and* is in principle knowable—are mind-dependent in a way that compromises their objective existence (in the sense noted in the Introduction). The reason for this is that the existence of the phenomena is connected with the obtaining of certain epistemic conditions which render them knowable (including thinking of them as being in space and time). Kant was quite upfront regarding the implications of this commitment:

> Thus the transcendental idealist is an empirical realist, and grants to matter, as appearance, a reality which need not be inferred, but is immediately perceived. (…) (I)n our system, on the contrary, these external things – namely, matter in all its forms and alterations – are nothing but mere representations, i.e., representations in us, of whose reality we are immediately conscious. (…) Empirical realism is beyond doubt, i.e., to our outer intuitions there corresponds something real in space. Of course space itself with all its appearances, as representations, is only in me; but in this space the real, or the material of all objects of outer intuition is nevertheless really given, independently of all invention; (O)bjects (…) are unknown to us as to what they are in themselves. (…) if I were to take away the thinking subject, the whole corporeal world would have to disappear, as this is nothing but the appearance in the sensibility of our subject and one mode of its representations A371-2; A375; A379; A383.

If the whole corporeal world depends on the thinking subject in the sense that is a network of appearances (in the technical Kantian sense which allows their reality *and* their conformity to synthetic a priori principles of knowledge), Kant's empirical realism can hardly honour the spirit of the objectivity condition.

3.2 Dummett's Semantic Anti-realism

Michael Dummett famously resuscitated verificationism by tying truth to justification or warranted assertibility. As he (Dummett 1982, 108) put it, a "statement cannot be true unless we know it to be true, at least indirectly, or unless we have means to arrive at such knowledge, or at least, unless there exists that which, if we were aware of it, would yield such knowledge". For Dummett a view of reality is not merely a view of what kinds of objects there are and of "what constitutes the existence of such objects". As he says, "it is necessary to say what kinds of fact obtain, and what constitutes their holding good" (Dummett 2006, 2–3). But talk of facts brings with it talk of truth, since the facts are the true propositions. Hence, for

Dummett, the concept of truth plays a central—and ineliminable—role in the realism debate.

In this setting, to *resist* realism is to resist the realist view that what makes a statement true is an independently given reality which renders our statements true when they are true and false when they are false; that is a reality which is "independent of our knowledge of it and of our means of attaining such knowledge" (Dummett 2006, 65). This kind of anti-realism fosters a notion of truth which is not evidence-transcendent; hence it denies the principle of bivalence. More specifically, it ties ascriptions of truth with the existence of an effective decision procedure which allows deciding whether a statement in a given domain is true or false. Famously, Dummett's model of anti-realism is intuitionism in mathematics.

The point relevant to our purposes is that this decision procedure is not connected to a certain type (or types) of object(s). Material objects and facts about them, insofar as statements about them are decidable by means of perception or other observational procedures, are parts of reality. Hence, the existence of a fact of the matter about statement S (e.g., that material objects exist or are real) depends on the existence of grounds for holding that there is a fact of the matter about the outcome of a decision procedure such that were it to be carried out, or had it been carried out, the truth of S would be, or would have been, decided. The subjunctive or counterfactual element in the foregoing clause is significant. The decision procedure need not be actually carried out; it is enough that it can, or could, be carried out. Those (and only those) statements for which the relevant decision procedure is available, that is those statements which are *decidable*, are such that the principle of bivalence holds. Where there is no possibility of knowing (where, that is, there is no relevant decision procedure) there are *gaps*. Dummett's anti-realism, then, consists in the claim that there can be *gaps* in reality, that is that reality is

> in some degree indeterminate, for we have no conception of reality save as that which renders true those true statements we can frame and those true thoughts we can entertain. If our statements and our thoughts are not all determinately either true or false, then reality itself is indeterminate; it has gaps, much as a novel has gaps, in that there are questions about the characters to which the novel provides no answers, and to which there therefore are no answers (Dummett 1991, 318).

The presence of gaps in reality (associated with verification-transcendent assertions) does not imply that if the truth of an assertion is verified, a gap in reality is filled with a fact popping into existence. Rather, Dummett's point is that if it is at all possible to verify a truth, then a gap in reality would not have been there in the first place. Note that this view is consistent with the claim that the facts that there are would have existed even if human beings had not evolved relevant to their recognition epistemic capacities. What, of course, could not have possibly existed for Dummett are facts for recognition-transcendent truths. To put the point differently, Dummett's anti-realism does not imply that the determinate part of reality (which is the whole of reality for Dummett's anti-realism) wouldn't be there if beings like us had not evolved relevant epistemic capacities to recognise truth.

But some extra care is needed here. Dummett insists that something can be a fact *only if* it can be known to be a fact, which is the same thing as saying that there is a fact of the matter about the truth of a statement S only if this truth can be in principle verified. As he (Dummett 2006, 92) put it:

> On a justificationist view, however, what we could have known extends only so far as the effective means we had to find out: the entailment is not from its being true to the possibility of knowing it, but in the opposite direction. It would be wrong to say that we construct the world, since we have no control over what we find it to be like; but the world is, so to speak, formed from our exploration of it.

This is a striking point. Obviously, it places significant constraints of what there is in the world. As Dummett put it: "Our world is thus constituted by what we know of it or could have known of it" (Dummett 2006, 92). There is no more content in the world than what can be in principle known of it. And it is there, (constituting the world), in some important sense, only insofar as *it can be known to be there*. It can be seen then that though Dummett honours the irreducible existence condition, he does not accept the objective existence condition.

To probe this point a bit more, let us note that when it comes to the philosophy of science, Dummett is an anti-instrumentalist. But he goes beyond this by taking for granted a lot that is posited by current science. Though he stresses that some of the theoretical entities of science ("black holes, quarks, hidden dimensions, anti-matter, superstrings") "seem bizarre", he takes it that "it is difficult to make a sharp demarcation between constituents of the everyday world and those of the physicist's world. Electric currents were not but now are part of the everydayworld; presumably radio waves must also be assigned to it" (Dummett 1991, 5). But here again, insofar as he advocates some form of realism about theoretical entities it is because and insofar as statements about them are recognisable as true. Dummett admits that "we cannot but view science, at least before it transcends some critical level of abstraction, as attempting to arrive at such descriptions [of an object as it is in itself]", (Dummett 1996, 410) but this is because and to the extent in which scientific statements are verifiable in roughly the same way as statements about common-sense objects.

Insofar as the resulting position is a kind of scientific realism, it has accepted the basic tenet of Dummett's anti-realism! The independence of the world has been compromised not by denying that theoretical entities are real and irreducible but by denying that truths about them are evidence-transcendent. Hence, theoretical entities are indeed real, but only insofar as truths about them can be known and precisely because (in *some* sense of because) they can be known.

As is well known, the middle Putnam came to accept Dummett's verificationism. Recounting this 'conversion', Putnam noted that his basic disagreement with Dummettian neo-verificationism was about leaving the possibility open of a gap between justification-by-present-evidence and truth: he took it that truth should be identified with "idealised justification" (Putnam 1983, xvii). Part of his thought was that even if truth is tied to justification, one should be careful how exactly this tie is effected. As he (Putnam 1983, 162) put it, he was looking for a position "which recognises a difference between 'p' and 'I think that p', between being *right*, and *merely thinking one is right* without locating the objectivity in either transcendental correspondence or mere consensus". Truth is not a property that can be lost—nor does it have a sell-by date. Hence, the verificationist notion that replaces (or captures) truth should be such that it honours this property of truth. The 'correctness' of an assertion is a property that can be lost, especially if it is judged by reference to current standards or consensus (these come and go). Because of this Putnam tied correctness to "the verdict on which inquiry would ultimately settle" (Putnam 1982, 200). This might suggest that Putnam adopted an account of truth that has been associated with Charles Peirce, but as he (Putnam 1990, viii) noted, this (Peircean) ideal-limit-of-inquiry theory truth is "fantastic (or utopian)".

In the midst of his conversion to verificationism, (or to internal or pragmatic realism, as Putnam tended to call it), he published a piece in which he did endorse scientific realism, suitably dissociated from both materialism *and* metaphysical realism (cf. Putnam 1982). He took scientific realism to involve commitment to the following theses: theoretical entities have irreducible existence (they exist in the very same sense in which ordinary middle-sized objects exist and are irreducible to either them or complexes of sensations); theoretical terms featuring in distinct theories can and do refer to the same entities (hence, there is referential continuity in theory-change); there is convergence in the scientific image of the world; and scientific statements can be (and are) true. Yet, the verificationist Putnam of the early 1980s took truth to be the "correct assertibility in the language we use" (Putnam 1982, 197). So scientific realism was dressed up in a verificationist garment. But is this scientific realism *proper*?

4 The Possibility of Divergence

The Dummett-(middle) Putnam verificationism is clearly anti-idealist. It entails (or at least it is consistent with the claim) that material objects are irreducibly real (be they the middle-sized entities of common sense or unobservable entities). It denies (or at least it is consistent with denying) that worldly objects exist insofar as they are perceived; or that they are collections of ideas which cannot exist without mental substances. The substantive disagreement between verificationist anti-realism and realism concerns the *sense* of existence. Anti-realism makes the world (or a set of entities) mind-dependent in a sophisticated sense: what there is in the world is exhausted by what can be known (verified, rationally accepted and the like) to exist. Hence, it forges a logical-conceptual link between what there is in the world and what is licensed as existing on the basis that it satisfies suitable epistemic conditions. As seen, verificationist anti-realism compromises the condition of objective existence, associated with the realist claim of mind-independence.

Opposing verificationist anti-realism, the realist claim of mind-independence should be understood as logical or conceptual independence: what the world is like does not logically or conceptually depend on the epistemic means and conceptualizations used to get to know it. But can we understand this kind of independence more precisely?

I think the best way is by endorsing the possibility of a *divergence* between what there is in the world and what is issued as existing by a suitable set of cognitive and epistemic conditions. Verificationist anti-realism a priori precludes this possibility of divergence by advancing an epistemic conception of truth. No matter what the details of this conception are, the key idea is that truth is conceptually linked with epistemic conditions in such a way that it is *not* possible that a theory is epistemically right (it meets the relevant epistemic condition e.g., being ideally justified or warrantedly assertible) and false.

It's then natural to think that honouring the possibility of divergence requires adopting a substantive non-epistemic conception of truth. Do we have to say more about this notion of truth? At this stage, this is not necessary. We might try to offer a theory about the nature of truth, e.g., in terms of correspondence. This would be useful when it comes to understanding truth, yet the main point is that the required notion of truth is non-epistemic, viz., that truth is cognition-independent (cf. Vision 2004, 15). Note that this last claim does not imply that necessarily truth is unknowable; (this very admission would defeat the realists' epistemic optimism). But it implies that whether or not a truth-maker is knowable is independent of its eligibility as truth-maker. An attraction of this broad way to understand the non-epistemic nature of truth is that the nature of truth-makers is left unspecified in the sense that any sort of thing (be it physical or mental or abstract) can be a truth-maker. As I have argued elsewhere (cf. Psillos 2005), realism in general and scientific realism in particular need not be driven by a fundamentalist conception of reality, where only an elite set of things or facts can be truth-makers. To put the point somewhat crudely, one does not cease to be a scientific realist about, say, psychology if one is not a physicalist, or if physicalism is wrong. Nor is scientific realism about psychology disqualified from the start because psychological entities may well be mental entities.

The key idea behind a non-epistemic account of truth is an *asymmetric dependence* of the theories (and beliefs) on the world. This asymmetry might well be captured (as the early Putnam suggested) by some general theses about what truth is *not*. One of them is that truth should not be conceptually equated with whatever logically follows from accepted scientific theories, even when these theories are empirically adequate and well-confirmed (cf. 1978, 34–5). The point here is not that theories are or tend to be false. Rather, it is that when truth is attributed to the theory, this is a substantive attribution which is meant to imply that the theory is *made* true by the world, which, in its turn, is taken to imply that it is logically possible that an accepted and well-confirmed theory might be false simply because the world might not conform to it. This is precisely what is captured by the realists' possibility of divergence: a sense in which the world is independent of theories, beliefs, warrants, epistemic practices etc. For some realists (e.g., the early Putnam) this possibility of divergence—which is supposed to capture a modal fact about the world—should be admitted even when the theory is *ideally warrantedly assertible*.

This modal fact, if accepted, has a certain significant implication: there is a way the world is *independently* of how it is being described by humans—even at the ideal limit of inquiry and when all evidence has rolled in. This is a view that has been associated with the so-called Metaphysical Realism. Many (including famously the middle Putnam) have come to reject it, arguing that it is a mistake to think there is just one (privileged) way the world is; hence that there is just not a single (privileged) way to describe the world. Putnam employed the so-called 'model-theoretic' argument against metaphysical realism. But as I have argued in my (2012) (following the lead of others, especially Lewis 1984) this argument offers far from compelling reasons for rejecting Metaphysical Realism.

5 Against Verificationist Scientific Realism

At the end of Sect. 3.2, we raised the question: is verificationist scientific realism, scientific realism *proper*? The worry is this: Why should scientific realism incorporate the claim of mind-independence (as elaborated by the possibility of divergence)? Why, that is, couldn't someone who accepted the reality of unobservable entities without also rendering them mind-independent (in the above sense) be a scientific realist?

Crispin Wright has aimed to show that a verificationist version of scientific realism is indeed possible and desirable. He claimed that "nothing in an intuitive scientific realism requires semantic realism", that is the kind of realism that Dummett has been arguing against (Wright 1992, 158). Taking scientific realism to be the position that "there are aspects of reality for the description and cognition of which we are dependent upon the vocabulary and methods of scientific theory", he claims that this kind of position is fully consistent with a conception of truth which is evidentially constrained. Of course, scientific realism is more than anti-instrumentalism. But this more is captured by Wright by adding that part of scientific realism is the claim that "statements formulated in scientific theoretical vocabulary are apt to be true or false in a substantial way, one associated with representation of or fit with objective worldly states of affairs" (Wright 1992, 159).

This further content of realism, Wright argues, is consistent with verificationism since verificationism need not entail "a thesis about the bounds of reality—the thought that, as it were, the totality of facts is conveniently (but mysteriously) trimmed to ensure that there is nothing there that outreaches human inquisitiveness" (ibid.).

Well, we have already seen Dummett claiming that "the world is, so to speak, formed from our exploration of it", which might well imply that there is nothing *beyond* human inquisitiveness when it comes to how far the content of the world extends. But of course, the reaches of human inquisitiveness can be taken in a strict sense, viz., the actual capacities for probing the world, or in an idealized sense, viz., the in-principle (or ideal) human capacities for probing nature. The problem with the first option is that it is too narrow: actual probing capacities might change over time; hence, what there is cannot be equated with what can be actually probed. It is natural then to think of an idealized version of human inquisitiveness. This could easily leave a gap between what humans can actually come to know and what humans can come to know in-principle, or in an idealised end of the inquiry or under idealised cognitive conditions. And then Wright could be right in saying that the totality of facts might well outreach *actual* human inquisitiveness. Still, the totality of facts would not thereby outreach *idealized* human inquisitiveness; and in this sense, the content of the world would be trimmed to whatever can in principle be known or verified.

Indeed, Wright takes it to be the case that this verificationist version of scientific realism would require that the "faithfulness" of scientific representation—that is, its representing things as they are—is "in principle detectable". And though he is right in stressing that, on this view, there is "nothing to foreclose on the thought that scientific theory deals in genuinely representational but irreducible contents, rendered true or false by a real external source" (Wright 1992, 159), this very claim only shows that verificationist scientific realism is *not* instrumentalism. The key point here is that Wright's verificationist scientific realism renders the world that science probes mind-dependent in precisely the sense we have been describing, viz., it forecloses the possibility of a divergence between what there is in the world and what is in principle known (detectable) to exist. It compromises objective existence even if it honours irreducible existence.

Note that an appeal to idealized (or in-principle) conditions for justification is a double-edged sword. On the one hand, it allows for a gap between what is currently justified and what is justified *simpliciter*. Hence, it leaves open the possibility that what we think that exists is not what really exists—since the latter is tied to the obtaining of idealized conditions of justification. On the other hand, however, if we were to take it that the conditions for 'ideal justification' are really *ideal*, that is, such that *no human could possibly survey them*, then, as Davidson put it, "they are so ideal as to make no use of the intended connection with human abilities" (Davidson 1990, 307).

This tension is seen in Wright's favourite version of an epistemic account of truth, which relies on a strong account of justification. As is well known, Wright (1992) introduced the idea of superassertibility as a candidate for an epistemic notion of truth. Superassertibility is "assertibility which would be endurable under any possible improvement to one's state of information" (Wright 1992, 75). Only such a strong epistemic conception rules out the possibility that a warrantedlyassertible statement will not be controverted by subsequent information and hence enshrines the stability of truth. Interestingly, Putnam (2001, 599) claimed that this was exactly the notion of truth that he had in mind in his verificationist years.

To see how strong this requirement of incontrovertibility of justification is, it is enough to state that the only way to ensure it is to claim that the available evidence *entails* the truth of a certain statement S; otherwise, it is *always* possible that there may be further or future evidence that will remove the warrant for the truth of S. This requirement then would end up implying either that scientific hypotheses are non-ampliative or that we are already in possession of a priori justifications of ampliative methods such as induction or inference to the best explanation. But we are far from this, if it is possible at all.

Superassertibility is meant to close the gap between truth and incontrovertible warranted assertion. And to achieve this it encodes within it the *denial* of the possibility of divergence. Hence, the realist reaction to this should be that it is possible that truth and superassertibility are extensionally divergent notions (that is there are truths which are not superassertible and/or conversely).

Note that the very possibility of a verificationist scientific realism requires the a priori falsity of the underdetermination of theories by evidence, as Wright himself recognises (cf. Wright 1993, 287). It is obvious that the very logical possibility of two or more mutually incompatible theories being empirically equivalent entails (on the assumption that only one of them can be true) that truth doesn't necessarily lie within our cognitive capacities and practices. Moreover, if it is possible that there are incompatible but empirically equivalent alternatives even to an idealised theory of the world, it cannot be the case that an idealised theory of the world is necessarily true; hence the world might differ from the way it is described to be even by an ideal theory of the world.

Though scientific realism must find ways to block the argument from underdetermination of theories by evidence in order to ground the epistemic optimism associated with the third thesis (see the Introduction), it is questionable that it should be committed to the a priori falsity of the underdetermination thesis (cf. Psillos 2015). The reason for this is precisely that by doing so, the independence of the world is compromised—in particular, the objective existence condition. As Newton-Smith (Newton-Smith 1978, 88) has nicely put it, presented with the possibility of underdetermination, realists face a dilemma. Realists can go for an 'Ignorance Response' or an 'Arrogance Response'. On the first horn, realists choose to cling to a realist metaphysics of an independent world, but they put at stake their epistemic optimism. On the second horn, they secure epistemic optimism, but sacrifice the independence of the world by endorsing a view which denies that there are 'inaccessible facts'. We will come to the Ignorance Response momentarily. For the time being, the point is that it's hard to see how a scientific realist, qua *realist*, can endorse the 'Arrogance Response'. For, 'trimming down' the content of the world so that it contains no inaccessible facts leaves two options available (both of which should be repugnant to realists). The *first* is to re-interpret the empirically equivalent theories so that they are not understood literally and the apparent conflict among them doesn't even arise. This option would compromise the second thesis of scientific realism (see the introduction). The *second* option is precisely to adopt an epistemic notion of truth which makes it the case that only one of the empirically equivalent theories passes the truth-test (cf. Jardine 1986). This option would compromise the first thesis of scientific realism.

So is the 'ignorance response' a compelling move for realism? Realism, to be sure, makes scepticism possible. This is the price that realism has to pay for respecting the thesis that the world is mind-independent. In fact, a key motivation for the anti-realist denial of mind-independence has been blocking of scepticism. But though ignorance is possible, realism does not have to argue that it is actual! In fact, the standard argument for scientific realism, known as the Putnam-Boyd argument from the success of scientific theories to their truth, is precisely meant to ground the realist epistemic optimism on the contingent fact that the (approximate) truth of the scientific theories is the best explanation of their empirical successes (cf. Psillos 1999, 2009).

In characterising non-verificationist scientific realism, Wright says: "If it is as likely as not, for all anyone can have cause to believe, that prosecution of best method leads to false theories as to true ones, what is the basis for believing that we can even so much as get lucky?" (Wright 1993, 292). This kind of question, and the implicit point made by it, totally disregard the fact that scientific realists have argued precisely for the point that the best explanation of pursuing successfully best method is that the relevant scientific theories (that inform best method) are approximately true. This, it should be stressed, can be argued for even though the best explanatory link between truth and the application of best method is contingent.

When it comes to the argument from underdetermination, the right realist answer is to argue that the tie between competing empirically equivalent rival theories is broken by looking for reasons—mostly explanatory reasons—to prefer one theory over the rest. In other words, look for the best explanation of the evidence. It is not the purpose of this paper to revisit the debates about the so-called 'no miracles' argument or about the underdetermination thesis (see my Psillos 1999, 2015). It is enough for our purposes to note the following. A key motivation for verificationist scientific realism is that it renders scientific truth achievable. As Wright (Wright 1993, 298) put it, it wields an essential connection between "the harvest of best scientific method and truth". What it adds to ordinary scientific realism is that "truth is essentially certifiable by best method: that for any true theory, sufficiently extensive researches must disclose an adequate, enduring case for taking it to be so". Beware though: on the verificationist approach, it is not that the method harvests an independently given truth about the world; the truth *is* the deliverances of the method (suitable and persistently pursued). Hence, there is no possibility of divergence; hence, the mind-independence of the world (in the form of the objective existence condition) is compromised.

What, ultimately, is at stake in the scientific realism debate is whether there can be a robust sense of objectivity, that is a conception of the world as the arbiter of our changing and evolving theoretical conceptualisations of it. The kernel of the metaphysical thesis of scientific realism is that science is in the business of discovering what a world that is not of our making is like. This thesis implies that if the natural kinds posited by theories exist at all, they exist objectively, that is, independently of our ability to be in a position to know, verify, recognise etc. that they do; and hence that it is they, if anything, that make scientific theories true. Verificationist scientific realism compromises this robust sense of objectivity.

Moreover, a moral that might be drawn from the debates around scientific realism is that the success of science—that realism is meant to explain—is hard won. It is neither trivial, nor in any way guaranteed. The heated debate over the pessimistic induction (see my Psillos 1999, Chap. 5) has driven the point home that if there is continuity in theory-change, this has been a considerable achievement, emerging from among a mixture of successes and failures of past scientific theories. A realist non-epistemic conception of truth and in particular the possibility of divergence do justice to this hard-won fact of empirical success and convergence. Given that there is no a priori guarantee that science converges to the truth, or that whatever scientists come to accept in the ideal limit of inquiry or under suitably ideal epistemic conditions will (have to) be true, the claim that science does get to the truth (based mostly on explanatory considerations of the sort we have already canvassed) is quite substantive and highly non-trivial. If, on the other hand, an epistemic conception of truth is adopted, hence if the possibility of divergence is denied, the explanation of the success of science becomes *almost* trivial: (long-run) success is *guaranteed* by a suitably chosen epistemic notion of truth, since—ultimately—science will reach a point in which it will make no sense to worry whether there is a possible gap between the way the world is described by scientific theories and the way the world is.

Wright has called 'absolute realism' the view that is anti-instrumentalist, "but drops the epistemological optimism with which scientific realism is usually associated (Wright 1993, 292). This position, he adds, makes the connection between scientific theoretical truth and the deliverances of best method "inscrutable". Absolute realism, in other words, is non-verificationist realism and the remedy that is suggested to bring back epistemic optimism is to adopt the view that all truth is detectable (and harvested, in principle, by the scientific method). I have already stressed that what Wright calls absolute realism is not the position associated with current scientific realism. But not because scientific realism adopts an epistemic notion of truth; but because it need not abandon epistemic optimism. All it needs to do (and defenders of realism have done) is to show that there is a hard-won contingent link between the harvest of scientific method and truth. Cognitive success might is a contingent matter; but it is no less success for this.

Let me grant, at least for the sake of the argument, that, as a matter of contingent fact, whatever is issued by an epistemically right theory of the world (that is a theory which meets some ideal epistemic constraints) is what *really* exists in the world. This is certainly a possibility. It is consistent with the possibility of divergence noted above, and does not compromise the mind-independence of the world in any sense. Nor does it commit scientific realism to an epistemic account of truth. The scientific realist can easily accommodate the envisaged possibility of convergence by taking the right side in the relevant *Euthyphro* contrast: is the world what it is because it is described as thus-and-so by an epistemically right (meaning: suitably justified, superassertible etc.) theory *or* is a theory epistemically right because the world is the way it is? At stake here is the order of dependence. Scientific realists should go for the second disjunct, while verificationist anti-realism goes, ultimately, for the first.

Hence, it is wrong to pose to realist the following dilemma: either the concept of truth should be such that cognitive success is guaranteed or else any cognitive success is a matter of pure *luck*. What the realism debate has taught us, to be sure, is that the success of the realist project requires *some* epistemic luck: if the world were not mappable, science would not succeed in mapping it. But the realist has a story to tell us to why and how cognitive success, though fortunate and *not* a priori guaranteed, is not *merely* lucky or a matter of chance. The realist story (see Psillos 1999) will have to be phrased in terms of the reliability of scientific method and its defence. But there is good reason to think that this story is both sensible and credible.

References

Berkeley, George. 2008. *Philosophical Writings*. Desmond M. Clarke (ed.). Cambridge: Cambridge University Press.

Davidson, Donald. 1990. 'The Structure and Content of Truth'. *Journal of Philosophy* 87: 279–328.

Downing, Lisa. 2005. 'Berkeley's Natural Philosophy and Philosophy of Science'. In Kenneth Winkler (ed.) *The Cambridge Companion to Berkeley*. Cambridge: Cambridge University Press, pp. 230–265.

Dummett, Michael. 1982. 'Realism'. *Synthese* 52: 55–112.

Dummett, Michael. 1991. *The Logical Basis of Metaphysics*. Cambridge MA: Harvard University Press.

Dummett, Michael. 1996. *The Seas of Language*. Oxford: Clarendon Press.

Dummett. Michael. 2006. *Thought and Reality*. Oxford: Oxford University Press.

Garber, Daniel. 1982. 'Locke, Berkeley, and Corpuscular Scepticism'. In Colin Murray Turbayne (ed.), *Berkeley: Critical and Interpretive Essays*. Minneapolis: University of Minnesota Press.

Jardine, Nicholas. (1986). *The Fortunes of Inquiry*. Oxford: Clarendon Press.

Kant, Immanuel. 1998. *The Critique of Pure Reason*. Paul Guyer& Allen Wood (trans). Cambridge: Cambridge University Press.

Kant, Immanuel. 2004. *Prolegomena to Any Future Metaphysics*. Garry Hatfield (tran). Cambridge: Cambridge University Press.

Langton, Rae. 1998. *Kantian Humility*. Oxford: Clarendon Press.

Lewis, David. 1984. 'Putnam's Paradox'. *Australasian Journal of Philosophy* 62: 221–36.

Musgrave, Alan. 1999. *Essays on Realism and Rationalism*. Amsterdam: Rodopi.

Newton-Smith, W.H. 1978.'The Underdetermination of Theory by Data'. *Proceedings of the Aristotelian Society*, Suppl. Vol. 52: 71–91.

Psillos, Stathis. 1999. *Scientific Realism: How Science Tracks Truth*. London: Routledge.

Psillos, Stathis 2005. 'Scientific Realism and Metaphysics'. *Ratio* 18: 385–40.

Psillos, Stathis. 2009. *Knowing the Structure of Nature: Essays on Realism and Explanation*. London: MacMillan-Palgrave.

Psillos, Stathis 2012. 'One Cannot be Just a Little Bit Realist: Putnam and van Fraassen'. In James Robert Brown (ed.) *Philosophy of Science: The Key Thinkers*, London: Continuum, pp. 188–212.

Psillos, Stathis. 2015. 'Evidence: Wanted, alive or dead'. *Canadian Journal of Philosophy* 45: 357–381.

Putnam, Hilary. 1978. *Meaning and the Moral Sciences*, Routledge and Kegan Paul.

Putnam, Hilary. 1982. 'Three Kinds of Scientific Realism'. *The Philosophical Quarterly* 32: 195–200.

Putnam, Hilary. 1983. *Realism and Reason*. Philosophical Papers Vol.3. Cambridge: Cambridge University Press.

Putnam, Hilary. 1990. *Realism with a Human Face*. Cambridge MA: Harvard University Press.

Putnam, Hilary. 2001. 'When 'Evidence-Transcendence' is not Malign: A Reply to Crispin Wright', *The Journal of Philosophy* 98: 594–600.

Vision, Gerald. 2004. *Veritas: The Correspondence Truth and its Critics*. Cambridge MA: MIT Press.

Wright, Crispin. 1992. *Truth and Objectivity*. Cambridge MA: Harvard University Press.

Wright, Crispin. 1993. *Realism, Meaning and Truth*. Oxford: Blackwell.

Structural Realism and the Toolbox of Metaphysics

Steven French

Abstract Recent developments in the scientific realism debate have resulted in a form of 'exemplar driven' realism that eschews general 'recipes' and instead focuses on the specific, 'local' reasons for adopting a realist stance in particular theoretical contexts. Here I suggest that such a move highlights even more sharply the need for the realist to incorporate a healthy dose of metaphysics in her position, particularly when it comes to the theories associated with modern physics. Turning to another set of recent developments, having to do with the relationship between metaphysics and science, I argue that the exemplar driven realist can appropriate certain current metaphysical devices to help make concrete her commitments. Specifically I focus on a kind of exemplar based structural realism and examine the adequacy of, first, the determinables-determinate relationship as presented by Wilson and, second, Paul's 'one category ontology, as such devices within this framework.

1 Introduction

This paper sits at the intersection of two recent debates: the first concerns the contrast between so-called 'recipe' realism and an exemplar driven form (Saatsi 2016), whereas the second has to do with the relationship between metaphysics (Callender 2011; Ladyman and Ross 2007). In essence it represents an attempt to delineate a more moderate, 'third' way in each debate, using the example of structural realism to give concrete form to this attempt.

I shall begin by outlining 'exemplar-driven' realism which has been offered as an alternative to the traditional, 'recipe-based' framework. As an example of a form of realism that adheres to the latter, structural realism has been held up for criticism and I shall argue that this criticism is either unwarranted or can be accommodated. In effect, I shall suggest that structural realism can be articulated as an exemplar-driven project. However, when it comes to the relevant examples, it is

S. French (✉)
School of Philosophy, Religion and History of Science, University of Leeds, Leeds, UK
e-mail: s.r.d.french@leeds.ac.uk99

E. Agazzi (ed.), *Varieties of Scientific Realism*,
DOI 10.1007/978-3-319-51608-0_12

metaphysical considerations that crucially motivate the shift to structures and this takes us into the second debate.

A number of commentators have noted the apparent divergence of much of current metaphysics from modern science. Metaphysicians have been admonished for failing to pay attention to developments in modern science, especially physics, and various metaphysical devices, principles and theories have been taken to be ruled out by these developments. However, French and McKenzie (2012, 2015) have argued that metaphysics may yet have instrumental value in providing a kind of toolbox that philosophers of science can use for their own ends. In particular, I shall suggest that metaphysics offers an array of tools that the realist can deploy to help make good on the claim that science offers a view of how the world is, and not just how it could be. And in particular I shall argue that these tools can help the structural realist further articulate her position and respond to various criticisms and concerns.

2 Exemplar-Driven Realism

In a recent analysis, Saatsi has argued that the realism debate is in the grip of what he calls 'recipe realism', where such a recipe is '... capable of distilling the trustworthy aspects of a theory, applicable to *any* good, predictively successful mature theory.' (2016 p. xx; see also Asay 2016). The example he gives (indeed, which leads the paper), is that of structural realism, according to which the structural realist insists that given any empirically successful, mature theory, it will 'get the structure right'. Of course other examples can be given—the entity realist will insist that such theories nail down the right entities, the dispositionalist semi-realist will insist that they get the relevant bundles of dispositional properties 'right' and so on—but they are all akin in spirit by virtue of proposing some abstract schema for capturing the truth-content of our best theories or, more generally, characterising the sense in which such theories 'latch onto' the world.

Thus, structural realists express their epistemic commitments in such general terms: 'all that we know is structure'. And in that spirit, the debate then focuses on the letter of how that notion of structure may be metaphysically characterised and represented. Some, such as Worrall (1989), have tended to highlight the relevant equations and deploy Ramsey sentences as representational devices, whereas 'ontic' structural realists (Ladyman and Ross 2007; French 2014) have emphasised symmetries and the associated group-theoretic structure and have used the semantic approach to represent these features of theories. But Saatsi argues that the underlying approach is the same: to give a general recipe that can be applied to any historical episode or any new scientific development (that meets the realists' general criteria of empirical success, maturity etc.).

Thus, the aim of recipe realism is '...to capture theories' epistemic commitments across a wide range of disciplines and different areas of scientific theorising in unified terms, offering recipes or algorithms that are first motivated by particular considerations and case-studies, and then optimistically projected well beyond those to the rest of science.' (Saatsi 2016, p. xx). However, Saatsi argues, this aim is thwarted for a number of reasons: first of all, the sheer diversity of science and the inhomogeneity of theorising suggests that there is really little reason to expect that one recipe will fit all disciplines. Of course the very abstract nature of these recipes may delude their advocates into viewing science as more homogenous than it actually is, but to expect that theories in high energy physics, on the one hand, and immunology, on the other, 'latch onto reality' in the same way is surely unwarranted.

Furthermore, Saatsi continues, the fact that we have so many different realist recipes should give us pause; perhaps, instead of thinking of them as competitors, we should regard them from a pluralistic perspective as capturing the different possible ways that a theory can 'get the world right'—i.e. get it right in its structural aspects, in its causal aspects and so on. But of course, the fact that each recipe fits certain cases gives no reason to think it can be 'projected' even within that particular science, much less across disciplines, nor that the other recipes are somehow ruled out across the board.

And the very abstract nature of such recipes means they actually say little about precisely how the theories concerned latch onto reality. Again, in the case of structural realism, to give content to the recipe the nature of the structure that is appealed to must be spelled out but this has proven contentious, or so it is claimed. Here we have the contrasting examples of Ramsey sentences and the semantic approach, and it is argued that '[i]t is entirely unclear why these recipes should in general be at all good for discerning something that will be carried over in various theory changes, something that furthermore genuinely accounts for the past theories' empirical success.' (*ibid.*, p. xx) Even giving illustrative exemplars such as the famous Fresnel case or phlogiston, is of little help, since they may pull in different directions, thereby watering down the content of the recipe; after all, in what sense can 'the structure' in the case of light and phlogiston be the same?

Nevertheless, Saatsi agrees that focussing on such exemplars is the way to go—except he recommends giving up on the recipe entirely: 'Recipe realists are right in leaning heavily on exemplars in explicating their realist commitments, but they go wrong in trying to generate a general recipe that captures the gist of those exemplars.' (*ibid.*, p. xx) Instead, we should regard realism in general in terms of adherence to the 'positive attitude' that theories' empirical success is due to their getting something right about the world and specific exemplars then give content to this attitude by specifying what that something is, with no expectation that it can be exported to other theories within that discipline, much less across disciplines. Thus we have a global attitude, applied locally and as Saatsi notes, this gives new meaning to the 'divide et impere' slogan that also underpins structural realism.

Now, there is much that is positive about this suggested reorientation of the realism debate. And indeed, despite being held up as an example of 'recipe realism', there seems little to prevent structural realism from being articulated within the framework of 'exemplar realism'. Although the epistemic form of structural realism was indeed originally presented as a general and perhaps abstract response to the

Pessimistic Meta-Induction (Worrall *op. cit.*), the ontic variant allies that response to an attempt to accommodate the specific metaphysical implications of quantum mechanics (Ladyman 1998). And as has been noted (French 2006), that alliance may in fact come apart. Consider: towards the end of his classic paper, Worrall speculated that his form of structural realism might be extended to quantum physics. Now, although attempts have been made to articulate an appropriate sense of continuity between classical and quantum mechanics (see Saunders 1993, French 2014 pp. 15–20), it might be argued that such attempts can only be deemed to be successful to the extent that we accept a certain 'plasticity' in the relevant structures (so that the Poisson brackets of classical mechanics can be deemed to be appropriately related to the Moyal brackets on the quantum side for example; see French *ibid.*). If one were to conclude that the bridge between the two is just too tenuous one might then be inclined to conclude that either one should acknowledge that the structures one should be realist about are different in the classical and quantum cases, with no relevant continuity between them (and hence convergent realism is in trouble, as Laudan famously suggested), or that if one has grounds to be a structural realist when it comes to quantum physics, one may have no such grounds in the classical context, where one should be an entity realist perhaps. This second option would certainly fit with exemplar realism.

And of course, setting aside the issue of responding to the Pessimistic Meta-Induction, the relevant grounds for shifting to structures will vary both within a particular discipline, such as physics, and between disciplines, such as physics and biology. Thus when it comes to quantum mechanics, part of the original motivation for ontic structural realism concerned the perceived metaphysical underdetermination between the views of particles as individuals and as non-individuals, both of which are supported by the relevant quantum statistics (see Ladyman 1998; Ladyman and Ross 2007; French and Ladyman 2011; French 2014). In response to van Fraassen's conclusion that such underdetermination undermined the realist's position, it was argued that it could be effectively dissolved by giving up on the underlying object-oriented metaphysics and claiming that, to repeat the slogan, 'all that there is, is structure' (Ladyman 1998). This is less a 'recipe' than a metaphysical commitment that of course then needs to be cashed out.

However, that cashing out of what is meant by 'structure' should not be understood in terms of Ramsey sentences or set-theoretic structures or category theory or whatever. As French (2014, Chap. 5) emphasises, these are the devices that we use as philosophers of science to *represent*, for our own purposes and aims, theories, data models, programmes and, yes, empirical and theoretical structures. But our realist commitments should not be to these devices in and of themselves. In deploying them we are not so much giving content to the relevant 'recipe' as using a meta-level tool. Thus it should come as no surprise that different philosophers of science, with their different meta-level commitments, should use different sets of such tools. Worrall, for example, is famously antagonistic towards the semantic approach and thus prefers the syntactic formulation of Ramsey sentences to capture the structural commitments manifested at the object level of the theories themselves in the form of the relevant equations. Others have opted for category theoretic frameworks (Bain 2013; Landry 2007), although these have been criticised for failing to appropriately represent the relevant structural features (Lal and Teh 2015; Lam and Wuthrich 2015). And, of course, the set theoretically based semantic approach has long been advocated as an appropriate means of capturing the inter-theoretic commonalities that are claimed to form the basis of the structural realist's response to the Pessimistic Meta-Induction (Ladyman 1998; French 2014, Chap. 5)

Furthermore, that cashing out at the 'object level' of the science itself will be specific to the relevant theoretical context. Thus in the context of the theory of light and, subsequently, electromagnetism, the relevant structure is presented by Worrall in terms of the equations of first, Fresnel, and then Maxwell (and beyond), taken to be interpreted of course. And although we find such features as Galilean invariance in the classical context, it is in the quantum physics that symmetries really come to prominence, beginning with the permutation invariance that lies at the heart of the quantum statistics underpinning the above metaphysical underdetermination. And just as the laws are presented mathematically via the appropriate differential equations, for example, these symmetries are presented via the mathematics of group theory. Hence in this case the 'structure' is cashed out in terms of the relevant laws plus symmetry principles (see French 2014),[1] where these are then clothed in an appropriate metaphysics.

Shifting to quantum field theory, we no longer have the original motivation in the form of the above metaphysical underdetermination (although we do have another kind of underdetermination in the form of fields-as-substantival versus fields-as-instantiated properties) but we still understand the structure in play through a combination of symmetries and laws, with the Poincaré symmetry of relativistic space-time playing a particularly significant role (French and Ladyman 2003; French 2014). And since it is quantum field theory that provides the framework for the Standard Model of high-energy physics, we can again cash out the relevant structure via laws and symmetries, with the gauge symmetries involved in representing interactions now added to the mix (French 2014).[2]

So, in one sense, we don't have the same recipe cashed out in each case, since the motivation for structural realism that is presented in the context of quantum mechanics is not present in that of high-energy physics, where the motivation has more to do with the way that fundamental properties such as spin and charge effectively 'drop out' of the relevant symmetries. But of course, in another sense it can be alleged that we do, insofar as it is the relevant symmetries that are focussed on in each case, as presented in the theoretical contexts by the appropriate groups.

---

[1]Actually, following Cassirer the structure of the world is understood in terms of a three-way interwoven complex of symmetries, laws and determinate phenomena (French *ibid.*).

[2]Curiously, Nounou (2015) suggests that ontic structural realism is almost exclusively focussed on quantum mechanics with very little attention paid to quantum field theory and hardly any at all to high-energy physics, although the former is covered in French and Ladyman (2003) and the other papers in the special issue of *Synthese* in which this appeared (in particular Cao 2003 and Saunders 2003), as well as in French (2014) which also tackles the latter.

However, one can insist in response that this is entirely driven by the relevant context not by some adherence to a particular realist recipe: it is because of the framework provided by quantum field theory that we find Poincaré symmetry also playing a fundamental role in the context of the Standard Model and it is because of the role of gauge symmetries more generally that we find the notion of structure cashed out in this context in this manner as well. In other words, what appears to be the same recipe applied again and again is in fact due to the features of the relevant physical theories.

But then of course we should not expect these same features to be exemplified either by other theories within physics or by the theories of other disciplines. So, no one of course would claim that when it comes to theories of light and phlogiston the structure is the same. Indeed, insofar as the latter example might be seen as falling under 'chemistry' (and here we might need to be sensitive to disciplinary boundaries), we would clearly not expect to encounter the same equations or laws much less any symmetries (see French 2014 Sect. 12.2). Likewise when it comes to biology, where we not only have no symmetries but no laws either, except perhaps for natural selection. Nevertheless, although we clearly no longer have the motivation for shifting away from objects that was articulated in the quantum context, the kinds of concerns with the nature and role of the notion of 'organism' and biological object more generally that have been articulated by Clarke, Dupré and others (see, for example, the papers in Guay and Pradeu 2016) have been taken to power a similar shift from understanding biological entities in object oriented ways to conceiving of them in terms of certain kinds of biological structures and processes (French 2014, Chap. 12; French 2016). Again it might be emphasised that it is reflection on the science itself rather than sticking to a particular realist recipe that is driving these moves.

All of which amounts to saying that in certain respects structural realism is already exemplar-driven and there seems to be no inherent barrier to rendering it explicitly so. Thus, from this perspective, the structuralist would acknowledge that, at the very least, the motivations and reasons for this shift will vary from context to context and discipline to discipline and indeed that in some cases there simply will be no such grounds. In other words, whether structural realism is the appropriate stance to adopt would have to be tested on a case by case basis.

However, if our realism is going to be exemplar based then there is even greater need to be clear on what it is we are going to be realist about. Consider: the structural realist has long pointed out that underlying the 'recipe' of standard realism is a certain kind of 'object orientation'. In effect this smuggles in an implicit metaphysics so that when the standard realist declaims 'I am a realist about electrons' and is then pressed on what these electrons are, she can then say 'they are objects, like tables and chairs, albeit subject to the laws of quantum physics which make them behave in weird ways …' As far as the structural realist is concerned, the object oriented standard realist gets away with a lot by means of this manoeuvre, since she never seems to face the equivalent to 'what is this 'structure' of which you speak?'; that is, she never seems to have to answer 'what are these objects that electrons are supposed to be?' In other words, the recipe masks the underlying metaphysics.

If that mask is then stripped away and we ground our realist stance in distinct exemplars, then we cannot get away with keeping the metaphysics implicit—it must be stated explicitly in each case. The alternative is to adhere to an entirely epistemic form of realism (or what Magnus calls 'shallow' realism; Magnus 2012) which would amount to pointing, if pressed, to the relevant features of the theory, as expressed in its equations or models or whatever, and insisting 'I am a realist about *that*!'. But as a response to the demand to say how the world is according to the theory, that hardly seems adequate. Hence we need to appeal to some appropriate metaphysics in each case. The question then is how to avail ourselves of that metaphysics.

3 Metaphysics as a Tool for the Realist

As I noted in the introduction, the relationship between metaphysics and science has recently come under scrutiny, with a number of commentators declaring the former not fit for purpose, given recent developments with regard to the latter. This rejection proceeds on (at least) two bases: first of all, many of the big debates in current metaphysics, such as monism versus pluralism or fundamentality versus gunk seem to proceed with little or no regard to the impact of the relevant science. At best, it is claimed, when science is dragged into the debate, it is in the form of a crude, long since discarded picture, amounting to little more than high school chemistry (Ladyman and Ross *op. cit.*). Secondly, a number of the concepts and principles that lie at the core of modern metaphysics appear to have been ruled out of court by developments in modern science.

Now some caveats are in order here. With regard to the first point, it has to be said that not all metaphysicians are ignorant of developments in science. Paul and Schaffer, for example, have both appealed to features of quantum mechanics in support of their different positions (a one-category ontology and monism respectively; see Paul 2013; Schaffer 2013). And when it comes to the second, this 'ruling out' is not always definitive (French and McKenzie 2015). Take Leibniz's Principle of Identity of Indiscernibles for example. Following French and Redhead (1988) it has long been held to be violated by quantum physics yet a 'Quinean' version has recently been constructed that is compatible with the physics (Muller and Saunders 2008; but for criticism, see Bigaj and Ladyman 2010). However, it might be felt that these are exceptions and that in general the dismissal of much of current metaphysics by philosophers of science is well justified on the grounds that it is simply out of touch with modern science (think, for example, of the way the notion of intrinsicality is usually understood in terms of 'lonely objects' and how this is discussed in the absence of any consideration as to whether physics can even accommodate a model in which there is a lone particle in the universe).

Now if one is a realist, exemplar based or otherwise, seeking to articulate a locally delineated view of how the world is, what are one's options given the above?

One, of course, is simply to eschew metaphysics entirely and in answer to the question 'what is the world like according to theory T?", to simply point to T, set out in all its glory on a whiteboard, say, and to declare 'That! It is like that!'. Now, that setting out will typically be—and certainly so in the case of physical theories—in terms of the relevant mathematics but only a radical Platonist will leave it at that. Physical realists will of course insist that the relevant terms must be interpreted, and those in eschewal mode will further insist that this interpretation will be 'purely' physics based. Now of course, purity is a slippery notion but one can imagine our eschewalling realist declaiming 'That! The world is like that! Where this term refers to the electron and that to the electro-magnetic field ….' and refusing to say anything more. But of course the door to metaphysics has already been opened via this interpretation, since it invites the further question 'Yes, but what *is* the electron? Is it a particle? Is it a wave? Is it even an object?' Of course, one could simply refuse to answer such questions, insisting that to do so would take us beyond what can be legitimately grounded in the relevant physics. But I suspect that many would feel that in so refusing the realist hasn't really lived up to the name and that our understanding of how the world is remains thin and impoverished.

And of course, even appealing to the 'pure' interpretation of T invites comparison to similar interpretations of both related theories and its predecessor. The term 'electron' for example is freighted with certain connotations associated with its deployment in, say, classical mechanics. There the electron is regarded as a particle and, further, as an object that possesses certain properties and, further still, as an *individual* object, assemblies of which can be statistically considered in certain ways which are dependent on permutations of those objects being counted. Here we see the door to metaphysics opening wider and wider. And the next obvious question would be 'Well, when our theory T is quantum mechanics, is the electron like *that*? Is it an individual object, permutations of which are counted in the appropriate statistical analysis of the objects' collective behaviour?' Now again the metaphysics eschewing exemplar based realist can maintain the line and simply utter the response 'No. It is not.' But that is going to invite obvious further questions and refusing to spell out in some metaphysical terms how the world is such that permutations of electrons do not count, or make a relevant difference, is again going to leave us with only the thinnest of understandings (indeed, one that is cast in largely negative terms).

An alternative is to eschew metaphysics *as it is currently formulated* and adopt some form of 'bespoke' metaphysics constructed to directly clothe the relevant features of modern science. We have been here before of course. One example is that of Whitehead, who drew on the early (or 'old') quantum theory and its apparent 'vibratory' features to motivate his process philosophy (see for example Whitehead 1926; for a recent consideration of this motivation see Epperson 2004). Another would be Eddington, who took the above feature of quantum statistics in particular (that is, its permutation invariance) to motivate a form of structuralism according to which objects are not prior to but on a par with the relevant relations and subsequently went on to articulate this structuralist metaphysics in the context of what can be considered to be an early form of quantum gravity (Eddington 1946). The obvious problem with such a move—which is evident in the later works of both Eddington and Whitehead—is that such a bespoke metaphysical framework must be elaborated via bespoke terms, concepts, principles and categories and runs the risk of descending into incomprehensibility.

Fortunately, there is a third and, I would argue, more reasonable alternative: treat current metaphysics as a kind of toolbox that although it may contain some devices that are not 'fit for purpose' may still contain others that the realist can use (French 2014; French and McKenzie 2012, 2015). So, although we might conclude that notions such as intrinsicality or principles such as Leibniz's are ruled out by modern physics, there may be others that we can adapt to fit. Let me expand on an example from (French 2014) and express it in the context of exemplar based realism.

4 Symmetries, Structure and Determinables

Consider the so-called Standard Model, which has been the subject of much discussion in the popular and philosophical literature, especially following the discovery of the Higgs boson. The overarching framework is quantum field theory. Here the non-counting of permutations of electrons, for example, is explicitly built into the theory via a fundamental symmetry known as Permutation Invariance, expressed mathematically by the permutation group. This yields the fundamental division of 'elementary particles' into the kinds fermions (to which electrons belong) and bosons (to which photons, for example, belong), corresponding to two of the irreducible representations of the permutation group. Quantum field theory is also relativistic, so it incorporates the symmetries of Minkowski space-time which are represented mathematically via the Poincaré group, the irreducible representations of which yield a classification of all elementary particles, with these representations indexed or characterised by mass and spin (the invariants of the group).

Furthermore, the Standard Model is a gauge theory, represented by the group SU (3) x SU(2) x U(1) via which further relevant symmetries can be captured within the theory. What this means, broadly speaking, is that the Lagrangian of a system—which basically captures the dynamics—remains invariant under a group of transformations, where the 'gauge' denotes certain redundant degrees of freedom of that Lagrangian. Thus, consider electrodynamics, for example, for which U(1) above is the relevant gauge symmetry group associated with the property of charge and the photon (a gauge boson) effectively drops out of this requirement that the theory be gauge invariant. Extending this requirement to the other forces, we obtain, for the weak nuclear force, the SU(2) symmetry group associated with isospin, a property of protons and neutrons, and for the strong nuclear force, SU(3) associated with the colour property of quarks. Mass is then accounted for via the

Higgs boson associated with the breaking of the isospin symmetry of the unified electro-weak force.

That, crudely sketched, is the relevant exemplar. Now, it has been argued that the appropriate realist stance that should be adopted towards this exemplar is that of the structuralist, where the metaphysical notion of 'object' is at best set on a par with that of 'relation' (Ladyman and Ross 2007) or removed from the picture altogether in favour of a fundamental conception of 'structure' (French 2014). The obvious question that has been asked (repeatedly) is 'What is that structure?', or putting it more generally, 'What is the world like, if it is structural?". Again, one answer would be to write out the details of the Standard Model on a whiteboard and pointing, insist 'It is like that!'. As before, this yields a thin sense of metaphysically informed understanding. An alternative is to attempt some form of bespoke account. Thus Eddington, before he went off the metaphysical deep end as it were, expressed such group theoretically described invariances in terms of 'patterns of interweaving', which at least is evocative if not perhaps very precise (and perhaps not really very bespoke, given the connotations associated with 'weaving'!).

Instead we might appeal to certain devices in the metaphysical toolbox to help capture the nature of 'structure' in this context. So, consider the way in which the fundamental properties from 'being a fermion' to charge and spin 'drop out' of the above symmetries. This is a core feature of this structuralist view: rather than considering the world as built from the bottom up, as it were, beginning with objects that have properties, between which there hold relations, which are expressed by laws, that are constrained, in some sense, by these symmetries, the structural realist inverts that order, and sees the relevant metaphysics as proceeding from the top down, so that we take the symmetries and laws as fundamental, and the properties to be derivative. How can we metaphysically express that inversion and capture the relationship between the above symmetries and the properties that drop out of them? One tool we can use is the determinable-determinates relation (French 2014, Chap. 10).

This has been extensively discussed of course (for an excellent overview of the various positions, issues and concerns that have been raised, see Wilson forthcoming) and the central idea is that determinables and determinates stand to one another in a certain specification relation, as the determinable 'colour' does to the determinate 'red', or the latter as determinable does to a particular shade of red, or as mass, qua determinable, does to a specific mass value. Part of the extensive discussion here has focussed on the nature of this relation but the crucial point is that it relates properties that are more or less specific, relative to one another; so, 'red' is more specific than 'colour' and a particular value of mass is more specific than 'mass'. 'Increased specificity' is just one of the features of the determinable-determinate relationship that Wilson helpfully lists (*ibid.*, pp. 8–9). Others that also motivate its deployment in this case include: 'determinate incompatibility', according to which if something has a certain determinate of a given determinable, then it cannot at the same time have a different determinate of that determinable (at least, not of the same or lower specificity); 'determinate opposition', according to which different determinates of the same determinable are not just incompatible but are relevant alternatives (so 'red' and 'blue' are determinates of 'colour' but 'red' and 'square' are not); 'requisite determination', which requires that anything that has a given determinable, must have some determinate of it; and 'asymmetric dependence' which states that for any determinable of some determinate, anything that has that determinate must have that determinable, but something could have that determinable without having that determinate (so anything that is red must be coloured but something coloured, may not be red of course).

Now, the suggestion is that we can apply this metaphysical tool to the case of the symmetries that the structural realist takes to be a fundamental feature of the structure of the world, in the sense that we regard such symmetries as relational determinables generating determinable properties and associated determinate values. Thus, consider the permutation group, mentioned above: this encodes a range of possible particle statistics, but in this world it appears that only two of those determinates are manifested, namely those corresponding to the kinds fermion and boson (yielding Fermi-Dirac and Bose-Einstein statistics and characterised by anti-symmetric and symmetric state functions, respectively). Likewise, the symmetry of relativistic space-time, characterised by the Poincaré group can be regarded as a determinable which also yields spin as a property-determinable, which in turn yields the property spin ½, associated with the electron for example, as a determinate. Again, it is through the determinable (with the emphasis on the -able) that the relevant possibilities are encoded (French 2014, p. 283). So, being a fermion, say, is more specific than being subject to permutation symmetry, so we have increased specificity; and being a fermion and being a boson are not just incompatible but are relevant alternatives; and anything that is subject to permutation symmetry must behave according to some particle statistics, whether fermionic, bosonic or, but not apparently in this world, parastatistical. Finally, of course, anything that is a boson is subject to permutation symmetry but something that is subject to the latter may not be a boson—it could be a fermion, for example.

Now applying this device to help flesh out the metaphysics of structure raises a number of issues. First of all, some have argued that increased specificity feature implies that determinates must be metaphysically prior to or more fundamental than determinables and if this were accepted, we could not take permutation symmetry to be prior to and more fundamental than bosonic or fermionic statistics. And hence we could not take that symmetry to be a feature of the fundamental structure of the world, in line with the core shift of structural realism from objects possessing properties to relations and structures. But this argument is problematic, not least because there is a lacuna that has to be filled: what has specificity to do with fundamentality? (French *ibid..*, p. 284) And this lacuna needs to be filled in a non-question begging way: so, it is not going to impress the structural realist to insist that maximal specificity corresponds to fundamentality because maximally specific determinates are the properties possessed by objects, such as elementary particles. Nor is it going to persuade the non-Humean structuralist by insisting that maximally specific determinate properties are categorical and only categorical properties can be in the fundamental base. Such a structuralist takes her structure to be modally informed and thus has no qualms about admitting modality into the fundamental base. Finally, it might be objected that reality must be maximally determinate else we allow a form of ontic vagueness to enter the world (Wilson *op. cit.*, p. 14) and as Lewis reminded us, '[t]he only intelligible account of vagueness locates it in our thought and language.' (Lewis 1986, p. 212). But Lewis' claim is highly contentious of course, and quantum physics has again been appealed to in order to motivate arguments that the world is ontically vague, in a certain respect (French and Krause 2003). Note that this is still in accord with the weaker claim that there cannot be *only* determinable features of the world (Wilson, *op. cit.* p. 14). The structure of the world incorporates both determinable and determinate features, such as the distinct bosonic and fermionic kinds and the specific spin of the electron, which Wilson refers to as 'existential witnesses'.

Not only can we use metaphysics as a constructive tool, but we can also use it as a contrastive one. Thus to get a (hopefully) clearer picture of the view being presented, and of the way in which the determinable-determinate relation can help as a metaphysical tool in fleshing out that picture, let us compare it to Paul's recent development of a 'one-category' ontology (2012; 2013).

She begins with the core question that obviously resonates with the structural realist: 'What is the fundamental structure of the world?' (2012, p. 221). Answering this question is a partly metaphysical project, where that metaphysics is informed and constrained by science but not governed by it.[3] Thus, by 'fundamental structure' here she understands fundamental constituents, from which all else is constructed via some 'building rule', and the fundamental categories, which are determined by the fundamental kinds or natures of things. In the balance that has to be achieved between metaphysics and science, the latter will determine what we take to be the fundamental constituents of the world, in terms of the physical properties, structures and objects that should be regarded as 'perfectly natural', to use Lewis' phrase. But metaphysics will take the lead in determining both the rule by which things are composed out of these constituents and the nature of the latter, in the sense of determining the fundamental categories to which they can be assigned.

So, for the 'building rule' she takes composition, on the grounds that we have a direct, intuitive grasp of proper parthood which forms the heart of the composition relation. Here immediately the likes of Ladyman might object that such intuitions, based as they are on naïve view of 'everyday' objects or, at best, classical mechanics, fail utterly when it comes to modern physics where the notion of being a part of is much slipperier and harder to grasp.

However, when it comes to the fundamental categories, Paul does draw on certain features of quantum physics to argue, first, that we should reject what she calls the 'traditional spatiotemporal view' that runs throughout much of contemporary metaphysics and which '…takes some or all of the fundamental constituents of the world to be spatiotemporal parts, i.e., chunks of spacetime, many of which are qualitatively rich, and the building relation to be spatiotemporal composition.'

[3]Paul explicitly considers how metaphysical realism meshes with scientific realism (2012, p. 232).

(*ibid.*, pp. 233–234). And here she acknowledges that the fault of such a view is that it conflates the metaphysics of the everyday, or 'manifest image' with that of 'the real' (ibid., p. 239) Secondly, and more importantly for my purposes, she maintains that we can still retain 'the world-building relation' but now applied to a different set of fundamental categories.

Thus Paul collapses the category of property into that of substance (2013). On her view the world is built from n-adic properties via property composition, which effects a kind of fusion or bundling (2012, p. 242) and since this 'mereological bundle theory' does not require this fundamental category to include spatio-temporal properties, it can accommodate a much broader range of possibilities when it comes to the nature of the fundamental entities. And since '…every fundamental physical theory ever given, including all of those currently on offer, is or can be couched in terms of properties and relations, even if these properties and relations are extremely abstractly specified …' (*ibid.*, p. 245), such theories will mesh with this particular metaphysics.

This is certainly an attractive metaphysics and it is for that reason that it acts as a useful contrastive tool. The contrast, of course, comes from Paul's reading of physical theories as couched in terms of properties and relations and the concomitant insistence on a metaphysical 'bundling' or property composition relation. Although the former is obviously true to a certain extent, this reading omits the crucial role of laws and symmetries. Indeed, if we take this role seriously in the context of the Standard Model say, then it would seem that although Paul has gone some way in the right direction by dropping spatio-temporal composition, she still retains an overall 'bottom-up' approach. At the very least mereological bundle theory needs to be able to accommodate the relationship between symmetries and kinds, as in the case of permutation symmetry and the boson/fermion distinction, and properties, as in the case of Poincaré symmetry and spin, say.

Paul herself explicitly invites the structural realist to adopt her mereological bundle theory, on the grounds that, '… structuralists can make good use of an n-adic property mereology, since they don't need substances or even monadic properties in order to construct the world.' (2012, p. 248; see also 2013 pp. 110–111). Indeed, she suggests, such a marriage would lead to a 'super-sophisticated structuralism' (2013, p. 111) that avoids certain of the problems that its less sophisticated form is held to face.[4] The idea then is that we take relations as constituting our fundamental base and then apply bundling as the appropriate building relation, thereby effectively constructing the structure of the world via fusion (Paul 2012 p. 245), with putative objects as 'nodes' in this structure, or as Cassirer put it, as 'intersections' of these relations.

Understood this way, mereological bundle theory would be in effect a further metaphysical tool that the structural realist could use (see French 2014, pp. 186–189). However the issue of how to accommodate symmetries remains. If

[4]Her emphasis on n-adic properties as fundamental also resonates with Mertz's ontology, which has also been taken to be a suitable metaphysics for OSR (see Mertz 2016).

they are viewed as merely 'by-products' of laws, expressing certain features of the latter then with laws themselves expressing the relations that sit in the fundamental base, mereological bundle theory might seem the appropriate metaphysical device for accommodating the relevant structure construction. On this account, each such relation would exhibit a certain feature that when 'fused' to create the network of relations that the laws of physics describe manifest the global features that we describe via symmetries. Of course, further work is required to 'mesh' this metaphysics with the physics.

In particular it might be objected that in the practice of physics, symmetries act as constraints on laws, or as 'meta-laws', which suggests more of a 'top down' stance, in contrast with Paul's. Now of course, one could respond that this might be correct when it comes to the heuristic use of symmetries but that doesn't require that they be regarded as 'standing above' laws, metaphysically speaking. A third way between these two extremes is to follow Cassirer and take symmetries, laws and measurement results as being on a par and together constituting the structure of the world (French 2014). This removes the necessity of adopting either a 'bottom up' or 'top down' stance towards symmetries but of course the relationship between them and the properties typically taken to be monadic must still be accommodated. In particular, although it might be regarded as merely 'loose talk' to say that a property such as spin 'drops out' of Poincaré symmetry, the close relationship here on the physical side needs to be matched by a similar relationship on the metaphysical.

Consider again permutation symmetry and the distinction between bosons and fermions. One could begin with that distinction as fundamental, so that quantum entities possess 'being a boson' or 'being a fermion' as kind properties. Bundling such properties together yields the relevant feature of assemblies of such entities that is represented by either the bosonic or fermionic representation of the permutation group, respectively. And the fact that this group yields other representations, corresponding to paraparticle statistics, for example, is primarily of mathematical rather than physical significance—unless of course, such statistics turns out to be physically realized (as was suggested for a short time in the case of quarks), in which case the relevant group representation would be applied (so this comes down to an issue in the applicability of mathematics), but metaphysically, of course, according to mereological bundle theory we would still begin with 'being a paraparticle (of a certain order)' and build up from that.

The alternative is to begin with the symmetry itself as part of the fundamental base and take the 'dropping out' of bosonic and fermionic statistics to express the relationship between this symmetry, as part of the fundamental structure of the world, and these kind properties. In terms of the mathematics this amounts to no more than the relationship between the group and its representations but this obviously needs to be matched on the metaphysical side. Here the determinable-determinate relationship seems to do the job, so that 'being a boson' or 'being a fermion' are simply determinate aspects of that partly determinable structure. Note the further contrast with mereological bundle theory: instead of a 'building relation' we have something akin to a 'manifestation relation' and instead of thinking of the structure as built up from certain parts (even if these are properties and relations rather than objects and substances), we are invited to think of it as given holistically, as it were, and as manifesting certain determinate features.[5]

Another way of seeing the contrast between this and Paul's approach is to consider the question of what should be our attitude towards the other possible properties, such as 'being a paraparticle' for example. According to mereological bundle theory we begin our construction with the properties that we actually discover in the world, such as 'being a boson.' We represent those properties mathematically via group theory and we find that such mathematics includes alternatives that do not appear to be realized—from this perspective these are just so much 'surplus structure'. According to the alternative, this 'surplus' represents certain possibilities which may or may not be actualized and the determinable nature of the structure flags the point that it encodes such possibilities. Thus rather than beginning with the actual, and building up from that, we begin with what is modally allowed and show how the actual world fits into that, as a determinate manifestation of that modally informed structure.

There is more to say here, of course, but this is perhaps enough to highlight the differences between these metaphysical tools.[6]

5 Conclusion

I began by sketching recent moves towards a more local or 'exemplar based' form of realism and suggesting that such moves do not preclude the adoption of a structuralist stance. However, if we are to take this move seriously then we need to pay close attention to the relevant exemplars, one such, in the context of modern physics, being the Standard Model with its emphasis on certain symmetry principles. The exemplar realist is then faced with the issue of spelling out how the world is according to that model. One option is just to point to the relevant physics and insist 'it is like that!', but that is obviously unsatisfactory. The alternative is to treat current metaphysics instrumentally, as a kind of toolbox and apply certain devices in an effort to generate a sense of understanding how the world could be that way. Focussing on the issue of capturing the relationship between such symmetry principles and certain properties, I've presented two such 'tools': the determinable-determinate relationship and mereological bundle theory. The former, I think, does a better job in meshing with the physics, but the latter cannot be discounted. And there may be other devices that can be used as well. The point is, if we are going to 'go local' and focus on the particularities of a given set of

[5] In a sense still be spelled out, this stance sits somewhere between metaphysical nihilism and monism.

[6] Paul herself remarks that she finds Wilson's defence of determinables 'interesting and plausible' (2012, p. 245 fn 22).

exemplars, whether historical or current, then in adapting our realism to those particularities there must be an even greater emphasis on spelling out what this realist stance commits us to, in terms of, as Paul puts it, not only the fundamental constituents, but the categories they fall under and the kinds of relations that hold between them. With particular metaphysical tools adapted to particular exemplars, this overall approach may reinvigorate and strengthen the currently strained relationship between metaphysics and science.

References

Asay, J. (2016). Going local: a defense of methodological localism about scientific realism. *Synthese*. Online First (DOI 10.1007/s11229-016-1072-6).

Bain, J. (2013). Category-Theoretic Structure and Radical Ontic Structural Realism. *Synthese, 190*,1621–35.

Bigaj, T., and Ladyman, J. (2010). The Principle of Identity of Indiscernibles and Quantum Mechanics. *Philosophy of Science 77*, 117–36.

Cao, T. (2003). Structural Realism and the Interpretation of Quantum Field Theory. *Synthese 136*, 3–24.

Callender, C. 2011. Philosophy of Science and Metaphysics. In S. French and J. Saatsi (eds.) *The Continuum Companion to the Philosophy of Science,* pp. 33–54. London: Continuum.

Eddington, A.S. (1946). *Fundamental Theory*. Cambridge: Cambridge University Press.

Epperson, M. (2004). *Quantum Mechanics and the Philosophy of Alfred North Whitehead*. New York: Fordham University Press.

French, S. (2006). Structure as a Weapon of the Realist. *Proceedings of the Aristotelian Society 106*, 167–85.

French, S. (2014). *The Structure of the World*. Oxford: Oxford University Press.

French, S. (2016). Eliminating Objects Across the Sciences. In A. Guay and T. Pradeu (eds.), *Individuals Across Sciences*, pp. 371–394. Oxford: Oxford University Press.

French, S. and Krause, D. (2003). Quantum Vagueness. *Erkenntnis 59*, 97–124.

French, S. and Ladyman, J. (2003). Remodelling Structural Realism: Quantum Physics and the Metaphysics of Structure: A Reply to Cao. *Synthese 136*, 31–56.

French, S. and Ladyman, J. (2011). In Defence of Ontic Structural Realism. In A. Bokulich and P. Bokulich (eds), *Scientific Structuralism*, pp. 25–42. Dordrecht: Springer: 25–42.

French, S. and McKenzie, K. (2012). Thinking Outside the (Tool)Box: Towards a More Productive Engagement Between Metaphysics and Philosophy of Physics. *The European Journal of Analytic Philosophy 8*, 42–59.

French, S. and McKenzie, K. (2015). Rethinking Outside the Toolbox: Reflecting Again on the Relationship Between Philosophy of Science and Metaphysics. In *Metaphysics in Contemporary Physics*. ed. T. Bigaj and C. Wuthrich, 145–174, Poznan Studies in the Philosophy of the Sciences and the Humanities, Rodopi.

French and Redhead, M. (1988). Quantum Physics and the Identity of Indiscernibles. *British Journal for the Philosophy of Science 39*, 233–46.

Guay, A. and Pradeu, T. (eds.) (2016). *Individuals Across Sciences*. Oxford University Press.

Ladyman, J. (1998). What is Structural Realism? *Studies in History and Philosophy of Science 29*, 409–24.

Ladyman, J. and Ross, D. 2007. *Every Thing Must Go: Metaphysics Naturalized*. Oxford: Oxford University Press.

Lal, R. and Teh, N. (2015). Categorical Generalization and Physical Structuralism. *British Journal for the Philosophy of Science*. Advanced Access (doi:10.1093/bjps/axv002).

Lam, V. and Wuthrich, C. (2015). No Categorial Support for Radical Ontic Structural Realism. *British Journal for the Philosophy of Science 66*, 605–634.
Landry, E. (2007). Shared Structure Need Not Be Shared Set-Structure. *Synthese, 158*, 1–17.
Lewis, D. (1986). *On the Plurality of Worlds*. London: Blackwell.
Magnus, P.D. (2012). *Scientific Enquiry and Natural Kinds: From Planets to Mallards*. Palgrave Macmillan.
Mertz, D.W. (2016). *On the Elements of Ontology: Attribute Instances and Structure*. De Gruyter.
Muller, F., and Saunders, S. (2008). Discerning Fermions. *British Journal for the Philosophy of Science 59*, 499–548.
Nounou, A.M. (2015). For or against structural realism? A verdict from high energy physics. *Studies in History and Philosophy of Modern Physics, 49*, 84–101.
Paul, L.A. (2012). 'Building the World from its Fundamental Constituents'. *Philo- sophical Studies 158*, pp. 221–56.
Paul. L.A. (2013). Categorical Priority and Categorical Collapse. *Proceedings of the Aristotelian Society 87*, 89–113.
Saatsi, J. (2016). Replacing recipe realism. *Synthese,* First Online, 2016. DOI:10.1007/s11229-015-0962-3.
Saunders, S. (1993). To What Physics Corresponds. In S. French and H. Kaminga (eds). *Correspondence, Invariance, and Heuristics; Essays in Honour of Heinz Post*, 295–326. Dordrecht: Kluwer.
Saunders, S. (2003). Structural Realism, Again. *Synthese 136*, 127–33.
Schaffer, J. (2013). The Action of the Whole. *Proceedings of the Aristotelian Society 87*, 67–87.
Whitehead, A.N. (1926). *Science and the Modern World*. Cambridge: Cambridge University Press.
Wilson, J. (forthcoming). Determinables and Determinates.
Worrall, J., (1989) Structural realism: The Best of Both worlds? *Dialectica 43*, 99–124. Reprinted in Papineau, D., (Ed.)., *The Philosophy of Science*, Oxford: Oxford University Press, 139–165.

Retention, Truth-Content and Selective Realism

Alberto Cordero

Abstract Selectivism, the *divide et impera* approach, is arguably the most promising realist project today, despite lingering issues regarding the selection of truthful parts in successful theories. Second generation approaches address those issues by basing selection on detailed inferential analyses of impressive predictions drawn from the theory at hand. The leading reformed strategies advocate commitment to just theory-parts that are "absolutely indispensable" for advancing the derivations at hand. The minimalist emphasis involved seems reasonable at first, but it weakens the realist project, leaving many thinkers gloomy about the prospects of scientific realism. This paper has two parts. Part 1 examines the reformed strategies and traces their main difficulty jointly to the pledge they make to interpretive minimalism and a neglect of the epistemic (and realist) import of explanatory power. The result is selections of excessively modest content and vulnerable grounding that deliver less than is needed. Part 2 advances a critical revision and naturalist generalization of the basic strategy. The proposed alternative maintains the focus on inferential analyses of predictions and content-reduction, but without commitment to content minimalism. By moving in a naturalist direction, the suggested strategy brings the valuation of theory-parts in line with confirmational criteria that, in scientific practice, give salience to success and freedom from reasonable doubt. The ensuing proposal recognizes both prediction and explanation as necessary conditions for realist commitment to theory-parts.

A. Cordero (✉)
CUNY Graduate Center & Queens College, The City University of New York, New York, USA
e-mail: alberto.cordero@qc.cuny.edu

E. Agazzi (ed.), *Varieties of Scientific Realism*,
DOI 10.1007/978-3-319-51608-0_13

Part 1

1 Introduction

A prominent response to Larry Laudan's skeptical reading of the history of science rejects the view that empirical theories fall flat if either their central terms fail to refer or some of their assertions are seriously wrong. A false theory, realists argue, can still be approximately true in the sense of having parts that have high truth-content—we just need to identify those parts in some principled way, hence the labels "Selectivism" and "Selective Realism." In the 1980s and 1990s, selectivist efforts to identify truthful content tried to distinguish between working or essential posits on the one hand and idle or inessential posits on the other. Fresnel's theory of light was presented as an exemplar case. Embedded in now rejected assumptions (notably about the existence and character of the ether luminiferous), Fresnel's theory nonetheless contains reliable content at some theoretical levels of description. Light is not completely as Fresnel imagined, yet light is made of microscopic transversal undulations, as Fresnel thought, and these fluctuations follow the laws of reflection and refraction he derived. The general theory-parts obtained by abstracting Fresnel's original proposal from his account of the ether—the wave substratum—spell out a "thinned out" description (call it "Fresnel's core"). This core is of reduced specificity relative to the original theory, but it comprises content which in fact all subsequent theories of light have retained. Selectivists point to this comparatively abstract core as their paradigmatic recipient of likely realist content in successful theories.

In the 1990s and early 2000s, the leading strategy[1] run as follows: To identify what a successful theory gets right (a) look for parts that seem indispensable for making the theory's winning predictions; (b) in the case of past theories, check that in subsequent theories the parts in question have been retained and deemed approximately correct; and (c) propose those parts as both descriptive of theoretical aspects of the intended domain and responsible for the theory's success. Objectors voiced complaints along several lines. One focused on posits once regarded as central to a theory's success only to be rejected by subsequent science—like the ether, phlogiston, and caloric (Chang 2003). More corrosive complaints included Lyons (2006)'s argument that metaphysical and even mystical beliefs, weak analogies, erroneous calculations, and logically invalid reasoning often seem "centrally responsible" for the success of theories. Much in these objections is fair. In mature scientific disciplines, theories are typically constructs tight enough to make breaking them into independent parts exceedingly difficult. Breaking into parts is particularly troublesome in theorizing that proceeds under traditional metaphysical constraints (as virtually all pre-Einsteinian theorizing did). In the case

[1]Multiply advanced by Philip Kitcher (1993), Jarrett Leplin (1996), and StathisPsillos (1999).

of the ether of light, the proposals by Fresnel, Arago, Maxwell, Hertz, and Lorentz, all invoked the subsequently discarded ether. Contrary to Psillos (1999)'s suggestion, it seems that the predictions these thinkers derived from their ether-based theories could not have followed paths that actually bypassed the ether, not least because all the alternative theories then at their disposal partook of a metaphysical framework in which waves (and fields) stood as modes of being and as such something in need of a substratum (Cordero 2011).

The point is that, in advanced scientific disciplines, posits are very rarely "idle" or "inessential".[2] Responding to the challenge, over the last decade some reformed selectivists have sought to address the suggested complaints by sharpening up selectivism. The resulting proposals tilt the project towards the minimalism of structural realism. At the center of these contributions are recent works by Juha-Saatsi (2005, 2011), Peter Vickers (2013), and IoannisVotsis (2011) [SVV], whose focus is on the truth and falsity content of descriptive claims seemingly needed for deriving impressive predictions from theories.

2 Content Minimalism

Which parts are to be deemed "truthful" or even "reliable" in an empirically successful theory? SVV focus on the realist import of impressive predictions: If a prediction comes out right, it is because the theoretical claims invoked in its derivation have some significant truth-content, which they pass downstream to the prediction. The thesis here is this: A theory that is false but otherwise empirically successful lodges theory-parts that, although of weaker content, are still strong enough to make the derivation of impressive predictions go through. Spotting those parts *prospectively* (as opposed to just retrospectively) requires purging the derivation steps of content superfluous to making the predictions at hand. On this matter, SVV variously draw inspiration from the structural realist line revived by John Worrall (1989). In examining a theory T, they look for content that plays a substantial role in the derivation of impressive prediction. They begin by taking a representative set of remarkable predictions from T and then for each prediction pick the most cogent derivation available. The inferential steps are then subjected to critical scrutiny from the bottom (prediction-statement) up, looking for assertions involving "causally active" posits that seemingly cannot be further abstracted away without blocking the prediction. Because of the emphasis on purging content, the theory-parts that gain selection will often be thinned down versions of richer and more detailed theory-parts available in the full theory, as Fresnel's Core exemplifies.

One key question is how aggressive a purge one should attempt here. To Votsis (2011, p. 1230), the only parts dependably involved in the logical deduction of impressive predictions are "mathematical parts," chiefly equations and

[2]For an overview of early selectivism see e.g. (Cordero 2011).

mathematically structured concepts. However, since these cannot generate predictions without interpretation, they must be given one, which opens the door to superfluous content that—SVV agree—realists must strive to reduce to just the minimum needed to generate the impressive predictions of the theory at hand.

Thinning down derivations can be problematic, however. Minimalism pushes realism towards antirealism. Where is one to stop? A major shortcoming of bare bones trimming is a lack of clarity regarding what content is "absolutely necessary." Wary of the perils of minimalism, Vickers (2013) looks for a "natural" differentiation between the posits invoked in the derivation of predictions. A prediction typically resorts to premises that include claims drawn directly from the theory at hand ("internal") as well as ideas that, although "external" to the theory-proper, have guided the beliefs and judgments of the scientists who developed the theory. Vickers assesses these two varieties regarding their realist import. Once an approach (e.g. Fresnel's) is articulated, he urges, many of its surrounding posits (e.g. the ether) become "optional" supplements no longer indispensable for deriving the prediction in question. Being optional, "external" posits do not directly gain epistemic strength from the successful predictions of the theory, Vickers argues. Nor does realist commitment extend to all the internal claims invoked in the derivation: Commitment, he stresses, goes only to the "working parts" of the *internal* claims involved.

So, how well does this strategy handle the old ether? The proposed distinction assumes that the conceptual relations between internal and external claims are "optional," but this can be denied, as the already cited critics of Kitcher, Leplin and Psillos noted. Vickers presents the ether as a posit external to Fresnel's Theory, yet it seems implausible that Fresnel's 'internal' tenets could be affirmed without claiming also every claim they clearly presupposed. This is especially serious in theories born with conceptual links to higher-level ideas (Cordero 2011). In the ether example, arguing in the 19th century that something (say, X) was a wave amounted to claiming that it was a traveling perturbation, which presupposed that X required a supporting medium (since, metaphysically, theorists then widely regarded waves as modes of being). If so, using the external/internal distinction to bracket the ether required an extra step. The conceptual network associated with the predicate 'being a wave' needed first to have the noted conceptual "necessities" broken up, i.e. turned into contingent relations. But attempting that seemed "nonsensical" in the 19th century. It is only in the early 1900s that theorists such as Einstein, thinking against the current, proposed the required fragmentation. Einstein did this in his Special Theory, in a move that proved far from popular in the early 1900 but which subsequently gained wide acceptance in the natural sciences. The point is that Vickers' proposal can work only if one places it within a "modern" naturalist framework in which all conceptual links rest on empirical generalizations (indefinitely open to the possibility of revision).

Other hurdles stand in the way of the reformed selectivist approach. Vickers is aware of two of them: Pursuing interpretive minimalism pushes theory-part selection into enemy territory—at each step the "absolutely necessary" might seem no more than the step's constructive empiricist version, i.e. the step freed of commitment to non-empirical content in the original version. Secondly, no matter how one conducts the sought purge, part of what remains *may* still be idle. One question is thus whether one can make "reliable enough" selections along the proposed lines. Vickers pessimistically concludes "the realist is still some distance from prospectively identifying (even some of) the working posits of contemporary science." Indeed, in his view, the realist should commit to just some unspecified parts of what remains once the posits acknowledged as idle are removed. Vickers exudes pessimism:

> In all this, we find ourselves—even 30+ years after Laudan (1981)—unsure of the extent to which the *divide et impera* strategy can succeed. Even if the 'working posits' of contemporary science cannot be prospectively identified, it remains possible that we might develop a recipe for identifying at least certain idle posits. That would be a significant achievement, even if not quite what the realist originally had in mind. Vickers (2013: 209).

The above gloom harks back to Vickers's way of seizing on content separation in terms of his internal/external distinction, which is problematic on yet further grounds. For example, although conceptual fragmentation makes the external/internal separation possible, it is not enough for realism. Abstraction keeps out of sight some of the original background of a claim, but for a claim to have realist import the parts left out must be left out for *specific reasons*—ones explicitly backed up by the extant confirmational arrow. Otherwise, the selection will be *arbitrary*.

The strategy just reviewed improves on earlier selectivist strategies for identifying theory-parts suitable for realist commitment, and is both explicitly focused on abstraction and "prospective" (as opposed to merely retrospective) identification. But now there is this additional hurdle: The gloomy conclusion in the above quote clashes with the realist expectation of warranting commitment to explanations that make the theory's observable domain more *intelligible* in terms of unobservable structures and mechanisms underlying the intended empirical domain. It also leaves underappreciated much of the array of theory-parts that the best current confirmational practices in the natural sciences sanction as empirically successful and beyond reasonable doubt. Recognition.

To overcome the pessimistic trap, I suggest, selectivists need to do better at both discarding posits seriously off the mark (e.g. the ether) and identifying posits worthy of commitment. Otherwise, the resulting realist stance will be disappointingly bland, something like "In every successful theory, somewhere some abstract versions of some of its claims about unobservables get some things somewhat right about the intended domain."

More is expected of selective realism—a criterion both aimed at identifying substantive specific theoretical content and found reliable against the history of modern science.

3 Selectivism Without Minimalism

I suggest the pessimism expressed by Vickers is an artifact of particular selective emphases in SVV's works—especially regarding content minimalism, individual derivations taken by themselves, and their conditioning realist commitment to just predictive success. Such stresses tend to neglect some highly plausible content that successful theories invoke to make their respective observable domains intelligible. In the case of Vickers' pessimism, arguably there is also an overreaction to the tentative character of realist selections, compounded by a contentious reading of historical cases, notably Kirchhoff's Theory (see e.g. Cordero 2016).

Still, Vickers' worry about minimalism is on target: The equilibrium point of interpretive minimalism seems just too close to antirealism (Constructive Empiricism, Ramsey sentence non-realism, and such). But Vickers' specific complaints rest on *global* rather than specific misgivings: Subjecting a theory to the selectivist strategy he endorses strengthens the theory's epistemological credentials, but—Vickers urges—it still leaves open the possibility that some of what remains may be idle, and the minimalist stance behind the strategy pushes towards antirealism. Lamenting the lack of certainty and the vulnerability of theoretical claims in empirical science are old complaints, however. Non-skeptical responses to them are also old. In 1581, confronted with general doubts about the status of astronomy, Christopher Clavius judiciously argued that it is not enough merely to speculate that there may be some other way or method of accounting for the celestial appearances than the one then accepted. For a challenge to have any force, he maintained, an opponent must produce a specific alternative; only then can we profitably decide whether there is a reason to worry about a given option[3]. A related reaction occurs in Newton's Fourth Rule of Reasoning:

> In experimental philosophy we are to look upon propositions inferred by general induction from phenomena as accurately or very nearly true, notwithstanding any contrary hypotheses that may be imagined, till such time as other phenomena occur by which they may either be made more accurate, or liable to exceptions. (My emphasis).

Applying these suggestions to the selectivist approach, then, we should not trim content that is presently both successful and free of compelling specific doubts against it—leave in place content that faces merely possible doubts. This recommendation runs against minimalisms that declare inflationary all content that remains open to "possible" doubts, but the point is that those minimalisms lack compelling reason in their support. Unless explicit reasons are given for either keeping or weakening content, the resulting theory-parts will rest on "preferences" and feel rationally vulnerable. This fear of subjectivism is a factor behind Vickers' gloomy conclusion. The good news is that the needed reasons are frequently available, at least in disciplines that thrive in predictive power, or so I will argue.

---

[3]The case is presented in Gardner (1983) in connection with the debate between realists and instrumentalists in pre–Newtonian astronomy.

In the rest of this paper, I propose a naturalistic generalization of the SVV strategy that maintains the focus on inferential analysis of predictions and content-reduction, but lowers the commitment to content minimalism. One general difficulty with SVV's concentration on minimalism and individual derivations is the way this concentration calls attention away from significant realist resources, particularly confirmational relations that become unavailable when one goes for minimalist interpretation or takes derivations in isolation. Particularly pertinent among these resources are representative diachronic records of the empirical and explanatory difference that specific parts make to success as a theory plays the field.

The proposal outlined in the remaining sections moves in two complementary directions: (a) It makes both predictive power and some explanatory power necessary conditions for realist commitment to theory-parts. (b) The proposed approach sides with naturalism in trying to bring the assessment of theory-parts more in line with leading confirmational criteria of success and trustworthiness found in scientific practice.

The starting point is a clarification of what makes a theory-part crucial, superfluous, or toxic for advancing the scientific study of a domain.

Part 2

4 A Naturalist Generalization

If the previous considerations are correct, then recognizing specific theoretical content as crucial (or not) to deriving successful predictions from a given theory should rest on compelling specific reasons. For this task the naturalist has no meta-scientific insight or grand a priori philosophy to rely on—naturalists lack resources beyond those available in actual science; they just add rigor and concern for overall coherence.

One appealing platform, invoked by naturalists of a previous generation, rests on scientific appraisals of scientific success and freedom from compelling specific doubts—in Dudley Shapere's vocabulary (1982, 1984). Can this "first approximation" platform help the inferential selectivist approach? No single prediction outcome taken in isolation can stamp much of a seal of reliability or unreliability on any functioning posit. To do better, one needs to analyze clusters of derivations of different predictions (and phenomena) in which the theory-part in question plays a role. Acting on this suggestion, the remainder of this paper tries to develop an alternative to minimalism, picking out for realist commitment only claims that are taken to be both predictively successful and free of compelling specific doubts by scientific standards. How might a selectivist convincingly identify theory-parts likely to have high truth-content, and do this with projective force? Two selection strategies common in scientific assessments seem especially relevant to the task at hand (Cordero 2015, 2016). One focuses on inference analyses applied to representative sets of the predictions a theory T makes as it gains applications. The other strategy looks for independent support (from sources external to the theory) that may be available for parts left by T as assumptions or postulates that then play a role in the derivations at hand. As these two lines do their jobs during the life of a theory, they may concentrate confirmational weight on some theory-parts; when that occurs there is a promising case for selective realism.

5 Confirmational Strategy 1

The first strategy (S1) looks for encouraging and discouraging posits in the theory T. Here one valuable resource lies in the discovery of apparent problems for T, usually by opponents who subject T's central tenets and auxiliary assumptions to hostile probing, hoping to expose those tenets as either wrong or superfluous. Consider, for example, the ingenious efforts in the 1810s by defenders of the corpuscular theory to show that the central ideas of Fresnel's wave theory were wrong. Those efforts reached a peak in the episode leading to the experimental demonstration of the so-called "Poisson Spot," a prediction Poisson and other corpuscularians thought was to be the ruin Fresnel's theory—to their surprise it crowned it. In a parallel set of battles, the supporters of T struggle to revise its auxiliary assumptions upon encountering difficulties in the theory's application. This kind of effort was apparent, for instance, when Thomas Young's discovery of double-slit interference in 1801 forced corpuscularians into convoluted auxiliary hypotheses to account for the phenomenon. Corpuscularians failed to convince, leading to the effective collapse of the particle camp.

Study of how these animated investigations play out in a theory's course can help a realist make a preliminary selection of theory-parts that seem either crucial or not for the theory's empirical success. Two lists, drawn from multiple applications of this confirmational strategy, can be developed at this stage: List (L+), made of parts implicated in cases of impressive predictive success; and (L−), made of parts found implicated in unfulfilled predictions.

Once these two lists are preliminarily drawn up, the next realist task is to assess the impact on T's empirical progressiveness of each of the items listed. This assessment can be done for each part by evaluating the effect of removing it from T while keeping the others. The proposal is to recognize as (a) probably "crucial" to T's success only those parts in L+ whose removal clearly leads to T's stagnation, judging from T's overall track record, and (b) "suspect" those parts in L− whose removal clearly improves T's predictive power and/or frees T from seemingly intractable conceptual conundrums. In Fresnel's theory, for example, the yield of impressive predictions plummets if one removes certain posits (e.g. the transversal character of the light wave and the abstract spatial structure given by Fresnel's equations).

How well does this strategy help realism? The historical record of manifest retentions across theory-change seems good for L+ parts. Good but not without blemish, for it would have left in place the ether of light, at least throughout the 19th century (Cordero 2011). Strategy S1 helps realists but more is needed.

6 Two Additional Confirmational Strategies

A second strategy (S2) looks for support available to assumptions made by T from sources initially *external* to the theory T at hand, especially independently successful empirical theories. Think, for example, of the time when Fresnel's claim about the transversal character of light became derivable from Maxwell's theory of electromagnetism. Elucidations like this have become common in the last 150 years. From the 1950s on, for instance, numerous aspects of cell biology have gained justificatory elucidation from biochemistry or molecular biology (Thagard 2007). Cases in point include initially postulated neural mechanisms subsequently explained by realizing that neurons consist of proteins and other molecules that organize into functional sub-systems such as the nucleus, mitochondria, axons, dendrites, and synapses.

Since elucidation springs from an independently supported theory T*, it raises the credibility of the assumptions and narratives it casts light on—hence its interestto realists. External support thus helps sharpen list L+. Strategy S2 has clear pluses. Since the ether posit never got any such support in the 19th century, references to the ether would have been filtered out (Cordero 2015). On the other hand, as a marker of probable truth, elucidation seems neither necessary nor sufficient for realism. Unsavory counterexamples give pause to granting any given theory-part high probability from elucidation alone. For example, when Kepler looked for broader theoretical support for his 2nd Law, he derived it from the Aristotelian laws of motion for sublunary bodies and some principles of optimal action. Kepler elucidated his law, but only by invoking as premises some of the most hopelessly wrong claims of Aristotelian physics. So, strategies S1 and S2 improve the realist project, although the job each can do seems less than ideal.

There is yet another policy relevant to the selectivist cause. A third evaluation strategy (S3) develops when a theory starts to wane, and continues for some time after it dies. It comprises efforts to explain why a discarded theory showed empirical success. The retrospective analyses involved are not automatically "self-serving." Some S3 analyses unveil previously unappreciated causal or structural justification for a theory's accomplishments. Recall, for instance, the explanation wave theorists gave for the success of corpuscularian optics regarding the phenomena of reflection, refraction, and polarization. Retrospective analyses frequently add precision to specifications of the parts a past theory gets right, as can be seen, for example, in recent attempts to show why erroneous fundamental posits in Kirchhoff's theory of diffraction led to correct predictions (e.g. Brooker 2008). Realists can sharpen L+ by turning to the materials yielded by this form of retrospective elucidation, as indeed practicing scientists routinely do.

Unlike the versions of retrospective reading denounced by Stanford (2006), in S3 the elucidation of past truth is both informative and non-trivial. Past proposals may or may not show epistemic gains by S3 lights. Pre-modern theories fare poorly in this regard. The geocentric accounts of a planet's motion, for example, lacked cumulative content in terms of epicycles, eccentrics, and equants. In Ptolemaic models content growth was basically limited to (a) the most directly observable structures and (b) a set of claims held to be beyond doubt—notably the Principle of Uniform Circular Motion for heavenly bodies and the Aristotelian arguments for the absolute fixity and sphericity of the Earth. The main reason why the resulting orbits were denied realist interpretation is not because they failed the intelligibility requirement. If anything, at many levels, Ptolemaic constructions went out of their way to honor intelligibility—then guided by the Aristotelian theses just noted. Rather, the realist reason for rejection was that the epicycles, deferents, and equants invoked were multiply-realizable by the lights of existing knowledge (i.e. available data and commonly accepted principles). Similar comments apply to numerous other theories (think, for example, of the medieval approaches to physiology based on the doctrine of the humors). Here the point is that strategy S3 is not self-serving: Finding truthful content about unobservables in past theories can be a meaningful achievement. Moreover, finding such content is even central to getting realist inductions off the ground.

7 Realism About What?

What historical generalization (if any) can we reasonably induce concerning posits licensed by the joint application of confirmational strategies S1 and S2? There seems to be a historically supported expectation of projectability of retention for theory-parts that make to L+ by passing both the S1 and S2 filters. Admittedly, compared to fuller counterparts in their mother theories, the parts that make into L+ are comparatively abstract (less precise), restricted and coarse-grained, as are the descriptive theoretical schemes built on them. On the cheerful side, the postulated entities, processesand accounts presently sanctioned by S1 and S2, far from making a meagre picture of reality, provide a remarkably thick and highly textured array of claims about the world beyond the reach if unaided observation. Examples range from detailed cosmological histories from 13,000 million years back to the present, to descriptions at various levels of generality of the composition, structure, and interactions of matter, to organic life and its diversity, to histories of the rise of humans and human nature, and more. The resulting theoretical corpus is not a haphazard quilt of dubious significance but a body of abstract and perspectival, finite-range, coarse-grained assertions that, nevertheless, display astonishing (and growing) levels of integration into a detailed and textured picture of the world.

If the suggestions made in this second part are on the right track, then, contrary to many thinkers (realists and antirealists), *more than thirty years after Laudan (1981), it seems reasonable to claim that the divide et impera strategy of selective realism can succeed*—or so I have tried to argue.

Taken separately, neither S1 nor S2 can accomplish the required task. Strategy S1 does a specially good job of filling list L− with theory-parts that face compelling specific doubts against them; also, S1 eliminates candidates marred by underdetermination or conceptual conundrums. In turn, S2 brings both external support and novel intelligibility to conjectures introduced by a theory. Strategy S3 sharpens the assessment of the epistemic success achieved by past accounts. Selectivism, therefore, should require prospective theory-parts to have passing grades from both S1 and S2. The ensuing naturalist realist proposal comprises two complementary criteria:

Refutational Criterion (R−): A theory-part will reveal itself as "doubtful" if multiple pieces of recalcitrant data converge inferentially in that particular part, and saving the part in question is consistently accompanied by degeneration of the whole system, as measured by current scientific criteria.
Positive Criterion (R+): A theory-part will reveal itself as appropriate for realist commitment (i.e. for being deemed "very probably approximately true") if it passes muster by *both* S1 and S2.

Note that these criteria can be applied while a theory is in full flight. The historical record seems to support both criteria well.

8 Concluding Remarks

If the above considerations make sense, to be a selectivist scientific realist is to accept as truthful only the theory-parts that Criterion R+ picks out. I conclude with a few remarks about the selectivist stance proposed in this section.

(i) Criterion R+ picks theory-parts prospectively. Each part selected is both backed by strong reasons for accepting it and free of compelling specific doubts against it.
(ii) The theory-parts that R+ picks lodge most commonly at inferential levels below those of the highest theoretical postulates, as in the example provided by Fresnel's Core. The relevant point is that the selected parts are both substantive and clearly theoretical, i.e. placed at levels significantly above the empirical ground level. Commitment applies most easily to intermediate and low-level theory parts—theory-parts at the highest levels of current theorizing qualifying for comparatively weak levels of realist commitment. This restriction seems a sensible outcome, namely a realist stance that progresses "from the bottom up," also one seemingly borne out by the history of science.
(iii) If, as it is now widely recognized, the contemporary realism/antirealism debate is primarily about the limits of ampliative inference in science, then Criterion R+ helps scientific realists affirm what antirealists deny, namely that scientific inference is robust enough to support claims about realities beyond the reach of the unaided senses.

(iv) Requiring approval by S2 enhances the explanatory import of realist commitment. The elucidation received by theory-parts in L+ flushes down to the ground level—to the phenomena that the theory in question predicts. In this way, the R+ selectivist (unlike the antirealist) makes the intended empirical domain less mysterious, more intelligible.

References

Chang, Hasok (2003): "Preservative Realism and Its Discontents: Revisiting Caloric." *Philosophy of Science* 70: 902–12.

Cordero, Alberto (2011): "Scientific Realism and the *Divide et Impera* Strategy: The Ether Saga Revisited." *Philosophy of Science* (Vol. 78, 2011): 1120–1130.

Cordero, Alberto (2015): "On Scientific Realism and Naturalism." *Journal of Philosophical Research* (40): 31–43.

Cordero, Alberto (2016): "Content Reduction for Robust Realism," in Laura Felline et al. (eds.), *New Directions in Logic and the Philosophy of Science*. Milano: Società Italiana di Logica e Filosofia delle Scienze (2016): 31–42.

Gardner, Michael R. (1983): "Realism and Instrumentalism in Pre–NewtonianAstronomy." In *Testing Scientific Theories*, J. Earman, (ed). Minnesota, pp. 201–65.

Kitcher, Philip (1993): *The Advancement of Science*. Oxford: Oxford University Press.

Laudan, Larry (1981): "Confutation of Convergent Realism," *Philosophy of Science*, Vol. 48: 19–49.

Leplin, Jarrett (1997): *A Novel Defense of Scientific Realism.* Oxford University Press.

Lyons, Timothy D. (2002): "Scientific Realism and the Pessimistic Meta-modus Tollens," in *Recent Themes in the Philosophy of Science: Scientific Realism and Commonsense*, Steve Clarke and Timothy D. Lyons (Eds.). Dordrecht: Kluwer, 63–90.

Psillos, Stathis (1999): *Scientific Realism*. London: Routledge.

Saatsi, Juha (2005). "Reconsidering the Fresnel-Maxwell Case Study." *Studies in History and Philosophy of Science* 36 (3): 509–38.

Saatsi, Juha and Peter Vickers (2011): "Miraculous Success? Inconsistency and Untruth in Kirchhoff's Diffraction Theory." *British Journal for the Philosophy of Science* 62: 29–46.

Shapere, Dudley (1982): "The Concept of Observation in Science and Philosophy," *Philosophy of Science* (49): 485–525.

Shapere, Dudley (1984): "Objectivity, rationality, and Scientific Change," *Proceedings of the Biennial Meeting of the Philosophy of Science Association* 1984: 637–663.

Stanford, P. Kyle (2006): *Exceeding Our Grasp: Science, History, and the Problem of Unconceived Alternatives*. Oxford: Oxford University Press.

Thagard, Paul (2007): "Coherence, Truth, and the Development of Scientific Knowledge," *Philosophy of Science* (74): 28–47.

Vickers, Peter (2013): "A Confrontation of Convergent Realism." *Philosophy of Science* 80: 189–211.

Votsis, Ioannis (2011): "Saving the Intuitions: Polylithic Reference." *Synthese*180 (2): 121–37.

Worrall, J. (1989): "Structural Realism: The Best of Both Worlds?" *Dialectica* (43): 99–124.

A Scheme-Interpretationist and Actionistic Scientific Realism

Hans Lenk

Abstract Any "grasping" of world factors or "the world is necessarily not only relative to presupposed interpretational perspectives, but only possible from within such a perspective. Thus, we have only the possibility of attending to or grasping states relative to interpretive perspectives. That means any actions and recognitions whatsoever are dependent on methodological schematizations and perspectives previously taken over or methodologically implied. All this also relates to the interpretative interconnection of action, action-orientation and formation, cognition and recognition, interpretation in representing, depicting or cognitional modeling or even abstract modeling—not to speak of active interventions, e.g., in experiments or everyday agency. Some of these schemata are primary interpretations or "*Urinterpretationen*" (original hereditary interpretation schemata) which are biologically, even possibly genetically fixed; others are variable as regarding the different levels and types of interpretation. Important however is to distinguish between what is accepted as ontologically basic and what is only methodological-epistemological. This is all the more decisive for the problems of realism—or different realisms, i.e., for the conception of reality and what is called "real" (be it "in itself" and independent of humans or be it "real" in a secondary, e.g., socio-cultural or even virtual, sense). Any direct recognitional "grasping" whatsoever is also interpretational. "Reality in itself" is then only indirectly recognized in the sense of methodologically "entangled" systems potentials.

1 A Kaleidoscopic Relational Reality

In quantum theory and its interpretations the classical concept of "object" would dissolve into a quasi interpretationist "rainbow reality" (Herbert 1985)—or even "kaleidoscopic" (Lenk 1995, 2003 Chap. 9) conception of reality comprising, to be sure, objective though interpretation-dependent phenomena or, rather, phenomenal

H. Lenk (✉)
Department of Philosophy, University of Karlsruhe, Karlsruhe, Germany
e-mail: hans.lenk@philosophie.uni-karlsruhe.de

E. Agazzi (ed.), *Varieties of Scientific Realism*,
DOI 10.1007/978-3-319-51608-0_14

descriptions. In the last analysis, it seems indeed that not only quantum "objects" ("quones", Herbert) have such a character of "rainbow" or "kaleidoscope" reality,[1] but also everyday objects of our usual perception of and action with things and by their respective projections regarding their kinds, forms and schemes. *Mutatis mutandis* the same applies to perceiving and conceiving events—indeed for any "grasping" whatsoever of "objects", be they real or virtual.. Even the description and grasping of dynamical and comprehensive whole systems are dependent on extant possibilities of action and interpretations (as, e.g., in quantum theory the preparation of situations, i.e., of measurements and observations, the decision regarding these preparations and actions on the respective manipulation and delineation of the situation). But holistic interpretations are indeed interpretations proper, i.e., human-made constructs necessary even for an objective description as for any agent-related or subjective projection in the first place.

Any recognition or identification of objects or events as "entities" confronting us is certainly interpretation-dependent, essentially co-constituted or in part influenced by schematic and structural modeling and the respective modeling activities, even though they are impregnated in the above-mentioned sense, i.e., co-determined by what can be called "world factors". Any recognition whatsoever is dependent on so to speak "having" and using constituted schemes and structures and ways of "grasping" the world or parts of it by the acting and designing human being. In acting, measuring, "grasping", applying structures and projections we usually have alternatives, possibilities of a deviating or modified design or decision—and yet we can get relatively good and objective, i.e., intersubjectively valid and testable, results.

In the following paragraphs I would like further to dwell on such epistemological generalizations of this insight about the preparation of situations and measurements as well as observation procedures in but not only in quantum mechanics. Primas (1983, 254ff) would even go much further and say that not only the Cartesian split of subject and object has to be given up in (a philosophy of) physics, but even the old dream of a monistic natural science consisting in explicitly describing and explaining something as occurring totally outside of the subject and being as well as changing independently of it—at least with regard to describing the engendering process of phenomena and any observational "grasping". Indeed, the dream of a purely mechanistic natural science being independent of any delineation, modeling, structuring, theoretical integration etc. seems to be outdated after Primas. (But this to my mind does not at all contradict in any sense the methodological(ly) indispensable objectivity and inter-subjectivity in natural science and also not the respective realism of sorts.)

---

[1]As everybody knows a rainbow is an objective phenomenon you may take a photographic picture of, but which disappears if you come closer. The metaphor of a kaleidoscope is in some sense better than the one of the rainbow because we are active in preparing the respective observation and measurement situation: We make and *actively* shake a kaleidoscope, whereas a rainbow is just *passively* being observed and not human-made.

"The theory of entangled systems leads us to a new holistic seeing" of the things of the world "which was alien to classical physics and its systems theory" (Primas, ibid. 258): "The world is no longer seen as an aggregation of particular mutually interacting things existing just for themselves, but as a unity in which objects only exist in connection with their relationships to the observer and his models and abstractions. There is no possibility of describing phenomena without prejudice" and (in this strong or strict sense) totally "objectively", we cannot "grasp" them as such without taking or having chosen a specific perspective. "Entangled systems would represent a whole or a totality. 'Whole' in this sense is any thing, "of which no other description can be given than a complementary one".

Rather interestingly Primas even infers: "Reality is a *relationship* between observer and the observed" (ibid.). (But what is that beast "the observed" without an(y) interpretative, if virtual, "grasping" again?) Instead, one can say that "grasping" reality is inevitably scheme-bound and thus interpretation-impregnated. Here, an impregnation is defined as necessarily being or at least implying loosely speaking a sort of "impact" of a "world factor" but is indeed, methodologically speaking, an interpretative approach, i.e., epistemologically as well as neuro-physiologically speaking, dependent on the activation of schemes. This is an important statement for pragmatic interpretative realism (of course of an indirect provenance). Sentences about matter would always be statements about relationships, about "our relationships toward the external world; *the laws of the natural sciences are (*basically, H.L.*) not (*just, H.L.*) laws of nature, but orders summoning (instructions for) actions to the natural scientists*" (ibid. 258). Phenomena are but "context-dependent" and interpretation-laden, i.e., indeed dependent on a special choice of a perspective required by any observation and measurement whatsoever. This choice of perspectives in some sense—at least methodologically speaking as an abstract structure of necessary predeterminations or prior or implied decisions—is certainly also (to be interpreted as) an *action*. In the same vein, performing, e.g., an observational measurement would also signify an (however minute) intervention into the system displaying more or less important consequences impacting or modifying the state of the system.

"Only the totality of all complementary descriptions would represent the unseparated reality" (Primas, ibid. 258). Whatever this "unseparated reality" may be and the term may designate or mean, in any case the conception is a construct, an idea to be conceived of integrating all wholes or totalities of potential descriptions in a unitary (metaphoric) picture. Any experience of world and reality is therefore in a sense also interpretation-dependent not regarding whether it is quantum states or macroscopic "graspings". *Always* we have to imply perspectival predeterminations about the scheme-interpretational approach to be taken, about the presuppositions, circumstances and context, about the instruments of measurement, the apparatus (es), and means of observing, about the delineation of the analyzed system as well as about the requirements and ways of action by which we would design and perform a specific representation, observation, measurement etc. of a system.

According to Lockwood the very consciousness itself would select these perspectives, would "designate" them; but I would not like to understand this in any ontological sense or supposition. Important is the methodological aspect: We necessarily need something like a starting base or take-off point given by a sort of interpretational perspective either pragmatically, at times tentatively given by our modes of seeing and understanding as mediated by our culture and language or assumed by a conscious choice: The "real" *is* (part and parcel, process and prior prompting of) *impregnation*. (More exactly, the "grasping" of the real is the result of impregnation processes.[2]).

All this corresponds in a certain sense—*mutatis mutandis*—with the scheme-interpretational perspective as developed here. There is only one important difference: I do not relate directly to consciousness and its faculties, but I would like to understand the connections rather in a methodological or even (quasi) transcendental manner—as a necessary interconnection of conditions. Lockwood so to speak understands this probably horizontally: The choice, e.g., to measure the momentum of one of pairwise entangled electrons would restrict and determine the correlated magnitude on the side of the other electron (may however be observed by another observer). This is so to speak a horizontal relativization of states. But I think that they might also be vertically understood in a more general sense, namely as the precondition of implying perspectivity in the first place and the necessary decisions about assuming one of the respective perspectives. This may in part be by an individual decisional choice (as exemplified in the EPR example of an electron pair) but it can also be determined by a more general interpretational prestructuring or preschematization by language, culture, history, by presuming the previous (observed or theoretically interpreted) interactions with the system etc.

The perspectivistic, interpretation-constructivistic understanding of the quantum-mechanical measurement problem does of course not lead to a solution of this problem on the object-level (in quantum theory itself) nor even just in an experimental-methodical manner, but nevertheless it would put the whole situation in regard to constituting and perceiving of objects and their delineations in everyday contexts in a new perspective. By this, of course, the interpretation of quantum-mechanical entities and of the measurement problem are also to be considered in a new light: The situation so to speak turns around as more generally illustrated. On the other hand, by this now changed relationship between macro- and micro-objects—amounting to a *veritable turnabout*—also the presuppositions, perspectives and methodological attention regarding micro-objects themselves will be modified in general, too. It is then not thus extraordinary any more if perspectival and interpretation- or construct-bound modes of "grasping" micro-entities are so to speak "read" into micro-entities, since a respective understanding is already at hand in mesocosmic interpretations of objects and reality. These insights therefore may be used from a double vantage point: First for a modified and novel

[2]One may proceed in analogue fashion to the EPR-correlated partial system of, say, a pair of electrons with their respective entanglements of quantum states relative to the respective fixation of possible measurement processes and the necessary interdependence by which the measuring intervention into one part (e.g. a measuring the spin of one entangled electron of a pair) the quantum state of the other is automatically decided upon ("designated", if you wish, or "fixed").

construct-interpretationist conception of micro-entities implying consequences regarding the overlapping problem in quantum theory and secondly for the generalization of respective topics engendering constructs under perspectival or scheme-interpretational preconditions in usual mesocosmic everyday world situations—or rather in a respective epistemological interpretation on a higher level.

2 "Grasping" in Perspectives and Interpretations

Any "grasping" of world factors or "the world" whatsoever is therefore necessarily not only relative to presupposed interpretational perspectives, but only possible from or within such a perspective. Thus, we have only the possibility of attending to or grasping states relative to interpretive perspectives. That means that any actions and recognitions whatsoever are dependent on methodological schematizations and perspectives previously taken over or being methodologically implied. All this also relates to the interpretative interconnection of action, action-orientation and formation, cognition and recognition, notably also involving scheme-interpretation in representing, depicting or cognitional modeling or even abstract modeling—not to speak of active interventions, e.g., in experiments or everyday agency. Recognition and action cannot be totally separated from each other, but can only be analytically distinguished in a *certain sense*. Any conception of "world" (or "world factors"), any experience of the external world, but also any recognition of other connections (say, e.g., of social origin) is certainly dependent on presupposed, predetermined or chosen perspectives of interpretation, on specific means of interpretation, like, e.g., forms and activation routines of schematization. *One can only "grasp" the so-called "reality in itself" in connection with and* via *dependence on the respective frameworks of previous perspectives shaped by such interpretation schemata, patterns* etc.

Some of these are primary interpretations or "*Urinterpretationen*" (original hereditary interpretation schemata) which are biologically, even possibly genetically fixed; others are variable as regarding the different six levels I have distinguished elsewhere (1993, Engl.: 2000, 2007,18).

These levels are also levels of flexibility and variability of schemes and interpretative constructs as one can easily see. Important however is to distinguish between what is accepted as "ontical" or ontologically important and what is only methodological-epistemological. This is all the more also decisive for the problems of realism–or different realisms (i.e., for the conception of reality and of what is called "real" (be it "in itself" or independently of humans or be it "real" in only a secondary, e.g. socio-cultural or even virtual, sense)—What does all this now mean for realisms and the respective conception of reality? Certainly, *grasping reality is only possible from or by using perspectives*, by taking some interpretational perspectives of maybe unconsciously implied primary ones or by using consciously developed habitual or other higher-level conventional or normative forms of interpretation. Thus one may indeed defend a hypothetical-pragmatic residual realism or minimal realism only being concretized in interactions within the forms of interpretations by contact with respective systems, so to speak in interactional interpretations, notably by our actions, by our theoretical and scientific representations, by choosing perspectives for descriptions and actions. With regard to external perception this is mediated by what I have called "impregnation in the narrower sense",[3] i.e., by being impinged by the so-called "world factors". However, even if we presuppose primary interpretations (biologically fixed, e.g.) with respect to direct sense perception such a recognition of "world factors" impregnating themselves on our sense organs and faculty of perception can only be conceived of and recognized by interpretatively colored concepts and processes (activities). Any so-called 'direct' recognitional "grasping" whatsoever is already interpretational. "*Reality in itself*"[4] is then only indirectly recognized in the sense of methodologically "entangled" systems potentials no matter whether strictly quantum-theoretically understood or in a classical interpretation. Impregnations are always results of schematic interactions and can thus be represented and conceived of only in action-dependent interpretation, in short: by interactional scheme-interpretations.

By our categorial schematizations and interpretations any "grasping" of reality is so to speak concretized in active, quasi active and *inter*active processes (activities in the wider sense) or procedures. If we would take a transcendental epistemological point of view, we would not like to talk about the activities as such, but about the basic dependence on such schematic activities and also the respective interactions.

The classical picture of an object as forwarded by determinism and classical physics—is but only a very simple special case, easy to understand and apparently easily accessible, but indeed dependent on an idealizing view and also relatively seldom encountered in the world.[5] Classical deterministic systems are a special case, since locality and separability of the entangled systems is presupposed or hypostatized there admitting of no unlimited generalization, but to be only approximately instantiated. Locality and separability of systems (notably entangled ones) is not generally guaranteed, but is to be presupposed to a certain degree and any individual concrete observation or measurement in the process of isolating a specific variable and measuring result. This is even true in quantum mechanics if we undertake a definitive process of measurement of, say, the location or polarization

[3]Even pure "impregnations" in direct perception may be considered in a weak sense to be interventions from the perspective of schematic ordering–at least in a rather restricted methodological sense.

[4]Even the concept of reality in itself is methodologically speaking an epistemological meta-or higher-level construction, but a practically and pragmatically well-founded one.

[5]This is true for classical deterministic (better: deterministically interpreted) systems with but very few, less than three variables. Even the classical example of the moon-earth system (embedded in the solar system) is certainly an idealized case in point eventually and finally admitting of unpredictable, so-called chaotic phenomena of development. Any system with three and more degrees freedom displaying a growing sensitivity on initial constellations of the respective system variables would in principle show such chaotic phenomena (the description of which necessarily has to use "chaotic" or "strange attractors"). In the long run, this is even true for our solar system.

fixing of a pair of entangled photons. In this sense classical deterministic and commonsense approaches seem to be the exceptional cases, whereas quantum-mechanical holistic entanglements of states are rather usual.[6]

In general, what we were or are still used to see as the *normal* conditions, are in some sense the *exception*. This seems to be a very important insight for any (re) cognition of reality whatsoever—in particular with the traditional classical projection of classical mechanistic descriptions to the external world of material objects and the respective systems.

Our attempts to structure and pattern the world by using concepts of objects and processes are certainly dependent on interpretative, perspectivist frameworks and (theoretical as well as activated) schematizations: They are dependent also on actions, interactions, forms of intervention in the narrower sense. (In a way, this even applies to construing the respective forms of ordering phenomena and observations). We can even say, *scheme-interpreting* is finally (under special circumstances very weak) *intervening and impregnating* (or being impregnated) as far as it refers to "external" objects, their "make-up" and constitution or classification. Theoretically speaking also usual objects and processes are even in everyday connections to be conceived of as consisting in some sort of states of overlapping—analogously the possibility of overlapping states or the respective linear combinations of state functions in quantum theory. At least conceptual ways of sorting out, classifying, developing and applying theories are generally speaking (at least methodologically) to be understood as the delimitation and analytical segregation of subsystems from the everyday-"entanglement" of phenomena.[7] This interpretation is but pragmatic, allowing the classical-objectifying (object-isolating) simplification of common sense as an approximation by ignoring the strong correlations (implying non-locality and non-separability of the respective entangled systems). The unitary holistic comprehensive description does not seem possible without respecting restrictions similar to Bohr's "complementarity"—even if generalized to an epistemological approach (as also Bohr would have it). In any case, one could understand such choices of one out of complementary perspectives in such a way that basically the possibility of relativized interpretations is acknowledged by which other possible approaches of different interpretations are for the time being "excluded" in the special situation, but in general being supposed as alternative possibilities. This seems to be relevant in everyday connections–even with a doctor approaching his patient as a biological-physiological organism on the one hand and complementarily—and hopefully, complimentarily!—as a person on the other. Usually then only both approaches in systematic relationship (although not compatible under a single complete description of the situation) are to be acknowledged

[6]Included is the chaos-theoretical inseparability and unpredictability of phase-state descriptions regarding chaotic systems characterized by three and more degrees of freedom and the respective sensitivity with regard to initial conditions.

[7]One need not adhere to the many worlds interpretation of quantum mechanics as developed by Everett, but one may claim for a relativized interconnection of state designations and combine this with a perspectivistic-interpretationist approach as developed here.

as representative (e.g., the patient is at the same time an organism *and* a person and to be treated according to both!).

To sum up, as an interim general result: *Interactional and scheme-interpretationism allows for and requires the choices of respective observation perspectives and makes room for integrating complementarities (complementary interpretations) and to overarch isolating perspectives and at the same time to use and approximate restrictive subclassifications of disciplinary model instantiations. This is true not only for the realm of quanta, but for reality in general (although the mutual exclusion of perspectives as regarded by Bohr may be different in the mesocosmic connections, they are only tentatively and preliminarily exclusive). Any perception and conception and "grasping" of reality is only possible by using the respective forms of interactions, impregnations* etc. *in terms of interpretations and schematizations methodologically presupposing or implying the respective choices of perspectives* etc.

3 Indirect Realism as Resistance

Even subjects as agents or epistemological recognizing instances can themselves only be understood as a certain kind of interpretative construct. Also the recognizing subject is an epistemological construct and already a result of interpretation procedures on different mostly higher levels. Already William James saw this, but from a methodological interpretationist point of view we have also to take this into account (Lenk 1992).

Nevertheless, we as persons and organisms may also run with our head against the wall: There is a certain *experience of resistance*, an impenetrability of real "objects" in the true sense ("throwing against": "*obicere*"). Such experiences of confrontation and opposition show that what we call "*reality*" does play an *influential role in impregnating our interpretations of experiences of, e.g., resistance*. Thus, the constructs are not only of our making, but mediated by a certain kind of *interplay between "world factors" and our patterns of interpretations and schematizations*. These latter need in application to the so-called "reality in itself" (to such experiences of external reality or resistance experiences) the "something" which is being interpreted as the "opposing entity or event" (although this need not only be constituted in a permanent object form). The "opposing other part" ("das Andere"/the *other* "thing"/referent) of interpretation has to be *pre*-conceived, *identified*, somehow (though interpretationally) "*distanced*" or already "constituted" in schematic form in order that interpretation may get "a grip" on that "something". In this sense, as in referencing, an "objectifying interpretation" as a kind of *"impregnation" presupposes something which it can "grasp"* in a rather active sense. (*Scheme-*)*Interpretation* therefore is *not just ideal production* (à la Fichte and a comprehensive action-oriented or even subjective idealism)—at least not with regard to the object world, to be experienced and represented. Theories of objects would presuppose something as already being in some sense "independently existent" though this can be "grasped", known or described only by interpretation again, if even by using higher-level schemes. The very representative constitution of experiential objects is certainly schematized, but not by just fictional productions on the side of the epistemological subject alone, but by schematized, interpretation-dependent processes interacting (in the sense of impregnation in the narrower understanding). "Impregnation" makes only sense with something which is presupposed as a "*fundamentum*" at which (scheme-)interpretation may somehow operate. There has to be this opposing "*Other*" of interpretation which in our reflection is hypostatized as being itself interpretation-free. As Röd (1991) turned Descartes' vision around: *I interpret, therefore there is reality*: "*interpretor ergo realitas est*". This is certainly the idea of a minimalized pragmatical hypothetical residual realism of an epistemologically indirect(ist) kind.

Here, this reality in itself may indeed be identified with Röd's "*residuum* of experiential analysis" (1991, 171, 174f, 178ff). This "elementary" *residuum* of reality has to be presupposed and may be somehow, if not identified, but intriguingly connected with reality in itself. Reality in itself and object reality are so to speak complementary modes of apprehending "things" (Röd 1995). I would not like just naively to talk of "things" here, but I would agree insofar as the perspectivistic choices and restrictions do shape our modes of apprehensions and recognitions if we attend to reality in itself or as such or to more specifically individual "things in the realm of appearance" à la Kant. It is even compatible with a quasi Kantian interpretation that the "*residuum* of experiential analysis", the *fundamentum reale*, so to speak, "the Other of the interpretation in a certain sense may be conceived as something, which again—from another or slightly modified perspective—may be interpreted" (e.g., epistemologically speaking) "as reality in itself". Indeed, something which is interpretation-free and no way interpreted or produced by interpretation would amount to a utopian limiting concept. It is, however, a concept having only *methodological* valence, a limiting concept being necessary because one can only interpret *something*. Any interpretation (as scheme activation) has to get a hold or fix at some entity or point to take off from. Interpretation—and in particular "*impregnation" as interpretation*—has to concretize at *something*. That means that such a limiting concept of what is not available to recognition and interpretation is at least methodologically speaking meaningful, if not even necessary.

The "Other" of interpretation could according to Röd also be identified with Kant's "thing in itself".[8] However, in Kant the foundation runs in the other direction: "The thing in itself" or presupposing "the world in itself" is considered as being a necessary condition for continuity, for the unity of the self and of the subject in the first place: The I, the self or transcendental subject—thus Kant's "rejection of idealism" (*CpR* B 275)—can only be constituted by presupposing something permanent ("*Beharrendes*"), being distanced or separated from itself:

[8]This sounds a bit too simple since Kant's concept of "the thing in itself" should be understood as an interpretative one! (See my 2007 chap. IV and Lenk-Wiehl, eds., 26 ff.).

The "thing in itself" would then in a certain sense be considered as interpretation-free, as only being hypothetically existent. This is for Kant necessary. But understanding all this from a higher level of interpretation certainly also the concept of "the thing in itself" or "the world in itself" can epistemologically speaking only be conceived of and "grasped" as and by an interpretative construct, if on a higher level, that is to say, from an epistemological point of view or perspective.

If we approach as a hypothetical realism of sorts, his theory is certainly compatible with this epistemological modeling of his concepts of "thing in itself" etc. from a higher level—and even this modeling from an ever-higher perspective or meta-level (e.g. IS_6) in my hierarchy of interpretative (meta-)schemes again. "The thing in itself" is so to speak—conceived on a higher or the highest meta-level (cumulatively understood)—itself to be interpreted as interpretation-dependent. (The separation and segmentation of "the thing in itself" from a totally entangled point of departure prior to epistemological specializing and concretizing may be understood in terms of complementarities not only between perspectives on the same level, but also between meta-levels.)

4 "Grasping" Is Interpretation-Laden

The relationships between the different levels and models regarding what is necessary for taking the respective perspective for interpreting the "entities" which are accepted as ontologically "real" is certainly shaped and schematized by what is methodologically necessary as a requirement of recognition. All this has so to speak to be again interpreted as model-dependent, as interpretatively qualified on a higher level or meta-level. There is *no direct access, no do directistic projection of structures into the external world without any interpreting from a perspective whatsoever*.

We might *and should* for practical and theoretical reasons hypostatize, even *realistically* postulate, "a reality in itself" but nevertheless this can only be conceived of and captured or referred to in an interpretative manner: Even the ideal limiting-concept of it would only be an interpretational concept as seen from a higher level. Still for the mentioned practical, pragmatic and even hypothetical-realistic reasons for existence and constitution of reality in itself may be and should be conceived of as the interpretation-independent "Other of interpretation"—to be sure in necessarily interpretative make-up. *In no activity or process of "grasping" we can transcend beyond the horizon of our interpretations; we cannot leave our "universe of interpretativity"* so to speak—*metaphorically*. We have to work with our interpretative forms, patterns, even with our biological inherited, pregiven schemes of, e.g., sense perceptions etc., lending themselves to primary interpretations. Indeed, we may mean and refer to a "reality in itself", nevertheless only in an interpretatively structured pattern or mode of shaping. We have, however, to be clear that even this is, methodologically and epistemologically speaking, a *model*, even though applied throughout in practical life to resistant realities. It is our interpretative constructs based on schematization, schematic-interpretive activities, rather fixed impregnations and even interventions which would in some sense function as a bridge between our "universe of interpretations" and the "world in itself" which however can only be referred to in an interpretative manner. We thus *metaphorically* speaking "*live*" therefore *in our "interpretation worlds"* (Abel 1993): we have to assume the perspective of interpretational recognition and can only in this way—built into these extant perspectives—perform a certain kind of "access" (which we pragmatically speaking have necessarily to assume) by a representing interpretational system to the interpretation-transcendent world in itself. However, from another, if you wish ontological, perspective a higher-level philosophical one, this can be called (interpreted as) "a real" though (quasi) "interpretative world".

This would sound relatively complicated, but the basic idea is very simple. It figures under the transcendental meta-perspective or—more generally—a methodological one along the lines of the question: What have we to presuppose as faculties of formation on our side and what schematization and scheme-stabilizations are to be activated or developed, respectively, in order to (re) cognize the (accessible "world factors" and "realms" of the) world and even to manipulate or influence part of it by our actions? This is certainly the question regarding necessary presuppositions as already focused on by Kant; a characteristic difference is only that Kant believed that there is a general *fixed* pattern of presuppositions built into the "Reason" ("*Vernunft*") and the "Understanding" ("*Verstand*") of any rational being, namely the categories which might (according to Kant) already be derived from the logical form(s) of judgment (i.e., of propositional statements, see my 1968, chap. I). That, however, cannot be upheld in the very specific Kantian presentation (ibid.). Not only are the forms modifiable (recognition may follow a probabilistic form or just empirical trends instead of logical exceptionless determination), and humans as active beings do not only avail themselves of some biologically fixed primary schematizations and interpretational patterns, but they have to develop many of them parallel or similar to the development of language and culture as well as enculturation, socialization etc. Secondly, these patterns of interpretation and schematization are not only presuppositions for any representation and description of theoretical-cognitive provenance, but of any form of action whatsoever, too. And, thirdly, pragmatically and (life-)practically we have to presuppose something interpretation-transcendent which is independent of our interpretative processing, but which we can only on our side "grasp", even designate and conceive of by interpretative models or constructs. (This is also agreed on by Kant in a way—particularly in his later phase where he apparently (according to Hossenfelder and Röd) thinks of "spelling out" or "interpreting" "objects in themselves" (*CpR* B 69, *Prolegomena* § 30) as "experiential appearances". This—and insofar both quoted authors are right—would render a non-contradictory possibility to abide by *a minimal or residual or rudimentary realism also in a Kantian approach. You cannot dissolve everything in a pure internalism of interpretations, although you cannot only "grasp" anything by using interpretation and by conceiving of it in interpretative constructs and forms.*—In addition, Kant's epistemological activism is even taken seriously here in a stronger sense than he himself had in mind! *Any recognition and cognitive process whatsoever is an activity* (not just the connecting representations ("*Vorstellungen*") in theoretical-cognitive judgments, but in a rather general sense), and *any action whatsoever is in turn schematically structured, dependent on interpretative constructs*, too - yet basically not only on them but (as in direct sense perception) also on the so-called "world factors".

In understanding ourselves as existing beings, we can do that only if we also have a counterpart, a world from which we would distance us, if by way of interpreting (i.e., using interpretative forms) again, or which epistemologically speaking we distance from our subject-dependent unity of representations and subject-constituting activities (schematizations). Indeed, we ourselves are compelled to understand us as part of this external world, to embed our body and in some sense also the subjective self—into this world. Indeed, again this is certainly dependent on internal models of the world and of (a representation of) the subject itself and its body within such a world model. (Also, the embedding of these processes of interpretation in social and cultural environments and other sign systems is a necessary precondition, to be sure.)

Natural-philosophically speaking it may be perhaps the most interesting point here to ask: How does it come to be that in nature a being did occur which would in a relatively reliable manner represent and recognize the world and is able to embed and recognize itself within this world attempting to bring to light "what connects the world in its most inner constitution" or at least entertaining a consciousness of this attitude towards nature and the modeling of "nature". Thus, the fact that humans develop theoretical consciousness, knowledge, cognition and recognition means that one part of nature, namely the sentient and cognitive human being is so to speak able as *pars pro toto* to recognize not only nature but also herself or himself. Only in humans in a certain sense (and this is a very basic natural philosophical idea) "*Nature*" would have developed a certain concrete living being which is able at least partially to recognize nature "in itself" and interpret it as "reality in itself". This is on the one hand a very high and valuable faculty as admired by Kant in the famous closing section of his *Critique of Practical Reason*: On the other hand, it also would impose the responsibility and obligation, to use this recognitional faculty in an adequate manner. Humans are therefore in a special situation within nature: They have special faculties, but also special obligations to develop these talents for recognition and knowledge—and also necessarily obligations and responsibilities for acting, say, from a moral point of view with regard to any other such beings or even other living creatures. (At least this holds from a specific and also as a generalized Kantian perspective).

Designing, acting, recognizing, knowing are indeed essentially interpretative and interpretation-dependent activities, at times interactionally impregnated ones. But the human being can certainly not only be characterized as the *cognitively recognizing* being as traditional theoretical philosophy would have it (*animal rationale*) and also not merely as the *acting* being as philosophical anthropology in the last century thought, but as the being of construing and constituting interpretations both in the cognitive as well as in the practical action-oriented realm. It is the being which does not only act goal- and aim-oriented, but which would also in turn judge and evaluate the goals, means and the procedural strategies—and finally also the very interpretations and interpretation products as well as interpretative constructs being used. Thus, the acting being is also the being evaluating actions and performances, indeed the valuating and normative as well as goal-oriented performing being capable of self-reflection: The human being is the actively designing, interpreting and meta-interpreting as well as reflecting and reflectively reflecting being encroaching necessarily by its constructed and reconstructed models into "reality", in part modifying, manipulating and reshaping part of it. It does not only structure the representation and interpretation of this reality, but also restructure and change it by manipulating, by interacting with the world always and by necessity with social partners in it. Anything procedural in all these connections is intricately, even intimately shaped by scheme-interpretations.

Humans are thus *the interpreting beings par excellence* (as Nietzsche already would have said). Beyond that they are also capable of distancing and reanalyzing reconstructing their interpretations from a higher level of interpretation or perspective: Therefore, they are not only the interpreting beings par excellence, but also the *meta-interpreting or supra- or super-interpreting* ones (cf. Lenk 1995b, 2007, chap. III).[9] We have to interpret the human being in a certain implicitly anthropological perspective as the meta-interpreting being par excellence, necessarily availing itself of constructions and reconstructions in recognition and action, i.e., interpretative constructs of descriptive or normative kind, reflecting not only the external world, but also society, culture and its own subjectivity in permanently ongoing interpretative activities. We cannot do and live without schematization and interpretation. Action, living, recognition, knowledge and any cognition are necessarily interpretation-dependent: *Interpreting is inevitable,* this being a fundamental insight for any topical epistemology and action theory. *We cannot* ***not*** *interpret.*[10] We cannot possibly evade from or avoid schematizations and interpretations. Any sort of experiencing, "grasping" and shaping as well as acting would figure under these necessary conditions of being interpretation-dependent, schematically structured or shaped according to the structures of the "graspable" for us. One could even speak of a "universe of interpretations" which we could not possibly leave or of a "horizon of interpretations" which we could not possibly transcend. But still the horizon is open or—to use the metaphor a bit further—recedes if we come closer to it: We may extend what we can grasp by our theoretical and interpretational means—as is obvious in any understanding of scientific progress and progressing knowledge.

[9]By necessity, the normative and judging as well as evaluating being has to be the meta-interpreting being, using not only cognitive-descriptive interpretations, but normative interpretative constructs, too.

[10]But equally well we cannot do and live *only* with interpretations and metá-interpretations, i.e. without actions and interactions.

In general however, it has to be taken into account, that the model of this scheme-interpretational (re)constructive realism of a pragmatic sort is itself an *epistemological model* on a meta-level which is in turn interpretatively (or for that: meta-interpretatively) shaped. The model and methods of applying interpretative constructs are themselves but interpretative constructs or even-products, they are so to speak interpretative meta-constructs. The interpretationist approach certainly has the advantage of possibly being reflexively applied to itself on the highest level IS_6 of the interpretation hierarchy. (Certainly the highest meta-level of interpretative types has to be considered as a cumulative category somehow, if potentially, containing the higher levels of meta-interpretations already.) Thus, methodologically speaking we do not end up in a contradiction or performative paradox or an epistemological circular argument. This is avoided by the very fact of providing the levels of interpretations in differentiating the perspectives.

By providing the levels of interpretation and the respective level-connected perspectives it is possible to avoid a vicious reflective circle of argumentation regarding the meta-interpretation of interpretations. Thus there is no trilemma of interpretations like Agrippa's skeptical dilemma in antiquity (later on called the "*Münchhausen Trilemma*" in critical rationalism (cf. Albert 1968)). Such a trilemma could only originate if we would conceive of a traditional foundationalist philosophy but not as a hypothetical methodology of epistemological constructions of a basically "prometheic"[11] discipline predesigning, constructive or anticipatory provenance. No trilemma of foundation would occur, possibly just a pragmatic indispensability and inevitability in the mentioned sense that we always can only recognize and act in an interpretation-dependent manner.

Thus, we do not end up in a vicious circle at all—not even in a virtuous circle (Vollmer), but rather in *a spiral leveling itself up the steps to higher planes of interpretation types*. Like acting also interpreting is a set of routines and customs or rules, routines, being anchored in society and shaped by cultural norms, institutionalizations etc. (at least this refers to interpretative levels of IS3 through IS_6). Any social phenomenon and regulation whatsoever is itself interpretative. Indeed and again, this does not lead to a circular foundation because the respective model of constructive and reconstructive philosophizing bound to levels of interpretations may be pragmatically interpreted—again and again—from a higher level respectively. This is true for a pragmatic shaping of actions as well as for the understanding of any cognition and recognition in science, philosophy and everyday circumstances. Indeed, philosophy should not operate on a remote plane separate from a common-sense understanding in everyday acting, even if it would critically reflect the extant hypostatizations of everyday conceptions, the respective objectifying, at times illusionary and skewed as well as manipulative ideological distortions or misrepresentations. Philosophy devotes itself to critically further developing constructive thought in a pragmatical context and feed-back. It should in this colloquial sense also remain "realistic". This is true for thinking, recognizing

[11]meaning "thinking ahead" in ancient Greek.

as well as for acting. We have to pragmatically and inevitably hypostatize a real world in which we act as against potential resisting events, "things" or processes. To reject such a world of objects in themselves would render a performative paradox analogous to the *petitio tollendi* in the foundation of logics.[12]

One might ask, where in the last analysis is "the real anchoring" of this interplay in the turbulence of interpretations? Do we need a fixed ground, so to speak an Archimedean point of philosophizing, from which to start off? Indeed, we don't—except the feedback to everyday language and our respective instruments of our theoretical, symbolic, socio-cultural constructions and media residing in a quasi Wittgensteinian manner in social practices and institutions. It is not necessary to have a fixed Archimedean point of action and interpretation as a *fundamentum inconcussum* of philosophizing. (This would indeed amount to a rationalist foundational philosophizing in a nowadays outdated absolute-rationalistic sense.) Instead, we know that we can successfully act and recognize as well as even, at least in part, anticipate "reality" in a relatively reliable manner. We know that without scheme-interpretation we can neither (re)cognize nor act. We may thus design a non-Archimedean philosophy of pragmatic *relative* foundational procedures without recurring to an epistemologically speaking ultimate and absolute, *non*-interpretative ground or a last fundamental security. Nevertheless, we can and must in "real life" connections pragmatically treat the "external world" as "real" (in a sense) constituting the context for our actions and as a counterpart to goal-oriented behavior even though we sometimes cannot absolutely and without doubt sever objective and interpretation-free "objects" independently of our interpretative constructs—which are in turn related to activities and potential actions. We cannot as we saw do without scheme-interpreting.

In a sense, we move in interpretation circles and even spirals, we usually do not consciously conceive of the respective levels and meta-levels: In acting and grasping we cannot get out of the "horizon" or "universe of interpretation". But again, this does not mean that only interpretations would exist or be conceived of as the only real processes. On the contrary, we could not think of a viable procedure and successful applying interpretative constructs without locating our actions and the very processes of interpreting in "real world" connections. As we saw, even

---

[12]As I called it more than 30 years ago independently of P. Strawson: You cannot reasonably–i.e. by arguments–reject certain strategies or principles in logics (like the Principle of Non-contradiction) without using it or functional equivalents of it on a higher level. Without rules of criticism it is not possible to reject a strategy of criticism or a rule (so, you have to have meta (level)-rules).Cf. Lenk 1970, reprinted 1973.-K.-O. Apel has later on (1973) used this argument for what he calls a "transcendental-pragmatic ultimate foundation" of rules of argumentation etc. It is however problematic whether or not such a circular structure of this *petitio tollendi* must or need be used to as sort of ultimate foundation: It is only a methodical-reflective interpretation of a methodological inevitability in order to render or illustrate the indispensability of specific rules like the Principle of Excluded Contradictions or a respective functional equivalent in logics. It is much more a normative postulate for the purity of construction and methodical progressing in developing and regulating the strategies of structuring arguments and constructions as well as interpretations than an ultimate rational foundation in the traditional absolute sense.

cognizing and recognizing as a sort of action is always embedded in and bound to the world, even a "world in itself". To repeat, interpretation is not everything, but without interpreting and world embedding nothing could be possibly grasped and/or understood. (Even formal procedures of thinking and representations are in the last analysis dependent on the development of action capabilities of a living being, i.e., on the embedding in a real world constellation and in social connections, e.g., in society, culture, language, institutions etc.)

5 Action, Interaction, and Deep Socio-Genic Impact

Of course we have always to start from everyday experience and everyday acting. We know that even this is deeply interpretation-laden. We might with Abel modify Wittgenstein: Thus we interpret as we are used to (or accustomed to) interpret: we are in a deep sense the designing, interpreting, meta-interpreting and acting as well as valuating beings—even if we know that this self-interpretation is again interpretative in an anthropological, (according to Wittgenstein and Kripke) "deeply social" and epistemological context. To be true, the later Wittgenstein would say: Thus we interpret in and how we act (we are accustomed to act). This does not exclude an insight about the interpretation-dependence of this very model and of all activities including everyday (re)cognitions and theoretical constructions in science (and also philosophy).

Such conditions and restrictions would be valid for any conceptual and linguistic foundations in Wittgenstein's deeply socially entrenched sense. Also language as a quasi "transcendental"-epistemological basis is interwoven with conceptual and factual potentials of "grasping" and acting. Thus, we have to go beyond Wittgenstein's transcendental lingualism not only in digging deeper to the very forms and requirements of *acting*, but also more basically to the forms and requirements of schematizing and non-linguistic interpreting in the first place. In my book on "*Schema-Games*" (1995) I extended the Wittgensteinian model of "language games" to the schematic forms of "grasping" or shaping any representations and actions whatsoever. We have to go beyond Wittgensteinian restrictions to just ordinary language formations. Transcendental lingualism (as E. Stenius interpreted Wittgenstein's philosophy) has to be superseded. Language is not the last and only basis for everything. Even the usage of language is necessarily dependent on the forms of actions and schematizations as well as non-linguistic interpretations in the elementary sense of IS_1 and IS_2 as well as IS_{3a}. Language only comes in later, though as a *very* important means of additional interpretative differentiation. Language is itself however actualized, it does only exist in acting and interpreting and resides, as the later Wittgenstein indeed saw, in socially conventionalized institutions, societal structures and customs, in rules, norms, symbols (as conventional signs or gestures etc.). "World" cannot be dissolved in or reduced just to language and signs and also not, as Nietzsche had it, to an ontologically hypostatized interpretative "happening". As we saw, we cannot just produce from our interpretativity anything existing at all. Not everything is a total result of interpretation, although anything whatsoever can only be "grasped" in an interpretation-dependent manner—or even indeed be conceived that way.

Would all this only be valid for world representations or the processes of "grasping" world versions? Indeed projections of meanings, hypostatizations are themselves interpretative, in some sense they are "world-producing" insofar as the manipulation of linguistic and symbolic signs as well as the respective systems of applications and embeddings in socio-cultural contexts are dependent on such interpretations. But this is only a projective, "*secondary*" relationship of constituting not a really extremely radical one as Goodman for instance would postulate: To be sure, we "have" only world *versions,* i.e., we can only refer to "the world" in the light of our interpretational perspectives and interpretative constructs. Any world concept whatsoever is an interpretative construct. However, the world does not totally consist in such constructs; it is not disjunctively separated into incompatible "worlds" (or the "world versions" the late Goodman had in mind, misleadingly calling them "worlds"), i.e., special restricted world perspectives or related interpretative constructs are to be taken seriously instead of a global talk about "incompatible" and "disjunctive", if not "many worlds". Any "grasping" of "the world" or "worlds" (or, for that, "*world versions*") is/are certainly interpretation-dependent; therefore, any world version in Goodman's sense is interpretatively constituted. This does not exclude that we do, for pragmatic and practical reasons, hypostatize a common social world of actions and interpretations: We act, to be sure, in one and the same world like our neighbor or partner—even at times including an interaction partner from another culture. However, any world versions we would avail ourselves of are indeed to be embedded—at least in practice and practical interacting—in a common world (to be represented in a comprehensive model of suchlike). Even the Indians of the recently discovered tribes in Bolivia and Brazil which have never to date been confronted with the so-called (Western) "civilization" would necessarily act (thus we are obliged to hypostatize) in one and the same world—"our" common world as we have to stipulate—when (and even before) the first encounter took place. This is true even regarding the fact that no common language or symbolic or representational world version does thus far exist. In spite of distinct and different world versions we necessarily have to hypostatize one common and "real" world. We yet know that "grasping" world versions of it would always be interpretative—and that would *a fortiori* also apply to the hypostatized basic common world. We really—in terms of actions and interactions—do not live in totally different worlds: There are overlapping zones of confrontation, action and interaction contacts in a situation which has to be located in one and the same world despite all differing perspectives with regard to differing projections, languages, cultures and different modes of interpretations—maybe from both sides. Even if inhabitants of different world regions, cultures or even remote planets never would encounter each other or get into contact living in or under totally distinct world versions à la Goodman, nevertheless they are as acting and interacting beings to be located as existing in the same world. (At least thus we have to understand it, and similarly the other side has correspondingly *mutatis mutandis* to conceive of it also.) (The many-worlds interpretation as of Goodman's is practically equally absurd as the many-worlds interpretation developed in interpreting quantum mechanics.)

6 Resistances in Ropes and Nets but Without a Rigid Fix-Point

It's a long way and walk to Tipperary indeed! The surveying of the epistemological and methodological as well as anthropological areas of recognition, cognition in general and acting as well as deciding, valuing etc. from the vantage point of an interpretative pragmatic realism and methodological schema-interpretationism leads to a rather multi-leveled and manifold picture: We have no last, ultimate foundation which cannot be doubted at all, which would render a conceptual or linguistic formative basis to build a safe intellectual construction on it. We however do not operate like a rope artist without net, but we ourselves—on the basis of biological fixed genetic dispositions and formal-operational necessities (for example involved in the fundamental rules of logics as methodologically interpreted by Lorenzen (1955)) we ourselves would knit or construct our nets in which we try to catch or capture elements and parts of the world. Thus, we elaborate our own net including the rope on which we try to balance ourselves. These nets and ropes may be extended and modified. We work to a large extent with self-constructed or culturally "given" classifications, shapes, symbols, representational instruments and in most (not all!) cases rather flexible possibilities of "grasping" external phenomena and objects we are confronted with—and also reflecting ourselves as subjects, bodies and persons. We know that the nets are means and instruments of schematizing and ordering as well as of structuring and reshaping; they are interpretation-engendered as representative media and instruments, constituted on different interpretational levels, in a last analysis "deeply" socially conventionalized and linguistically or symbolically differentiated. Any form of "grasping" the world is unavoidably and indispensably deeply per se interwoven with interpretations—including not only elementary and refined schematizations, but also theories, everyday theoretical—and practical!—pre-suppositions as well as conceptual and linguistic "colouring", if not even soaking. Nevertheless, from any necessarily interpretation-laden perspective it is practically inevitable (in order to avoid pragmatic performative paradoxes and contradictions) to hypostatize "the world" independent of us as "real"—even if we may not be able to objectify and identify elements in it independently of any pre-schematization or interpretation. Any identification of objects is per se already interpretative. To repeat the obvious for a last time: *Any "graspability" whatsoever is interpretation-laden. The world is real, but (any description and action of) "grasping" the world is always interpretative, i.e. only conceived of and formed by scheme-interpretation. It is furthermore internally action-bound and deeply societal.*

We have to reject all full-scale interpretational idealisms, absolutisms or even imperialisms as well as the so-called 'direct', allegedly interpretation-free realistic objectivisms of, say, naive naturalistic or other provenance. We have good practical, pragmatic and theoretical reasons for this rejection. The argumentation in this respect can–as any possibility of "grasping" and representing—of course only be performed within the "universe of interpretations" and meta-interpretations (over interpretations) themselves.

Moreover, this is also valid for our *subjectivity* proper: We have to understand ourselves as acting, as "real" beings, as responsible and even causally manipulating beings always by way of working out conceptions and interpretations. We are necessarily and all the time interpreting and meta-interpreting beings. In any case, even the subject as such—in experience and as a center of agency as well as in its being "grasped" as an epistemological subject from a methodological point of view—is to be conceived of, methodologically speaking, as an interpretative construct: Even here we cannot evade interpretativity. Any "grasping", action and understanding—even of our own selves or subjects—is (scheme-)interpretation-dependent, can only be performed and designated in and by interpretative constructs. However, as repeatedly stressed this does not mean that everything being interpreted would *merely* be a product of our interpretations. "Reality in itself" and "objects" or "things in themselves" would exist independently of our interpretations, even if we know that we can only "grasp" them within the realm of our interpretative nets and constructions of conceptual and theoretical provenance and in (as well as under) action constellations. We finally gained the insight that even this process of interpreting does not capture reality in a sufficient and absolute manner. We also found out that the very model of realistic and pragmatic interpretations from an epistemological perspective is itself an interpretative conception. This would in a sense even apply to the self-understanding of man by the philosophical anthropological model of man as the ever-interpreting and meta-interpreting being (Lenk 1995b, 2007 chap. III, 2010, 2013).

References

Abel, G.: Interpretationswelten. Frankfurt/M: Suhrkamp 1993.
Charles, D.—Lennon, K. (eds.): Reduction, Explanation, and Realism. Oxford: Clarendon 1992.
Child, W.: Causality, Interpretation, and the Mind. Oxford: Clarendon 1994.
Giere, R. N.: Explaining Science: The Cognitive Approach. Chicago - London: Chicago U. P. 1988.
Giere, R. N.: The cognitive structure of scientific theories. In: Philosophy of Science 61 (1994), 276–296.
Giere, R. N.: Science without Laws. Chicago: Chicago U P. 1999.
Giere, R. N.: Scientific Perspectivism. Chicago: Chicago U P. 2006.
Hacking, I.: Representing and Intervening. Cambridge, New York: Cambridge U P. 1983.
Harré, R.: Varieties of Realism: A rationale for the natural sciences. Oxford: Blackwell 1986.
Heelan, P.: Space Perception and the Philosophy of Science. Berkeley: U. of Calif. P. 1983.
Herbert N.: Quantum Reality: Beyond the New Physics. New York: Doubleday, 1985.

Hossenfelder, M.: Kants Konstitutionstheorie und die transzendentale Deduktion. Berlin: De Gruyter 1978.
Ihde, D.: Technology and the Lifeworld. Bloomington-Indianapolis: Indiana U.P. 1990.
Ihde, D.: Instrumental Realism: The interface between philosophy of science and philosophy of technology. Bloomington-Indianapolis: Indiana U.P. 1991.
Kuhn, T.: The Structure of Scientific Revolutions (1962, postscript 1969): Chicago U. P. 1970^2.
Kutschera, F. v.: Die falsche Objektivität. Berlin - New York: De Gruyter 1993.
Lenk, H.: Philosophie im technologischen Zeitalter. Stuttgart: Kohlhammer 1971, 1972^2.
Lenk, H.: Pragmatische Philosophie. Hamburg: Hoffmann & Campe 1975.
Lenk, H.: Handlung als Interpretationskonstrukt. In: Lenk, H. (ed.): Handlungstheorien interdisziplinär, II, 1. Munich: Fink 1978, 279–350.
Lenk, H.: Zwischen Wissenschaftstheorie und Sozialwissenschaft. Frankfurt/M: Suhrkamp 1986.
Lenk, H.: Zu einem methodologischen Interpretationskonstruktionismus. In: Zeitschrift für allgemeine Wissenschaftstheorie (J. for General Philosophy of Science) 22 (1991), 283–302.
Lenk, H.: Interpretationskonstrukte. Frankfurt/M: Suhrkamp 1993.
Lenk, H.: Schemaspiele. Frankfurt/M: Suhrkamp 1995.
Lenk, H.: Interpretation und Realität. Frankfurt/M.: Suhrkamp 1995a.
Lenk, H.: Realistischer Realismus als ein methodologischer und pragmatischer Interpretationismus. In: Lenzen, W. (ed.): Das weite Spektrum der analytischen Philosophie. (FS Kutschera) Berlin 1997, 149ff.
Lenk, H.: Einführung in die Erkenntnistheorie. Interpretation–Interaktion–Intervention. Munich: Fink 1998.
Lenk, H.: Kreative Aufstiege. Frankfurt/M: Suhrkamp 2000.
Lenk, H.: Grasping Reality. An interpretation-realistic epistemology. Singapore: World Scientific 2003.
Lenk, H.: Bewusstsein als Schemainterpretation. Paderborn: mentis 2004.
Lenk, H.: Global TechnoScience and Responsibility. Berlin: LIT 2007.
Lenk, H.: Methodological remarks on dynamic functional representation. In: Agazzi, E. (ed.): Representation and Explanation in the Sciences. (Special ed. Of Epistemologia). Milano: Angeli 2013, 122–140.
Lenk, H.–Maring, M. (Eds.): Advances and Problems in the Philosophy of Technology. Münster: LIT 2001.
Lenk, H.–Skarica, M.: Wittgenstein y el giro pragmático en la filosofía. Cordobá/Arg.: Del copista 2005.
Lenk, H.–Skarica, M.: Wittgenstein und die pragmatische Wende. Berlin. LIT 2009.
Lenk, H.-Wiehl, R. (eds.): Kant Today-Kant aujourd'hui-Kant heute. Berlin: LIT 2006.
Leplin, J. (ed.): Scientific Realism. Berkeley CA: U. of Calif. P. 1984.
Primas, H.: Chemistry, Quantum Mehanics und Reductionism, Berlin, Springer, 1983
Rescher, N.: Methodological Pragmatism. Oxford: Blackwell 1977.
Rescher, N.: Process Philosophy. Pittsburgh: U.P. 2000.
Röd, W.: Erfahrung und Reflexion. Munich: Beck 1991.
Röd, W.: Das Realitätsproblem in der Transzendentalphilosophie. In: Lenk, H.–Poser, H. (eds.): Neue Realitäten. Herausforderung der Philosophie, Berlin: Akademie Verlag 1995.
Wittgenstein, L.: Philosophical Investigations. Oxford: Basil Blackwell 1953.

Part III
Scientific Realism in Particular Sciences

Semantic Definition of Truth, Empirical Theories and Scientific Realism

Jan Woleński

Abstract The paper applies formal logical semantics to analysis of empirical theories and scientific realism. Empirical theories are ordered sets of propositions and, under the assumption of consistency, they have models. On the other hand, semantic models as abstract algebraic structures. cannot be regarded as parts of the real world. In order to define links between theories, their models (understood as mediators) and the real world, the concept of empirical valuation is introduced. This innovation allows to base scientific realism on the semantic theory of truth.

1 Introduction

This paper tries to show how semantic theory (**STT** for brevity) of truth can be used as a tool for analysis of the scientific realism (**SR** for brevity). In particular, I will argue that **STT** with some additions provides a good foundation for the realistic account of empirical theories. Roughly speaking, **SR** regards such theories as true or false due to their relation to the external (real, actual, etc.) world (see Psillos 1999 for an extensive analysis of **SR**). One can say using a very fashionable semantic and ontological jargon (I will not employ this way of speaking in what follows) that real facts function as truth-makers for statements made by scientists.

2 General Remarks on Theories, Truth and Scientific Realism

However, the above general characterization requires further qualifications. What is a theory? The simplest account see theories as ordered sets of universal (possibly universal generalizations) propositions (laws, hypotheses, etc.) related to a specified

J. Woleński (✉)
University of Information, Management and Technology, Rzeszów, Poland

E. Agazzi (ed.), *Varieties of Scientific Realism*,
DOI 10.1007/978-3-319-51608-0_15

domain of reality. If T is a theory (empirical or mathematical), its internal order qualifies some of its elements as basic, primary, initial, fundamental, etc., but other as derived or secondary. Although I do not claim that empirical theories need to be axiomatized or formalized, it is convenient to accept the equality $T = CnX$, where X is a subset of T (consisting of axioms, postulates, principles, etc.) and Cn refers to the consequence operation. What is truth? Scientific realists, like Popper, for instance, insist that the concept of truth in **SR** should be based on the idea of correspondence (see Agazzi 2014, Chaps. 4 and 5): truth consists in the correspondence of truth-bearers (propositions, sentences, beliefs, statements, etc.) with reality (facts, states of affairs, etc.). I follow this view and consider **STT** as a modern version of the correspondence theory (see also Sect. 4 below). A further possible constraint on **SR** postulates (Chakravartty 2007; Ellis 2009, Chap. 2), that if T is a theory, its ontology accepts theoretical entities as well as denotations of observational terms.[1] I will not focus on this question, because it is not particularly relevant for my tasks. What is the world? If ontology of **SR** is rich, it must admits abstract entities. Since my own ontological preferences are rather nominalistic and naturalistic, I am inclined to see the reality as the mereological unity. Hence, so-called abstract items exist in language eventually as constructs. Finally, following Agazzi, I think that **SR** appears as the best possible justification for objectivity as perhaps the most essential and socially required feature of science.

3 Semantic Definition of Truth

SDT was introduced by Alfred Tarski in 1933 (see Tarski 1933, 1956). I will present its modern version, that is, based on the concept of model. More specifically, we define truth-in-**L**-**M**, where **L** is a language and **M** is a model. **L** is a set of sentences, **M** consists of **U**—the set of individual objects, subsets of **U** (set-theoretical counterparts of properties of objects, subsets of $\mathbf{U} \times \cdots \times \mathbf{U}$ (set-theoretical counterparts of n-termed relations defined on **U**) and, eventually, fixed individuals and functions. Thus, **M** is a structure defined in set-theoretical terms. **L** has a well-defined syntax prescribing the ways of constructing sentences from simpler expressions (constituents), that is, individual constants, predicates, variables or quantifiers and propositional connectives. **L** can be formalized or not. In particular, it possible to understand **L** as a technical scientific language or as a specialized (for instance, by adding symbols) part of natural language. To fix the scenario, any theory is a proper consistent subset of **L** (in symbols, $T \subset \mathbf{L}$ (see below why **L** is not a consistent theory).

[1]Since I do not enter into a detailed polemics with opponents of **SR**, that is, anti-realists, idealists, instrumentalists, conventionalists, etc. I resign from a characterizations of these views. To illustrate the point, I do not discuss (or even allude to the question whether anti-realists (to use this label as general) treat observational.

L and **M** are connected via a semantic interpretation *V*. Omitting various formal details, *V* is a function, which ascribes semantic values to sentences of **L** and their constituents. More specifically, *V* values sentences of **L** by truth-values (truth, falsehood), individual constants (if any) of **L**—by distinguished objects from **U** as references, predicate letters—by subsets of **U** or relations defined on **U**, function symbols (if any)—by functions, and quantifiers as ranging over **U**. In consequence truth-in-**L**-**M** is defined under an interpretation generated by *V*. The concept of interpretation is just semantic in its very essence, because expressions take extra-linguistic items as their values (I omit here the problem of the nature of truth and falsehood).[2] The first step toward **SDT** consists in defining the concept of satisfaction. Let *P* be an one-place (unary) predicate letter; by definition, it refers to a subset of **U**. We say that the formula *Px* is satisfied by an object *o* if and only if $o \in V(P)$, that is, *o* is an element of the set being the value of *P*. The full definition of satisfaction proceeds by induction on complexity of formulas of **L** and can be skipped here without any loss of generality. I only mention that since formulas can be of an arbitrary finite length, it is convenient to introduce infinite sequences of objects in order to obtain a general scheme for all possible syntactic cases.

Px as an open formula is neither true nor false. On the other hand, *Pa* (*a* is the individual name for the object *o*) has a definite logical value dependent on *V*. For instance, the formula '*x* is a prime number' is true for $V(x) = 2$, but false for $V(x) = 4$n. This fact shows an intuitive connection between satisfaction and true. Tarski's ingenious idea consists in considering truth as a special case of satisfaction. Let *A* be a sentence (closed formula, that is, not having free variables). One can prove that *A* is either satisfied by all (infinite) sequences of objects from **U** or is not satisfied by any sequences of objects (satisfied by no such sequence object). Tarski proposed to define truth as satisfaction by all sequences of objects and falsity as satisfaction by no sequence of objects object. The satisfaction by all sequences of objects is equivalent to satisfaction by at least one such sequence or to satisfaction by the empty sequence of objects. These facts motivate:

[2]It is literally true for **L** being an object language. If **L** is a metalanguage its expressions are valued by items taken from the object language. Generally speaking, if $\mathbf{L_n}$ is a language of *n*-rank, *v* ascribes to its expressions either extra-linguistic objects. If $n = 0$; $\mathbf{L_0}$ refers to the object language) or items belonging to $\mathbf{L_{n-1}}$. The basic feature of semantics as contrasted with syntax consists in the fact that the former is not completely reducible to the latter. Hence, semantic values of (at least some) expressions of $\mathbf{L_n}$ must be external with respect to this language. Practically, the interplay of $\mathbf{L_0}$ (the object language; **L** for brevity) with $\mathbf{L_1}$ (= $\mathbf{ML_0}$; **ML** for brevity) suffices to explaining what is going on in semantics. One can identify $\mathbf{L_0}$ with the first-order language and **ML**—with the second-order language. This means that ranks of languages are interpreted as orders in the sense of predicate logic.

(**SDT**) A sentence n(*A*) of a language **L** is true in a model **M** under an interpretation *V* if and only if is satisfied by all infinite sequences of objects from **U** (or at least one such sequence or the empty sequence; otherwise *A* is false in **M**).[3]

The symbol n(*A*) refers to a name of the sentence *A* and belongs to **ML**, although the sentence *A* itself is an element of L. (**TD**) entails

(**TE**) A sentence n(*A*) is true if and only if A^* (in symbols, **T**n($A \Leftrightarrow A^*$),

for any *A* belonging to **L** (A^*—refers to a translation of *A* into **ML**). Tarski claimed that every reasonable truth-definition should entail (**TE**) (frequently called **T**-scheme); this requirement is called the convention **T** ((**CT**) for brevity), which establishes the condition of the material adequacy for a truth-definition. The use of symbols n(*A*) and A^* underlined that the construction of (**SDT**) proceeds in **ML**. On Tarski's account, this way allows to eliminate semantic paradoxes.

Two important metamathematical results have an utmost relevance for my further remarks. These are (**PA**—Peano arithmetic; **VER**(*T*)—the set of truths of *T*):

(**GM**) A set $X \subset$ **L** has a model if and only if *X* is consistent;
(**TU**) If *T* suffices for capturing **PA**, the set **VER**(*T*) is not definable in *T*.

(**GM**) (the Gödel-Malcev completeness theorem) and (**TU**) (the Tarski undefinability theorem) have significant consequences. The former supplements the earlier assumption that *T* is consistent. If so, *T* has a model and, intuitively speaking, concerns something being its subject-matter. One can observe that the assumption of consistency of *T* is at odds with the reality of science, because inconsistency of scientific theories appears as a notorious facts. However, another fact, also notorious, is that if a theory is demonstrated as inconsistent, steps toward its repair to achieve consistency appear immediately. Other strategy used in such situations tries to isolate inconsistencies until a new, typically more general, theory is formulated (see Vickers 2013 for a historical and methodological analysis of inconsistent science).[4] Thus, we can assume that scientists tacitly consider theories as consistent, at least in majority of cases. Still one consequence of (**GM**) deserves attention. **L** in its integrity has no model, because for every sentence *A*, the formula $\neg A \in$ **L**. Hence, **L** is inconsistent. On the other hand, **L**, unless it is purely formal, possesses an interpretation *V*. Consequently, **M** and *V* should be very carefully

---

[3]I distinguish the semantic theory of truth, denoted by **STT**, and the semantic definition of truth, denoted by (**SDT**). The former is the richer than the latter, because (**SDT**) is a part of **STT** which contains several other elements, for instance, theses about the relation between **L** and **ML**. The distinction in question leads to some inconveniences in my further remarks, because I say sometimes about applications of **STT**, but about uses (or abuses) of (**SDT**) in other cases. I hope that this dualism does not lead to misunderstandings.

[4]Claims that inconsistencies motivate paraconsistent logic as basic for science are not well-justified in my opinion. This paper assumes classical logic as the ultimate formal foundation of scientific theories.

distinguished. In fact, models are algebraic structures, but interpretations act as functions. (**TU**) justifies the view (see footnote 2) that semantics cannot be reduced to syntax.

4 Semantic Theory of Truth as a Correspondence Theory

As I earlier indicated that **STT** is the correspondence theory of truth in modern setting. Presumably, this qualification can be justified either by appealing to (**SDT**) or by reference to (**TE**). The latter option seems fairly correct and justified by Tarski's own explanations. In particular, his famous example that the sentence (i) 'snow is white' is true if and only if snow is white, points out that because (i) says that snow is white and it is so as this sentence says, (i) is true. What about the intuitive content of (**TD**)? We have two possibilities, firstly, that it is a mathematical trick, and, secondly, that (**SDT**) brings some intuitions concerning the correspondence theory of truth. Clearly, sequences of objects cannot be identifies with facts. Moreover, the satisfaction by the empty sequence appears as an artificial construction (see Tarski 1933, p. 195; page-reference to Tarski 1956). On the other hand, if (**SDT**) is a special case of the definition of satisfaction and the latter is based on explicit intuitions, it suggests that perhaps some philosophical intuitions are behind (**SDT**) as well. I am inclined to take this option. That an open formula is satisfied or not by an object, depends of valuation of free variables. Such valuations are irrelevant in the case of sentences, because they do not contain free variables. Consequently, every infinite sequence of objects can be ascribed to bound variables or individual constants. As far as the second case is concerned, we can eliminate individual constants via identity and existential quantification or say that if $V(a) = o$, every infinite sequence of objects with o as the first member, satisfies the formula Pa, if o satisfies this formula. The same can be expressed by saying that the empty sequence satisfies a sentence, because no free variables occur in it. What remains? The answer is that being true depends on how **L** is interpreted and, metaphorically speaking, how things are in **M** associated with **L** interpreted by V. And it precisely expresses what is established by the **T**-scheme. Informally speaking, truth depends on the domain sentences say about. If one wants connects this explanations with facts, we could perhaps say that a given sentence A of **L** cannot be demonstrated as false in **M** if it is the satisfaction by the empty sequence (or by all infinite sequences or by one infinites sequence) excludes, provided that V is fixed. Putting this in other words, if A is true, no fact defined on **M** can be A-falsifier.

5 Semantic Definition of Truth and Empirical Theories

Tarski himself explicitly emphasized **STT** applies to empirical theories (see Tarski 1944). In particular, he proposed to explain conditions of acceptability of theories by constraining that they do not imply false sentences. On the other hand, Tarski did not discuss the problems pertained to realism, because he considered them as too metaphysical.[5] In what follows I will comment various objections advanced in order to demonstrate that **STT** has no application to scientific theories and, hence, it has nothing to do with scientific realism. Consequently, it cannot justify this view or its opposition, that is, anti-realism. The main points can be classified in the following way:

(1) **STT** blurs the difference between logical and empirical truths;
(2) **STT** can be applied to formal languages, but empirical are not such;
(3) **STT** can be applied to theories understood as set of sentences, but so-called non-statement view is more proper;
(4) Even if we agree that a kind of truth in the semantic sense can be attributed to empirical theories, we should apply the concept of partial (approximate) truth;
(5) The world as the subject matter of empirical theories cannot be identified with semantic model.

I will argue that these objections can be met from the point of view of **SDT**, eventually enriched by further constraints.

Ad (1) This objection is to be find in O'Connor (1975, p. 109) and (Haack 1978, p. 113). Both authors maintain that if truth of a sentence consists in its satisfaction by all sequences, this conditions holds for logical as well as empirical sentences. However, this argument rests on entirely mistaken (see above) views on sequences as such. Once again, sequences of objects are not facts, states of affairs, etc. We can define logical truths as true in every model (or under any interpretation) and empirical truths as holding in some models only. If someone, as, for example, Tarski himself did, maintains that the borderline between logical and empirical truths is vague or fuzzy, still employs the definition of logical truth as valis (true in all models). If so, the division of truths into logical and anti-logical has nothing to do with (**SDT**) as such.

Ad (2) Tarski never said that (**SDT**) applies to formal languages. It is clear, if we consult the following proclamation (Tarski 1933, pp. 166–167; page reference to Tarski 1956):

> It remains perhaps to add that we are not interested here in 'formal' languages in sciences in one special sense of the word 'formal', namely sciences to the signs and expressions of

[5]Two remarks are in order in the context of Tarski's views. Firstly, acceptability of theories can be defined by recurring to other truth-definitions. Secondly, Tarski used the term 'realism' as related to the philosophy of perception (direct realism, critical realism). Hence, one should be very careful in saying that scientific realism in the present sense would be qualified by Tarski as too metaphysical.

> which no material sense is attached. For such sciences the problem here discussed [the problem of truth] has no relevance, it is not even meaningful. We shall always ascribe quite concrete and, for us, intelligible meanings to the signs which occur in the language we shall consider. The expressions which we call sentences still remain sentences after the signs which occur in have been translated into colloquial language.

Thus, a language **L** for which (**SDT**) applies is always interpreted, even if it is formalized. Consequently, interpretation of **L** always precedes definitions of semantic concepts including the notion of truth. Thus, we arrived at the problem of how formal is related to formalized. The answer is that formal language do not need to be equipped with meaning, contrary to formalized languages.

A common misunderstanding of Tarski's views consists in attributing to him the opinion that **STT** applies to formal (formalized) languages only (this objection goes back to Black 1948). This mistake neglects that Tarski explicitly explained that truth-bearers are correct syntactic units of the propositional category and having meaning. Yet it does not mean that Tarski's view about language and meaning have no weak points. In particular, he did not define the concept of meaning. In fact, he intentionally avoided this question and deliberately preferred to speak about interpreted languages as semantic items.[6] Yet such languages (see above) can be formalized or not. Now the question arise whether formal semantics can be applied to non-formalized languages or which amount of formalization suffices for the exact semantic analysis. A particularly interesting questions concern natural language as a subject of logical semantic constructions. Tarski pointed out (see Tarski 1933, Sect. 1) that natural languages are closed in the sense that they mix **L** and **ML**, and thereby generate semantic paradoxes. This feature of ordinary parlance motivated Tarski's skepticism concerning applicability formal semantic constructions to natural language. In Tarski (1944), this skepticism is weaker, because Tarski explicitly says that considerable parts of natural language can be analyzed by tools of logical semantics. He even introduced the category of languages having a specified structure, that is, not formalized but syntactically well-defined. Although Tarski illustrated (**SDT**) by an example from formalized theory of classes and exposed mathematically his definition of truth, there is nothing in its construction which would have to preclude its application to the language of physics. In particular, there is nothing in the definition of satisfaction which restricts this concept to formalized formulas. If we way that Warsaw satisfies the predicate 'is a city in Poland", it would be difficult to find a difference with saying that the number $ssss(0)$ satisfies the formula '4 $\in$ **N**", where **N** is a set o natural numbers and is an odd number' and $ssss(0)$ has the meaning 'the successor of the successor of the successor of the successor of 0').

To conclude this point, **SDT** is fully consistent with the view that the language of empirical theories is semantically interpreted (see also Popper 1963, pp. 398–399). It can be formalized or not, but cannot be purely formal.

[6]I do not suggest that Tarski's remarks on meaning are sufficient and agree that they should be supplemented by more positive assertions. On the other hand, (**SDT**) does not depend on any concrete theory of meaning. Hence, it is quite enough that expressions of **L** are meaningful.

Ad (3) Some authors (see Morrison and Morgan 1999, p. 3) contrast the syntactic and semantic understanding of empirical theories. Roughly speaking, the former view sees theories as ordered, for instance, axiomatized, sets of sentences, whereas the latter approach identifies theories with classes of models (see Suppes 2002 for the most comprehensive elaboration of this position).[7] I consider this distinction as very misleading. The simplest argument for that qualification stems from (**GM**). If being consistent (a syntactic property) has its exact counterpart in having a model, both ways of speaking, syntactic and semantic, are equivalent.[8] Hence, every consistent theory has a model, but also every model can be described by a set of sentences belonging to **L**. The characterization (see above) of *T* as *CnX*, although syntactic, due to the status of *Cn*, concerns interpreted languages, that is, involves semantics by definition. In fact, Patrick Suppes, the main proponent of the semantic view, had nothing against (**SDT**), although focuses not on it but on model-theoretic constructions as usable in analysis of empirical theories. Taking into account the history of contemporary methodology of science, it is easily to show that the purely syntactic point of view on theories can be attributed to the early Vienna Circle, but not to philosophers a la Tarski. If one agrees that scientists use interpreted languages, (**SDT**) and defining theories by *Cn* are perfectly compatible (see also Przełęcki 1969, 1977; Ruttkamp 2002).[9]

Ad (4) Many authors maintain that scientific truth is approximate or partial (see Wójcicki 1979; Wójcicki 1995/96; Psillos 1999; Costa and French 2003). This account does not need to be at odds with **STT**. In fact, all mentioned authors propose some modifications of this theory in order to capture 'the nearly true' (the phrase of Stathis Psillos) as a semantic property. Personally, I am very skeptical about conceptions of degrees of truth. On Tarskian semantics, sentences are either true or false under a given interpretation *V* ((**SDT**) implies the principle of bivalence; see Woleński 2003 for details) and there are no other or partial logical values. Of course, one can propose many-valued logic, fuzzy logic, probabilistic logic or account via the concept of verisimilitude, but they require a fairly fundamental revision of logic and metalogic. I prefer a more opportunistic approach consisting in keeping classical logic as defining formal properties of logical values with simultaneous admitting that we have various degrees of justification. To conclude, the is no need to revise (**SDT**) in its traditional model-theoretic setting.

Ad (5) At first, the following account of **SR** via **STT** could be attractive. Take an empirical theory *T*. By the assumption of consistency and (**GM**) it has model, say **M**(*T*). Call this model the fragment of the real world being the subject-matter of

[7]An alternative terminology consists in using the labels 'the statement view of theories' (the syntactic understanding) and 'the non-statements view of theories' (the semantic understanding). Perhaps one cautionary remark concerning Suppes' view is in order. He did not exclude understanding theories as sets of sentences, but considered model-theoretic approach (more precisely, defining theories by set-theoretical predicates as more general.

[8]It does not mean that we can define semantics in syntax.

[9]To avoid possible misunderstandings, this section as well as the next one report not objections but proposals intended to modify **STT** to make it more suitable for analysis of empirical theories.

T. By (**UT**), **M**(*T*) is not definable within **T**, provided that **T** contains **PA**; this assumption is legitimate for most empirical theories. Using a philosophical jargon, we can say that **M**(*T*) transcends *T*. And it is the fundamental thesis of any realism, including **SR** (see Woleński 2004). An additional advantage of this picture concerns theoretical and observational parts of theories, independently of criteria of their particular delimitation. Since, as the history of science points out, observational data are consistent with many alternative theories, information coming from observational procedures, can be captured by several different theories and their models. Contemporary cosmology provides a good example of this situation.

Unfortunately, the above picture is too simplified (see Grobler 2001). First of all, the real world is not an algebraic structure. This statement is true even on Platonic view on the reality that abstract objects belong to its ontological equipment. Independently of the chosen ontological view on the reality, (**SDT**) does not qualify per se any model as proper (intended, standard, etc.). Using the traditional (in fact, going back to the Schoolmen) way of speaking, we might distinguish formal and material objects of knowledge or cognition. According to this distinction, every consistent cognitive result refers to a formal objects.[10] It concerns, for example, completely fictional stories. This implies that some cognitive results have no material objects, that is, considered as the parts of the real world. Consequently, a formal object of cognition becomes a material object, if some additional conditions are fulfilled. Employing this idea, (**GM**) says that every consistent set of sentences has a model as a formal object of semantic interpretation, but this statement does not solve the problem of how identify the material object how it is related to formal one.

6 Semantic Models as Mediators

Presumably, one could say that semantic models represent fragments of the real world as mediators between them and theories.[11] This account of models became quite popular in the recent philosophy of science (see Morrison and Morgan 1999; Suárez 1999; Morrison 2015, Chap. 4). Mediating model have three principal properties:

(a) They are not directly accessible from theories;
(b) They are not forced by observational data;
(c) The replace investigated phenomena.

[10]There is a delicate problem concerning the statement that cognitive results are consistent. One can argue, as Meinong did, that we can think about inconsistent items, like round squares. I do not consider this question. Limiting attention to consistent cognitive results, I intend to be close to (**GM**).

[11]I remind that, on my view, the real world is the unity consisting of mereological parts. Yet nothing special follows from this assumption for my further considerations.

From a very abstract point of view, mediating models in the contemporary understanding can be presented as algebraic structure, but such an approach would be at odds with their functions. In fact, the proponents of the idea of models as mediators, do not apply this notion to semantic models. They speak about models as results of mathematical modeling. The same lack of references to algebraic structures or semantic models occurs in typical surveys, like Zeigler (1976), Meyer (1984), Weisberg (2013), Pohljolainen (2016). Mathematical modeling is a branch of applied mathematics (see Higham 2015, Part V).

Although theoreticians and practitioners of mathematical modeling do not use semantic terminology, this attitude does not result from a anti-philosophical prejudice. In fact, the representational strength of semantic models is very weak. Even if one were incline to say that the structure <the set of cities, Warsaw, is a Polish city> represents the fact that Warsaw is a Polish city, this would be a trivial observation. Similarly, the structure <the set of particles, the Higgs boson, completes the Standard Model>, although is not trivial, provides no interesting information, because is entirely ex post.[12] On the other hand, the actual mathematical modeling provides powerful tools for description of empirical phenomena (see Pincock 2012), for instance, in order to represent them as discrete or continuous, measurable by ordering, additive or quotient scales, subjected to various statistical distributions, etc. The Higgs boson would not be discovered without a proper simulation, the history of quantum and atomic physics is inherently associated with various models (see Cook 2006), etc. In general, a radically abstract character of contemporary physics makes it practically incomprehensible without modeling and simulating (see Falkenburg 1997; Lewis Peter 2016 for more philosophical reports about this situation).

Mathematical modeling is essentially based on inputs coming from genuine empirical investigations. Even if (mathematical) models are regarded as more or less fictional, their users can precisely explain what is fictional and what is non-fictional in a given construction. Observe that comprehending the word 'fiction' and its various derivatives, assumes that we are able to explain 'non-fictional' as its antonym. There is no other way to learn how to use 'fictional' and 'non-fictional' as appealing to the world and its cognition. Clearly, a skeptic or a radical anti-realist can, using general philosophical arguments, always deny that cognitive acts are directed to real objects, but if one says something like that, he or she announces an understanding of the word 'real' and 'unreal'. I do not claim that this observation refutes skepticism, but only that this view is much weaker than it is customary asserted. To conclude this paragraph, mathematical modeling could be difficult to understand it without asserting that it produces models responding to empirical inputs, which generate criteria for qualifying something as real or

[12]The situation is different in the case of model theory as a part of metamathematics, because model-theoretic constructions have important applications in mathematics, for instance, in algebra and geometry.

non-fictional and something else, if any, as just unreal or fictional. Mathematical models are mediators between theories and the world just in this sense.[13]

The scheme of mathematical modeling distinguishes the following main stages (see Pohjolainen and Heiliö 2016, p. 2): real world problem, mathematical model, a solution for the model, interpretation of the solution, return to real world problem. Empirical ingredients of this schematization are evident. Thus, mathematical modeling has its links with the world by definition. If mathematical models mediate between theories and empirically accessible facts, they must be also somehow related to the former, although this aspect does not occur in the quoted schematization. Yet ways of the interplay between theories and mathematical models are very different and depend on various circumstances. However, it seems that theories as generators of models are considered as correct or even true on a certain class of phenomena. On the other hand, qualifications of particular statement made on the occasion of modeling seems to be a secondary matter. Even if one were insist that, for instance, the sentence 'changes of velocity as modeled by calculus are continuous' is true, this qualification remains on the level of (**TE**) and can be skipped without any loss of content. As a matter of fact, mathematical modeling only rarely becomes a subject of philosophical reflection perhaps except a general question of applied mathematics and its relation to pure mathematics.

7 Semantic Models and Reality

Speaking on semantic models as mediators I extend the standard terminology and, what is perhaps more important, embed it into philosophy. The reason is that I consider the points (a)–(c) provide hints for a philosophical enterprise consisting formulating and defending **SR**. The main issue concerns the thesis (a) reformulated in the following way[14]:

(a*) Semantic models as representations of the world are not directly accessible from theories.

Using another terminology (see above), one might say that given *T*, its intended model **M**, that is, assumed to be a representation of the real world, is not directly derivable from *T*. In other words, the formal object of *T* does not entirely determine its material object. Hence, something must be add to **STT** in order to show in which circumstances semantic models of theories represent the real world.

---

[13]The role of mediators play also other kinds od representations, namely maps, diagrams, mechanical replicas of objects, for instance, cars, etc.

[14]I will neglect (b) and (c). Since semantic models have a small representational content small, (c) appears as pointless with respect to them. The point (b) is related to the problem of the relation between theoretical and observational parts of theories. This question possesses a secondary significance form the present paper.

(**GM**) suggests that **M**(T) and T are equivalent. However, if the issue concerns **SR**, it is more convenient to begin with T and its semantics. The valuation function V correlates expressions with their references in **M**(T). This machinery is too weak in order to capture the relation between **M**(T) and the world (**W** for brevity).[15] In fact, provided that V is already given, the pair **Π** = <T, **M**(T)> organizes the T-semantics. If **M**(T) is to be characterized as a mediator, **Π** has no second element to be mediated by the model in question. Thus, **Π** must be extend to the triple **Π**′ = <T, **M**(T), **W**>. Since V acts from the T (more precisely, its language) to **M**(T), it does not reach **W**. In order to fill this gap, we need a link between **W** and **M**(T) assuring that the latter is a mediator connecting T and **W**-items. Consider the function V': **W** → **M**(T) and its composition with V, that is, $\Phi = V' \bullet V$.[16] This construction displays the required link between T, **M**(T) and empirical data derived from **W** and the role of **M**(T) as a mediator. Informally speaking, V reaches the real world via t V'. The latter is not a valuation function in the strict model-theoretic sense. Consequently, although **W** is not a model of T, we can still say that **M**(T) semantically represents **W**. It is not a problem of semantics to demonstrate how V' acts. For instance, mathematical modeling or simulation can be helpful for deciding whether and how a given linguistic item applies to a piece of **W**.

In the ordinary practice of science and the daily life, V and V' are mixed, because, due to ways of learning and using language, learning V' automatically qualifies V. Hence, the material object of knowledge is usually identified with the formal one. Both are distinguished in special cases, for instance, searching new theories, considering psychic abnormality, comparing reality with fiction, etc. On the other hand, semantics provides tools for a general philosophical picture. Although V' is not a valuation function in the strict sense Φ can play this role. Consequently, we can introduce the concept of empirical satisfaction and empirical truth. Consider the following theoretical sentence:

(ii) If p is a particle completing the Standard Model, it should have such and such properties.

This assertion is true in **M**(ii) in the semantic sense. Also we can say that the open formula

(iii) If x is a particle completing the Standard Model, it has such and such properties, is satisfied by a hypothetical particle p possessing the properties in question, provided that (iii) is not empty-satisfied, that is, because its antecedent is false. Before discovering the Higgs boson, it was not known whether p really existed or not. although V valued the variable x in (iii) by p. In other words, p existed in **M**(ii), but its reality in **W** was an open question. The Higgs boson was just identified as the particle p. Clearly, V does not

[15]In fact, fragments of **W** are involved, unless T is a theory everything. Yet such a theory is still a dream.

[16]This construction is strongly motivated by Gåerding (1977, p. 170), particularly to the effect the the pair <T, MIT)> should be replaced by the triple <T, **M**(T), **W**>.

justify of substituting x by the term 'the Higgs boson'. On the other hand, Φ, if restricted to the considered case, allows the substitution operation in question.

Call Φ the empirical valuation function. We say:

(**SDT**$^{\mathbf{E}}$) A sentence A of a theory T is empirically true in a model $\mathbf{M}(T)$ under Φ if and only if A is satisfied by the empty sequence of objects.

The simplest example concerning satisfaction is as follows

(iv) An object o satisfies the formula Px under Φ if and only if $\Phi(x) = o$ and $o \in \Phi(P)$. Although the ordinary parlance admits to say that if A is true, under Φ, it is true about **W**, more proper would be say that A is true in $\mathbf{M}(T)$ with respect to Φ. However, the former way of speaking hast its justification in fact that empirical valuations map **W** on $\mathbf{M}(T)$ and help in setting which semantic model is standard or intended with respect to data.[17]

8 Semantic of Empirical Theories via (SDT$^{\mathbf{E}}$) and Scientific Realism

How (**SDT**$^{\mathbf{E}}$) is related to **SR**? Introducing empirical valuations meet two objections raised against the realistic account of science based on **STT**. Firstly, **SR** + (**SDT**$^{\mathbf{E}}$) can be interpreted as the view that empirical theories are about the real world and, secondly, truth accounted semantically allows to distinguish intended empirical models from other possible semantic constructions. Moreover, (**SDT**) does not require any further supplement, for instance, by the idea of partial or approximate truth. (**SDT**) formalized the traditional conception of the absolute truth (see Woleński 2012, 2015), because relativizations to **L** and **M** are consistent with the assertion that if A is true in **M**, its logical value does not change, for instance, over time. Yet this statement does not mean that cognitive subjects are omniscient, because empirical valuations depend on many circumstances and can be replaced by other. In fact, it happens notoriously. A general argument for realism as a philosophical view remains unchanged. To repeat, (**UT**) suggests the conclusion that since the set of truths of **PA** is not definable in this theory, the same concerns the definability of the class of **PA**-models. Hence, every empirical theory containing arithmetic, that is, every interesting empirical theory, cannot define its models. Consequently, if T is an empirical theory **W** cannot be defined within it. The situation will not change in the case of the theory of everything. That models exceed conceptual resources of theories constitutes the main feature of realism from the semantic point of view.

[17]For simplicity, I take into account single semantic models, but similar considerations concern classes of models. If **W** were replaced by a set of models, Φ could act as a tool for selecting the standard model from various possibilities.

Here remains still one point related to the distinction of internal and metaphysical realism (see Putnam 1987; Gardiner 2000). According to Hilary Putnam, the latter view maintains that we have access to one ready real world. On the other hand, model theory suggests that scientific theories are not categorical. Hence, every (interesting) empirical theory has many models and model-theoretic semantics per se does not offer resources for picking up the real world as independent of a conceptual system. In particular, **STT** has no relevance for the proper account of the relation between science and its subject matter (see Woleński 2004 for criticism of this view). To sum up, the real world must be defined within conceptual systems constrained by epistemic criteria. Unfortunately, Putnam's explanations are too vague in order to be interpreted either as a kind of realism or as a species of anti-realism. According to my earlier declarations concerning the scope of this paper, I do not like to enter into discussing realism and anti-realism in general. Hence, I restrict my further remarks to few points concerning **SR** in the context of Putnam's distinction. **SR** can be interpreted as accepting metaphysical realism as well as associated with internal realism. In my terminology, T is a counterpart of a conceptual systems in Putnam's sense. $\mathbf{M}(T)$ is obviously dependent on T, but it is controversial issue to which extent. A more concrete solution depends on the view about the theory/observation distinction, the role of terminological convention, etc. As far as the matter concerns **W**, **SR** says only that the real world cannot be reduced to T or/and $M(T)$. The empirical valuation function acts mostly epistemologically as linking T, $\mathbf{M}(T)$ and **W**. This construction is entirely consistent with internal realism. Recall that T has a model, if it is consistent. Thus, the existence of a model appears as a direct consequence of semantics. Semantics actually does not suffice for proving that **W** exists. In other words, the existence of a formal model of knowledge does not entail the existence of its material counterparts. On the other hand, epistemic constraints imposed by Φ are difficult to be comprehensible without assumption that **W** exists. Although internal realism provided a basis for a sound **SR**, its main defect consists in a vague mixing epistemology (cognitive access to **W**) ontology or metaphysics (the existence of **W**). **SR** based on **STT** avoids this confusion.

References

Agazzi Evandro (2014) Scientific objectivity and its contexts. Springer, Berlin.

Black Max (1948) The semantic definition of truth. Analysis 8, 49–63; reprinted in M. M. Black, Language and philosophy, Cornell University Press, Ithaca 1949, 89–107.

Charravartty Anjan (2007) A metaphysics for scientific realism. Knowing the Unobservable. Cambridge University Press, Cambridge.

Cook Norman D. (2006) Models of the atomic nucleus. Springer. Berlin.

Da Costa Newton C. A. and French Steven (2003) Science in partial truth. A unitary approach to models and scientific reasoning. Oxford University Press, Oxford.

Ellis Brian (2009) The metaphysics of scientific realism McGill-Queen's University Press, Montreal.

Falkenburg Brigitte (1997) Particle mechanics. A critical account of subatomic reality. Springer, Berlin.
Gardiner Mark Q. (2000) Semantic challenges to realism. Dummett and Putnam. University of Toronto Press, Toronto.
Gårding L. (1977), Encounter with Mathematics. Springer, Berlin.
Grobler Adam (2001) Pojęcie prawdy w językach nauk empirycznych (The concept of truth in langugaes of empirical sciences). In A. Grobler, Pomysły na temat prawdy i sposobów uprawiania filozofii (Ideas on truth and ways of doin philosop0hy). Aureus, Kraków, 166–179.
Haack Susan (1978) Philosophy of logic. Cambridge University Press, Cambridge.
Higman Nicholas J. (ed.) (2015) The Princeton company to applied mathematics. Princeton University Press, Princeton.
Lewis Peter J (2016) Quantum ontology. A guide to the metaphysics of quantum mechanics. Oxford University Press, Oxford.
Meyer Walter J. (1984) Concepts of mathematical modeling. Mc Graw-Hill, New York.
Morgan Mary S. and Morrison Margaret (eds.) (1999) Models as mediators. Perspectives on natural and social Science. Cambridge University Press, Cambridge.
Morrison Margaret and Morgan Mary S. (1999) Models as mediating instruments. In Morgan and Morrison (1999), 10–37.
Morrison Margaret (2015) Reconstructing reality. Models, mathematics and simulations. Oxford University Press, Oxford.
Morrison Margaret and Morgan Mary S. (1999) Introduction. In Morgan and Morrison (eds.) 1999, 1–9.
O' Connor David (1975) The correspondence theory of truth. Hutchinson, London.
Pincock Christopher (2012) Mathematics and Scientific representation, Oxford University Press, Oxford.
Pohlolainen Seppo (ed.) (2016) Mathematical modeling. Springer, Berlin.
Poholainen Seppo and Heillö Matti (2016) Introduction. In Pohlolainen (ed.) 2016, 1–5.
Popper Karl (1963), Conjectures and refutations. The Growth of scientific knowledge. Routledge, London.
Przełęcki Marian (1969), The logic of empirical theories. Routledge, London.
Przełęcki Marian (1977) The concept of truth in empirical languages. Grazer Philosophische Studien 3, 1–17.
Psillos Stathis (1999) Scientific realism. How science tracks truth. Routledge, London.
Putnam Hilary (1987) Many faces of realism. Open Court, La Salle.
Ruttkamp Emma (2002) A model-theoretic interpretation of science. Kluwer, Dordrecht.
Suppes Patrick (2002) Reperesentation and invariance of scientific structures. CSLI Publications, Stanford.
Suárez Maurizio (1999) The role of models in the application of scientific theories: epistemological implications. In Morgan and Morrison (eds.) 1999, 168–196.
Tarski Alfred (1933), Pojęcie prawdy w językach nauk dedukcyjnych, Towarzystwo Naukowe Warszawskie, Warszawa; English edition Tarski 1956.
Tarski Alfred (1944), The semantic conception of truth and the foundations of semantics, Philosophy and Phenomenological Research **4**, 341–395; repr. in A. Tarski, *Collected Papers*, v. 2, Birkhäuser, Basel, 665-–699.
Tarski Alfred (1956) The Concept of Truth in Formalized Languages. In A. Tarski, Logic semantics, metamathematics. Papers of 1923 to 1938. Clarendon Press, Oxford 1956, 152–278; 2nd ed., Hackett Publishing Company, Indianapolis 1984.
Tarski, A. (1944), The semantic conception of truth and the foundations of semantics, Philosophy and Phenomenological Research 4, 341–395; reprinted in A. Tarski, Collected papers, v. 2, Birkhäuser, Basel, 665–699.
Vickers Peter (2013) Understanding inconsistent science. Oxford University Press Oxford.
Weisberg Michael (2013) Simulation and similarity. Using models to understanding the World. Oxford University Press, Oxford.

Woleński Jan 2003 Truth and satisfaction by the empty sequence. In Philosophical dimensions of logic and science. Selected contributed papers from the 11th International Congress of Logic, Methodology, and Philosophy of Science, Kraków, 1999. Ed. by A. Rojszczak, J. Cachro and G. Kurczewski, Kluwer Academic Publishers, Dordrecht 2003,. 267–276; reprinted in Woleński (2011), 17–24.

Woleński Jan (2001) In defense of the semantic definition of truth. Synthese 126(1–2), 67–90.

Woleński, Jan (2004) Logic, semantics and realism. In *Sémantique et épistemologie*. Ed. by A. Benmakhlouf, Editions Le Fennec, Casablanca 2004, 135–148; reprinted in Woleński (2011), 51–60.

Woleński Jan (2011) Essays on logic and its applications in philosophy, Peter Lang, Frankfurt am Main 2011.

Woleński Jan (2012), Truth is eternal if and only if it is sempiternal., In Studies in the philosophy of Herbert Hochberg, ed. by E. Tegtmeier, Ontos Verlag, Frankfurt am Main 2012, 223–230.

Woleński Jan (2015) An analysis of logical determinism. In Themes from ontology, mind and logic. Present and past. Essays in honor of Peter Simons. Ed. by S. Lapointe, Brill, Leiden, 423–442.

Wójcicki Ryszard (1979) Topics in formal Methodology of empirical sciences. Ossolineum/ Reidel, Wrocław/Dordrecht.

Wójcicki Ryszard (1995/96) Theories, theoretical Models, Truth. Part I: Popperian and non-Popperian theories in science. Foundations of Science 1, 337–406. Part II: T Tarski's theory of truth and its relevance for the theory of science. Foundations of Science 4, 471–516.

Zeigler Bernard (1976) Theory of modelling and simulation. Wiley, New York.

Underdetermination, Realism and Objectivity in Quantum Mechanics

Dennis Dieks

Abstract Underdetermination of theories by empirical data is a central theme in debates surrounding scientific realism. Underdetermination undermines epistemological optimism: if empirical evidence cannot decide between theories, skepticism about the progress of science seems justified. Philosophical defenses have been developed against this skeptical threat. Typical themes in these defenses are that significant scientific examples of empirical equivalence (as opposed to imaginary armchair cases) are virtually non-existent, as it is already difficult enough in scientific practice to develop one single satisfactory theory; that in the rare instances where empirical equivalence can be maintained to occur it is defeasible and only temporary; and that there usually will be substantial differences in empirical *support*, even if theories are empirically equivalent. Examples are usually constructed cases within classical physics that have not played an important role in actual history. In this article we draw attention to the present-day situation in quantum mechanics, which we think is very relevant to the issue. There exist several realist interpretations of quantum mechanics, each of which depicts a quite distinctive physical world, and each of which has its own circle of devotees in the scientific community. Most of these interpretations are empirically equivalent in a quite strong sense: they predict the same results for all experiments that can be expected to be feasible. The usual arguments against the significance of theoretical underdetermination seem to lose a great deal of their effectiveness here. One may wonder whether non-uniqueness of theories is not part and parcel of the practice of modern science after all, and much less threatening than often thought.

D. Dieks (✉)
History and Philosophy of Science, Utrecht University, Utrecht, The Netherlands
e-mail: d.dieks@uu.nl

E. Agazzi (ed.), *Varieties of Scientific Realism*,
DOI 10.1007/978-3-319-51608-0_16

1 Introduction: Underdetermination and Empirical Equivalence

It is a truism that empirical data, as actually collected in scientific experiments, do not deductively fix scientific theories. For one thing, scientific theories are universal with respect to space and time, whereas empirical results are finite in number and associated with 0. the locations and instants at which they were gathered. For another thing, measurement results and outcomes of experiments belong to the domain of observable phenomena and do not immediately inform us about the "theoretical parts" of theories. Both considerations illustrate that theories contain more than what is logically implied by empirical data. The amount of information contained in empirical data is consequently insufficient to make deductions leading to one unique theoretical scheme. The difficulty is comparable to the mathematical problem of solving equations when there are more unknowns than equations: in such cases solutions will generally not be unique, and usually there will even be infinitely many solutions. Similarly, one may expect that there will be many scientific theories that agree with any given amount of experimental results. However, these theories may well lead to very different pictures of the parts of the physical world that are not directly visible in experiments.

A distinction should be made at this point. The empirical equivalence that we just discussed consists in agreement with the same finite set of experimental data available at a certain instant of time. This is a limited equivalence, which may be broken by any new incoming piece of evidence. One may also consider the more drastic case of equivalence between different theories with regard to all data that will ever be collected, or with regard to all data that could *possibly* be obtained (even including those that require experiments that will never be actually done). Even the existence of the latter, very strong kind of empirical equivalence seems a priori plausible, since the equivalence is still only on the "surface of observable phenomena", which does not exclude that the theories may say very different things about unobservable objects and processes. The underdetermination argument therefore still applies.

It is no wonder that empirical equivalence of theories has often been discussed in the philosophy of science literature, as it appears to pose a problem for evidence-based choice between theories. Writers on the subject usually perceive the danger that the objectivity of science may be called into question and that this in turns threatens the very rationality of science. Accordingly, much effort has been directed at down-playing the significance of underdetermination in scientific practice. In a seminal paper, Laudan and Leplin (1991) (see also Laudan (1990)) have provided what has become the paradigm of pertinent argumentation, concluding that the ordinary practice of science will have no problem of principle with breaking any deadlock of empirical equivalence that might arise. They argue that empirically equivalent rivals will often be artificial and not genuinely scientific; and in the rare cases of contenders that are to be taken seriously, they will generally be *supported* differently, via relations with other theories or background information. Moreover, if new evidence comes in it should be expected that hitherto empirically equivalent theories will soon cease to be equivalent.

We shall argue, however, that these Laudan-Leplin style arguments underestimate the significance of empirical equivalence and underdetermination within present-day science. Indeed, differently from what is often assumed in the philosophy of science literature,[1] theoretical underdetermination and empirical equivalence of a non-trivial kind actually occur in modern scientific practice and are frequently discussed in the foundational scientific literature. It will be interesting to compare these cases with the philosophical arguments for the practical insignificance of empirical equivalence, and to see whether these cases pose a threat to scientific rationality.

2 The Standard Arguments Against Empirical Equivalence

Laudan and Leplin introduce the themes of empirical equivalence and underdetermination in the following way Laudan and Leplin (1991, p. 449): "By the 1920s, it was widely supposed that a perfectly general proof was available for the thesis that there are always empirically equivalent rivals to any successful theory. Secondly, by the 1940s and 1950s, it was thought that—in large part because of empirical equivalence—theory choice was radically underdetermined by any conceivable evidence". As they continue Laudan and Leplin (1991, p. 450):

> The idea that theories can be empirically equivalent, that in fact there are indefinitely many equivalent alternatives to any theory, has wreaked havoc throughout twentieth century philosophy. It motivates many forms of relativism, both ontological and epistemological, by supplying apparently irremediable pluralisms of belief and practice. It animates epistemic skepticism by apparently underwriting the thesis of underdetermination. In general, the supposed ability to supply an empirically equivalent rival to any theory, however well supported or tested, has been assumed sufficient to undermine our confidence in that theory and to reduce our preference for it to a status epistemically weaker than warranted assent.

Laudan and Leplin identify two pillars on which the doctrine of empirical equivalence and theoretical underdetermination rests. *First*, the doctrine relies on the supposed fact that for any theory T and any body of observational evidence E, there is another theory T' such that T and T' are both in agreement with E. The support for this, they say, is twofold. Some authors claim that there exist general algorithms that are able to generate an empirically equivalent theory T' given any theory T. In addition, there is the Duhem-Quine thesis that a hypothesis can only entail observable consequences with the help of auxiliary assumptions. Thus, it is imaginable that any conceivable evidence can be accommodated by any hypothesis by means of suitable auxiliary assumptions. Accordingly, one might suppose that if a hypothesis H–along with class of auxiliary assumptions A–entails the observational consequence

[1]For example, Musgrave (1985, p. 200) confesses to know of only one serious example of empirical equivalence, which he takes from the writings of van Fraassen (1980, p. 46), namely that all Newtonian cosmological models that differ from each other in the absolute velocity they assign to the center of gravity of the solar system are in accordance with the same evidence.

E, there exists another hypothesis H' that also entails E by using an alternative class of assumptions A'.

Second, the doctrine of empirical equivalence and underdetermination of theory choice is said to depend crucially on the premise that only observational statements that are directly logically linked to a theory can legitimately count as empirical evidence for it: an observational report provides evidence for a hypothesis only if its content can be derived, explained or predicted from that hypothesis.

From the first premise it follows that there exist empirically equivalent rivals to any theory; and from the second premise that all these rivals are equally supported. It then appears that the objectivity and even rationality of scientific theory choice come under severe threat (see for a more extensive discussion, also of the other material in this section, Acuña and Dieks (2014)).

A straightforward way out of this problem would be to have recourse to non-empirical features of empirically equivalent theories. For example, if one of the theories is easier to understand, leads to a simpler world picture, or is easier to use than its rivals, one certainly has a *reason* to prefer this theory. However, theoretical virtues of this kind are controversial—they will often have a pragmatic flavor and depend on subjective preferences. Although the consideration of such non-empirical virtues can certainly provide reasons to prefer one of the theories, and in this sense dissolves the problem of the rationality of theory choice, the *objectivity* of theory choice still seems endangered—something to which we shall return.

Laudan and Leplin therefore choose a more robust way of countering the threat of relativism, by directly attacking the two premises of the above argument. First, they argue against the significance of empirical equivalence in the practice of science. As they say, since observational techniques improve all the time, and auxiliary assumptions may change as well, the class of observable consequences of any theory is relative to a particular state of scientific knowledge. Therefore, in practice empirical equivalence is always a feature that is relative to a certain state of scientific development Laudan and Leplin (1991, p. 454): "Any determination of the empirical consequence class of a theory must be relativized to a particular state of science. We infer that empirical equivalence itself must be so relativized, and, accordingly, that any finding of empirical equivalence is both contextual and defeasible". In other words, if two theories completely coincide in their predictions now, it does not follow at all that they will remain so and will be empirically equivalent in any strong sense. Scientific progress will probably break both the equivalence and the underdetermination.

A natural response to this manoeuvre is the observation that any new theory may again have countless empirically equivalent rivals so that the underdetermination problem has not been solved but only shifted. Kukla (1993, 1996) in fact argues that given any input theory, one can always formulate an alternative and empirically equivalent theory by applying simple transformations to the form of the theory. But this criticism is rejected by Laudan and Leplin, on the grounds that such artificially formed rival theories T^* are parasitic on T and fail to provide genuine alternatives; they do not satisfy "theoreticity criteria" that would make them serious contenders.

From Laudan's and Leplins rebuttal of Kukla's criticism Leplin and Laudan (1993) three such criteria can be extracted: non-superfluity, plausibility and testability.

A hypothesis is *superfluous* if it can be dispensed with without any loss of empirical content, i.e., if it does not contribute to the derivation of observational consequences.

Plausibility comes in because we should not consider theoretical schemes as rivals if they contain far-fetched and artificial characteristics Leplin and Laudan (1993, p. 14):

> Provisions that fly in the face of what we have good empirical reason to assume must claim some offsetting rationale if they are to be admitted as part of a theory. It would be different if the course of nature were known to exhibit such vast and mysterious ruptures or bifurcations as T' envisions, if natural law did not exhibit isometry, at least. One might then be willing to entertain wild, unexplained and unconfirmable scenarios as genuine possibilities. But the world is not known to be like that.

So in order to be considered as genuinely scientific, a hypothesis must possess a minimum degree of plausibility—which can be judged on the basis of a background of already well-confirmed empirical knowledge.

Finally, if rival theories contain superfluous additional hypotheses, in the sense that these hypotheses do not play a role in the entailment of observational consequences, these hypotheses will be untestable Leplin and Laudan (1993, p. 13):

> Because the purpose of theorizing is, at least in part, to gain predictive control over the subject matter under investigation, a theory must, at least in principle, be open to test. A 'propositional structure' that is not even in principle confirmable, that could not logically be an object of epistemic evaluation, is not a theory; for it could not in principle impart understanding nor advance practical interests.

The second part of Laudan and Leplins argument is directed against the premise that only the confirmation of observational statements that are derivable from a theory can count as empirical support. As they argue Laudan and Leplin (1991, p. 460), "significant evidential support may be provided a theory by results that are not empirical consequences of the theory". Accordingly, "the relative degree of evidential support for theories is not fixed by their empirical equivalence". Laudan and Leplin illustrate their point by the following scheme of empirical confirmation Laudan and Leplin (1991, pp. 461–462): if a theory T entails two logically independent theoretical hypotheses H_1 and H_2, and if in turn these hypotheses entail the classes of observational consequences E_1 and E_2, respectively, then the truth of any member of E_1 will support H_1 and also H_2, even though H_2 does not entail any statement in E_1 (the same holds, mutatis mutandis, for the truth of the statements in E_2). This demonstrates that the class of the observational consequences entailed by a theory does not coincide with the class of observational statements that can support that theory. So even if two theories were to prove empirically equivalent by themselves, it could well be that one and not the other is indirectly supported via a relation with other empirically supported theories or with a body of empirical background knowledge. This would solve the underdetermination problem.

When we take stock at this point, we see that Laudan and Leplin have pointed out that empirical equivalence, if it occurs in practice, does not need to be of a strong kind—it need not be permanent. Moreover, they have argued that not any arbitrary formal concoction should be regarded as a genuine rival to a scientific theory; there are certain requirements of scientific respectability that have to be satisfied. The three requirements that they list all relate to the way a theory connects to empirical evidence, and in this sense are meant to be objective. Finally, Laudan and Leplin have called attention to the fact that even if there were serious cases of persistent empirical equivalence, this would not necessarily mean that no evidence-based choice between them could be made.

However, the critical remark can be made that Laudan and Leplin have not actually *disproved* the possibility of respectable rival theories that are empirically equivalent, nor demonstrated that the progress of science is *bound* to break any such equivalence.

3 Quantum Mechanics, Instrumentalism and Empirical Equivalence

It may strike strange to search for cases of empirical equivalence in quantum theory, of all scientific theories. Indeed, there is a well-entrenched account in the philosophy of science literature according to which quantum theory is the paradigm example of explicit instrumentalist thinking.[2] If this were accurate, there would be no point in being worried about empirical equivalence and underdetermination in this area of scientific research. Indeed, from a strict instrumentalist viewpoint the content and meaning of an empirical theory is fully fixed by its class of empirical consequences, so that two empirically equivalent theories by definition possess identical content and are the same. In this case there could be no philosophically interesting problem of theory choice, because the choice would be between two alternate conventions for writing down one and the same theory. Of course, no serious problem for scientific rationality arises if the choice for one *convention* over another is determined by pragmatic considerations or personal preferences.

However, the thesis that instrumentalism may be considered the "official philosophy" of quantum physics does not sit well with scientific practice. Of course, it *is* true that this practice focuses on predictions and the exploitation of new phenomena, especially in applied research. But this focus on empirical results should not be mistaken for a *philosophical position* implying that it does not make sense to inquire into

[2]For instance, Fine (1984) describes Heisenberg , Schrödinger and Bohr as philosophically aware and convinced instrumentalists—he calls Bohr "the arch-enemy of realism" Curd et al. (2013, p. 1203). According to Fine the founding fathers of quantum mechanics succeeded in establishing nonrealism as an official philosophical doctrine of modern physics, "part of what every graduate physicist learns and practices" Curd et al. (2013, p. 1200). As will be discussed in the main text, in the latter case Fine confuses pragmatic considerations with philosophical principles, and in the former case misconstrues the positions of the founding fathers.

mechanisms behind the phenomena or that it is inappropriate to wonder about the properties of the underlying physical reality. Quite the contrary, physicists have been fiercely debating such interpretational issues, especially in connection with effects that resist explanation by processes known from classical physics; and they have done so since the beginning days of quantum theory. The realization in the nineteen sixties that quantum systems exhibit features of holism or non-locality that are completely foreign to classical physics has given a boost to such discussions, and has led to many investigations in which the non-locality of submicroscopic systems like electrons and photons becomes visible.[3] The advent of quantum information theory during the last couple of decades, which makes use of non-locality and other typical quantum properties of matter, has given an additional impetus to foundational and interpretational research. Another example is the large research effort presently devoted to quantum gravity, which connects in a natural way with questions like whether properties of space and time on very small scales may "emerge" from the properties of quantum objects like "strings" (see, e.g., Dieks et al. (2015)).

This attention for the nature of physical reality is not a new development: already the founding fathers of quantum mechanics were wrestling with the question of how quantum reality should be described and understood. It is true that Heisenberg in his ground-breaking papers on the new quantum theory announced that he would only allow quantities that were open to observation—but this should not be interpreted as a basic resolution in favor of instrumentalism. Rather, at the time the breakthrough to quantum mechanics took place, it had become clear that the classical world picture was deeply inadequate for the explanation of atomic phenomena and that one could not rely on the actual existence of things like "the orbit of an electron in an atom". Heisenberg therefore took the methodological decision to start anew, almost from scratch, by allowing only concepts whose applicability could nearly directly be verified by observation. It is a serious misunderstanding to think that by this he confessed to philosophical instrumentalism: Heisenberg spoke freely about submicroscopic entities like electrons and atoms, although taking into account that their properties were still obscure and should certainly not be thought of as classical. In fact, Heisenberg devised thought experiments involving such quantum objects (like his famous X-ray microscope). This by itself already shows a conflict with the idea of instrumentalism.

Bohr attempted in a more focused and systematic way than Heisenberg to characterize salient features of the new quantum domain. For this purpose he developed his notorious doctrine of complementarity, whose central idea is that quantum objects cannot possess properties like position and momentum at the same time; according to Bohr such properties are only well-defined if the quantum system interacts with quite specific environments. In particular, in a context in which a particle interacts with a measuring device that is sensitive to position, "position" becomes well-defined; and *mutatis mutandis* for momentum. Bohr has frequently been criticized for being

[3]In 1964 Bell (1964) showed that correlated submicroscopic quantum systems display behavior that is incompatible with a classical "local-realistic" description of these systems; this incompatibility with the classical world picture has since been verified in many experiments.

obscure in his explanations of complementarity and in how he proposed that properties should be attributed to quantum particles like electrons—sometimes they can be thought of as point masses and then again as if they were waves (e.g., Cushing (1994), Beller (1999)). But this very criticism testifies to the fact that Bohr did not propose a simple instrumentalist reading of quantum mechanics. Indeed, what could be simpler than explaining the instrumentalist position that the formalism of quantum mechanics should not be taken as possessing physical meaning and that only empirical predictions possess "cash value"?[4]

The present-day situation is that there exist two versions of the quantum formalism, which differ slightly in their dynamics and the mathematical representation thereof. Traditionally, a special dynamical principle has been assumed for measurement interactions: the core idea is that in a measurement the wavefunction suddenly "collapses" in order to represent the single outcome that materializes. Indeed, before the measurement there generally are many outcomes possible, each with its own probability (the wavefunction predicts only possible values and probabilities for measurement results, this is the notorious indeterminism of quantum theory). After the measurement only one of these possibilities has actually been realized, and in collapse versions of quantum mechanics one requires the wavefunction to represent this unique actuality. This then motivates the collapse of the wavefunction.

A more modern alternate approach is to treat all interactions, whether measurements or not, in the same way. This means that in all cases an evolution equation like the Schrödinger equation is applied. This has the consequence that the evolution of the wavefunction is always continuous, so that no collapses can happen. Conceptually speaking, all possibilities that were present before the measurement are still represented in the post-measurement wavefunction, but now including the result indicated by the pointer of the measuring device. If before the measurement there was a probability p for a certain outcome, the non-collapse approach says that after the measurement there is the same probability p for the device being in the state indicating the relevant outcome.

These two theoretical schemes (collapse and non-collapse) make the same predictions in all ordinary circumstances, but could in principle give rise to empirical differences in very peculiar, not yet practically realizable, situations.[5] We are thus dealing with two theoretical schemes that are not empirically equivalent in the strong sense, and an empirical decision between them should be possible in delicate experiments (although not yet in present-day practice).

---

[4]See Dieks (2017), and references contained therein, for an extensive discussion of Bohr's philosophy of quantum mechanics and the way in which it is realist rather than instrumentalist.

[5]The difference is that in non-collapse interpretations "superpositions" remain intact, whereas these superpositions disappear during measurements if there are collapses. This difference is detectable in principle, and in fact recent progress has made it possible to detect superpositions of quite big systems.

These two versions of the theory, collapse and non-collapse quantum mechanics, have their own classes of interpretations. We shall here concentrate on non-collapse interpretations.[6]

4 Realist Interpretations of Non-collapse Quantum Mechanics

It is sometimes thought, in line with the earlier-mentioned notion that quantum physics is an inherently instrumentalist field of research, that there are no realist interpretations of quantum mechanics. Part of the background of this idea may also be attributable to the literature surrounding the so-called Bell inequalities Bell (1964); the violation of these inequalities as verified in many experiments of the last decades demonstrates that no local-realistic interpretations of quantum mechanics are empirically adequate. Obviously, however, the latter conclusion does not rule out realism altogether, but shows that the quantum world must possess a distinctly non-classical character: it cannot be *local*, in the sense of consisting of localized objects that interact with each other via forces propagating at subluminal speeds.

In fact there are quite some realist interpretations of non-collapse quantum mechanics: hidden variables interpretations à la Bohm (1952), Goldstein (2016), many worlds interpretations Vaidman (2016); Wallace (2012), modal interpretations Dieks and Vermaas (1998), Bub (1997), Lombardi and Dieks (2016) and consistent histories interpretations Griffiths (2014) are only the more obvious ones. Whether the Copenhagen interpretation of Bohr belongs to this class of non-collapse realist interpretations is more controversial Dieks (2017), Faye (2014), as is the question of the status of versions of the Copenhagen interpretation proposed by others, like Heisenberg.

In most interpretations the mathematical structure of quantum mechanics is construed as a description of one single physical world, in which we find ourselves, but the many worlds interpretation is notorious for its assumption that in a measurement interaction all possible measurement results are actually realized in non-interacting different worlds—so that in a measurement interaction the world can be imagined to branch into many copies that differ from each other in the result of the measurement.

The just-mentioned difference (many versus one world) is already well-suited to illustrate how different interpretations of the quantum mechanical formalism may agree in their predictions of measurement results and still give rise to very different pictures of what the universe is like. In the many worlds interpretation *all* possible

[6]This formalism is the more consistent one in the sense that only standard quantum evolution is assumed and no special status is attributed to measurements. It seems that recent experimental results, which—as just said—verify that superpositions of semi-macroscopic states are possible, inductively support the hypothesis that standard quantum evolution, of the Schrödinger equation type without collapses, is generally valid. The view that "collapses" should be seen as "bookkeeping devices" rather than as physical processes is therefore gaining ground. However, the matter remains controversial.

outcomes of a measurement are realized when a measurement is performed; but *each one* is realized in only one world. Because the epistemic access of observers is confined to a single world, each observer will empirically find only one single outcome. Of course, because of the branching that took place in the measurement, there is a copy of the observer in all other worlds, and each one of these "clones" will find a different outcome—unbeknownst to our original observer. The single worlds by themselves are governed by the usual laws of quantum mechanics, and each observer-copy will only make contact with his own world.

Modal interpretations, by contrast, are based on the idea that there exists only one actual world. This world has a quantum character and therefore should be described by means of non-classical properties; the task is to define these properties in such a way that on the level of macroscopic observations the standard quantum mechanical predictions are reproduced. As in the many worlds interpretation, these standard predictions are not supposed to appear as subjective impressions of observers but are realized as objective properties of measuring instruments (pointers indicating a certain position on a dial, etc.). These macroscopic properties are part of a network of properties that extends also to the micro-realm, so that all physical systems at all times possess well-defined albeit nonclassical physical characteristics. It is for this reason that the scheme may be called realist. The exact details of the property assignment depend on the version of the modal scheme that one considers Bub (1997), Lombardi and Dieks (2016). One well-known and much discussed example of the modal ideas[7] is the hidden-variables theory of Bohm (1952), in which all physical systems are assigned definite *positions* at all times, and in which the wave function is used to define a probability distribution over these positions.

It is instructive to look at Bohm's theory in some detail to illustrate why, and to what extent, this theory can be said to reproduce the standard predictions of quantum mechanics. According to Bohm, quantum objects like electrons are localized particles that at each instant of time possess a definite spatial position, and therefore also a definite velocity. Here we already recognize that the Bohmian world is drastically different from the world pictured in other quantum accounts: most of these other accounts abide with Bohr's complementarity, and Heisenberg's uncertainty relations, according to which a quantum system cannot possess a precise value for both velocity and position. In addition to the particles the Bohm'scheme works with the wavefunction, whose evolution is always governed by the Schrödinger equation, in exactly the same way as in all non-collapse schemes. This wavefunction, according to Bohm, defines the probability that a particle will find itself at a certain position, via the equation $P(x) = |\Psi(x)|^2$; $P(x)$ is the probability that the particle is at position x and $\Psi(x)$ is the value of the wavefunction at that spot.[8] Particles will move in

[7]The observation that Bohm's interpretation can be seen as a special case of the modal interpretation is due to Bub (1997).

[8]Note the difference with the quantum textbook rule that $|\Psi(x)|^2$ is the probability of *finding* a particle at x, in an experiment: here there is no mention of any measurement and the standard realist assumption is made that the particle just *is* somewhere, even if it is not observed. Whether $\Psi(x)$ should be considered a real physical field guiding particles or rather a mathematical quantity

such a way that the probability law $P(x) = |\Psi(x)|^2$ will be satisfied at all times.[9] The wavefunction evolves in the standard Schrödinger way, which means that in a measurement an entangled state will be formed of object plus measuring device, including the part of the device that eventually indicates the outcome (the pointer). As in all non-collapse interpretations, the final state will be a superposition in which $|\Psi|^2$ will have non-vanishing values for all position configurations of the combination of particle and device-plus-pointer, in which the pointer indicates one of the possible outcomes. As we have seen, in the many-worlds interpretation this would be taken to mean that all these outcomes are actually realized, albeit in different worlds—but here, in Bohm's theory, there is only *one* incoming particle with which the device interacts, and at the end of the interaction this particle can only find itself in *one* of the "wavepackets" corresponding to the different possible pointer positions. At the end of the measurement we therefore have a situation in which the pointer has taken on exactly one of its possible position values, with a probability given by the usual quantum expression $|\Psi(x)|^2$—exactly as the standard rules of quantum mechanics tell us.

This illustrates the way in which the Bohm scheme reproduces the standard predictions of quantum mechanics. The crux here, as in all non-collapse schemes, is that the evolution of the wavefunction is taken over, in unmodified form, from the standard quantum formalism. So $\Psi(x)$ and the probabilities $|\Psi(x)|^2$ are always the same as in the standard formalism. But the associated picture is quite different from the usual one: in Bohm's theory all objects always possess definite positions and follow well-defined trajectories through space. The probabilities in Bohmian quantum mechanics represent our ignorance about the actual positions of particles; but this ignorance also reflects an objective "chaos" in the world that makes it impossible to improve on the probabilistic quantum predictions. Indeed, according to Bohm there is a basic lack of controllability and repeatability in the world, which makes the use of probabilities unavoidable. For example, when we repeat a measurement of the kind just described, we cannot arrange for it that the incoming particle starts at the same position as in the previous run of the experiment. In fact, in many repetitions of the experiment the initial particle positions will be distributed according to the values of $|\Psi(x)|^2$. So although in each run of the experiment everything happens according to a deterministic law of motion, which fully determines the outcome given the initial conditions, we cannot fully predict that outcome because the initial conditions themselves are uncontrollable and only open to a probabilistic treatment.

(Footnote 8 continued)
occurring in the particle laws of motion is a matter of debate between different adherents of the Bohm theory, which gives rise to a bifurcation of Bohmian world pictures.

[9]In order to guarantee this, the motion of the particles is posited to obey the deterministic "guidance equation" $\vec{p} = \nabla S(x)$, with $\vec{p}$ the particle's momentum and $\nabla S(x)$ the gradient of the phase $S(x)$ of the (complex-valued) wavefunction $\Psi(x)$.

5 Empirical Equivalence of Interpretations of Quantum Mechanics

To what extent are these different interpretations empirically equivalent? All the non-collapse interpretations that we have been considering share the standard mathematical formalism, with a wavefunction governed by the same evolution equation. They differ in what this wavefunction is supposed to stand for in physical reality. However, they have all been constructed with the aim to make definite at the macroscopic level exactly those quantities that are relevant for the observations we actually can make, so as to make their predictions for outcomes of experiments and the associated probabilities the same.[10] At the same time, as we have seen, these different interpretations say very different things about the world *behind* the scenes of measurement. So we appear to be facing typical instances of empirical equivalence. But is this an empirical equivalence in the weak sense of a present and probably temporary lack of discriminating evidence, a gap that most probably will soon be filled; or is it an equivalence in the very strong sense of equal predictions no matter what future experiments might be performed?

The very fact that the description of the physical world is different in these interpretations already shows that they cannot be empirically equivalent in the strongest possible sense, with regard to all *logically* possible observations. Indeed, one might imagine scenarios in which human beings develop new senses by means of which they are able to immediately "see" what physical reality is like, on all levels of complexity and on all scales. Such new sensory powers would open up the possibility of directly deciding between different hypotheses about physical reality, even if these hypotheses agree with respect to what is the case on the usual macroscopic scale. For example, if we developed an awareness of the values of (sub)microscopic momenta, we could immediately observe whether the Bohm theory is true or not: we could simply check whether particles always possess the precise values for their momenta the Bohm theory prescribes, and we could see whether they indeed follow trajectories of the predicted sort. Similarly, if we evolved the ability to make contact with other worlds, we could verify directly whether the many worlds interpretation is right.

But this argument for the drastic defeasibility of all empirical equivalence, on the basis of logical possibility, is obviously irrelevant to scientific practice.[11] Empirical equivalence by definition refers to equal status with respect to observations by us, human beings who as a physical fact have a restricted repertoire of ways of interacting with the surrounding world. These possible interactions are physical, and therefore amenable to physical description and explanation; our best physical theories should be able to deal with them. For example, we communicate via electromagnetic radiation (light) and sound, and our theories describe the impact these signals have on our sensory organs, conceived as measuring devices. The best theory in principle to

[10]This is a criterion of adequacy: If a proposed interpretation is not able to reproduce our "ordinary world of experience", it has certainly to be rejected.

[11]Compare van Fraassen's arguments Fraassen (1980, p. 17) against Maxwell's thesis that nothing is "unobservable in principle" Maxwell (1962).

deal with these things is quantum mechanics, of which, as we have seen, there are diverse versions. But these different versions agree with each other on the macroscopic level, which is the level on which we live and function. So if the different versions of quantum mechanics satisfy the adequacy requirement that they all be able to recover our classical world, they all must say the same with regard to the sensitivity of our sense organs. Moreover, because the relevant facts of sensory physiology are largely independent of quantum mechanics and were mostly established in the context of classical research already quite some time ago, they can be expected to be robust and remain the same even if quantum theory is going to be superseded by a better theory. It follows that in order to create new observational possibilities that go beyond the macroscopic surface of physical reality, new principles of physical interaction or a change in the formalism of quantum mechanics is needed, or both. But such changes would mean that *all* current versions of quantum mechanics, with their shared macroscopic predictions, are wrong.

We may conclude that the different interpretations of quantum mechanics that we have been considering are empirically equivalent in the sense that no future measurements that make use of the well-established interactions on which experimental practice is based will be able to destroy it. Of course, it is logically possible that macroscopic regularities of the past will discontinue at some future point in time; and that novel laws will come to replace them. But such contingencies will spell trouble for *all* interpretations.

This gives us an empirical equivalence that is pretty strong. However, we have to proceed with some caution. As mentioned above, it is an adequacy requirement for interpretations that they provide us with the usual classical characterization of the macroscopic world. But this will only be achievable to some degree of approximation, of which it is usually assumed that it is irrelevant for observation. For example, although the Bohm theory tells us that all objects always possess definite position and velocities, other non-collapse interpretations will characterize objects with quantum properties that obey the uncertainty relations, so that there will be a finite spread ("latitude", Bohr called it) in the values of the quantities characterizing them. In the latter case the classical world, with definite positions, momenta, etc., will not be completely recovered. However, the spreads in question will be enormously small once we are dealing with macroscopic bodies with appreciable masses, and this usually is taken as sufficient justification for discarding them ("in the classical limit"). But in principle these quantum spreads never vanish completely and one could imagine that however tiny they are, they could play a role in some situation. Thinking back of our characterization of the human sensory organs as physical detection instruments, this would require that it could make a difference for our awareness of what we observe whether molecules in our brains, neurons, etc., have exactly definite positions à la Bohm or positions with a tremendously tiny "vagueness" or spread as predicted by other interpretations. In view of what we know about the workings of our sensory organs—most of it is chemistry at fairly high temperatures—the latter hypothesis seems very implausible. The situation resembles what we discussed a moment ago, in connection with novel observational possibilities. There is presently no theoretical clue what such observational differences could be, nor any empirical evidence for their existence. So again, we are dealing with a speculation that is remote from scientific practice.

To sum up, the different interpretations of quantum mechanics that we have sketched can be taken to be empirically equivalent in a strong sense: no future macroscopic observations can be expected to be able to distinguish between them. This is so in spite of logically possible scenarios according to which we might develop direct veridical insights into the (sub)microscopic world and in spite of the fact that quantum theory will only be able to recover the classical macroscopic world of experience to a high degree of approximation. In the former case the possibilities are merely logical and not physical, in the latter we have to expect that the differences are much too small to be significant for observations.

6 Comparison with the Arguments of Laudan and Leplin

Empirical equivalence thus does occur in the practice of modern science and is not a mere harebrained philosophical scheme. How does this observation compare to the arguments we rehearsed in Sect. 2, to the effect that empirical equivalence should *not* be expected to play a serious role in scientific practice?

The first criticism Laudan and Leplin brought against the significance of empirical equivalence was that it necessarily is defeasible: because new experimental techniques are constantly being developed, not yet verified empirical differences between theories will come to light sooner or later. We have already dealt with this objection in the previous section: in the quantum case, the discovery of such empirical distinctions is not plausible and is not expected to occur. Therefore, a strong form of empirical equivalence obtains here.

This may raise the question of whether we are dealing with different theories at all; isn't the designation "interpretations of quantum mechanics" by itself already a sign that we are facing *one single* theory (quantum mechanics), although in multiple formulations? This reaction, however, fails to appreciate that the different interpretations draw very different pictures of the submicroscopic world. The problem of empirical underdetermination of theory choice is therefore certainly present here, in the form of the empirical underdetermination of the choice of one picture of the world over another—which is the problematic kind of underdetermination in the context of the realism debate.

In cases of persistent empirical equivalence, as we are facing, Laudan and Leplin suggest that the contending schemes will likely be artificial variations on a standard form of the theory; and that these artificial rivals will not satisfy certain standards of "theoreticity" so that they can be discarded. These theoreticity criteria are of two sorts: artificial pseudo-theories will not pass the test of testability because they contain empirically unverifiable parts that do not contribute to the derivation of empirical results; and/or they will contain hypotheses that are highly implausible given our empirical background knowledge of what the physical world is like.

In our case of different quantum interpretations it not clear whether it can be said that new physical hypotheses are *added* to "standard quantum mechanics" in order to create the various interpretative schemes. The mathematical formalism that is vital for the derivation of empirical results is the same in all interpretations, and it is first of all the *physical meaning* of this formalism that is different. It is true, though, that depending on the nature of the interpretation additional symbols and equations will be necessary to make the theoretical scheme complete. For example, in the Bohm theory symbols are introduced for the definite particle positions and momenta that are central in this theory, $\vec{x}_i$ and $\vec{p}_i$, and these symbols are obviously not needed in interpretations that do without Bohmian particle properties—although other symbols will be necessary to denote the "elements of reality" of these rival interpretations. Staying with the Bohm theory, we need a dynamics for the particles in order to make the total system, including the Schrödinger equation and the wavefunction, complete and self-consistent; one way of implementing this dynamics is by specifying the "guidance equation" we mentioned before, $\vec{p} = \nabla S$.

The particle properties in the Bohm theory, and the equation defining their evolution, may be said to embody physical hypotheses but these are not gratuitous extensions or embellishments added to the physics of other interpretations or to "quantum mechanics proper". They play an integral part in the physics of the Bohm scheme. Moreover, they do not at all appear artificial or implausible given our background knowledge. Indeed, the notion of a particle with a well-defined trajectory is part and parcel of the usual conceptual toolkit of physicists and plays a major role in explicating the meaning of causality in classical physics. Accordingly, adherents of the Bohm theory often argue that their scheme succeeds in providing a clear causal picture of how experimental results come about, whereas other interpretations propose more unusual and rather nonintuitive pictures of reality in which the causal connections remain unclear. It can therefore not be maintained that the particles and their trajectories are a superfluous part of the Bohmian scheme and do not contribute to the derivation of empirical results; nor that there is a conflict with our empirical background knowledge. So the Bohmian hypotheses score adequately with respect to the criteria of integration and testability.

This is not to say that Bohmian quantum mechanics is clearly superior to other interpretations. In fact, the history of quantum theory has taught us that the classical world picture of localized particles and local fields is problematic—think of the well-verified phenomenon of particles that "interfere with themselves", as in the double-slit experiment. In Bohm's theory these unexpected non-classical phenomena are handled by the introduction of a couple of unorthodox features of the particle dynamics; in particular, the dynamics is non-local, so that changes in the position of one particle will be immediately felt by all other particles (cf. what was said in note 3), and there is the nonclassical guidance by the Ψ field. Other interpretations obviously also have to accommodate interference, wave-particle duality, non-locality and similar symptoms of non-classicality, and claim to do so in a more natural and simpler way by renouncing the concept of a classical particle from the outset, and setting themselves the task of explaining the appearance of an "almost classical" world as the macroscopic limiting case of a description that overall is typically quantum. In all these interpretations the postulated ontology plays a direct role in the explanation of empirical results, so the Laudan and Leplin criterion of no superfluity does not disqualify any of them.

It should be recognized that all interpretations have their own unusual features, which make them all distinctly non-classical in their own way. As we have seen, there is the strange non-locality of interactions in the Bohm theory,[12] whereas there is the seemingly metaphysically extravagant multiplicity of worlds in the many worlds interpretation. However, the latter interpretation does not need to introduce particles with a definite position, nor a guidance equation, and therefore can be said to reflect in a more direct and simpler way the quantum formalism with its linear Schrödinger evolution; the multiplicity of worlds mimics directly the multiplicity of terms in the entangled superpositions that—because of linearity—follow from the Schrödinger equation in the case of interactions.

The methodological desideratum that superfluity, artificiality and testability are to be avoided is thus too ambiguous and weak to come to an objective decision about which interpretation is to be preferred. But there is also the second prong of Laudan's and Leplin's analysis, according to which different empirically equivalent theories may well be differentially supported by empirical evidence because they relate differently to other theories or background knowledge. If one theory T_2, of a pair of empirically equivalent theories T_1 and T_2, can be subsumed under a more encompassing theory T, it may happen that evidence E comes in that supports T, although it does not directly support T_2 (in the sense that E is not a prediction of T_2 and may even pertain to a domain of reality T_2 does not speak about).

In order to apply this line of reasoning to quantum mechanics, we should find a more general theory T into which the quantum formalism can be imbedded, in such a way that the imbedding goes well for one (or some) interpretations and does not succeed, or does not go too well, for other interpretations. Empirical evidence for T will then count against interpretations that cannot easily be imbedded. Now, quantum mechanics can be seen as a special case of a more general quantum theory, namely quantum field theory. In quantum field theory more complicated physical interactions can be handled than in quantum mechanics, in particular processes in which quantum systems are created and annihilated.

However, the use of quantum field theory as the theory T in the just-explained scheme of empirical support does not lead to unambiguous results. The reason is that quantum field theory has different interpretations *itself*, analogous to our earlier interpretations of quantum mechanics. That is, also here there is a core mathematical quantum field formalism that is shared by different interpretations and to which different "rules of correspondence", which stipulate the physical meaning of the mathematical symbols, are applied. Where needed, new symbols may be added to denote quantities that are specific to an interpretation (like $\vec{x}$ and $\vec{p}$ for definite particle position and momentum). In this way, for example, a Bohmian version of

[12]There is also the odd double role of the wavefunction, which on the one hand specifies the particle dynamics via the guidance equation and on the other hand determines the probability distribution of the particles.

quantum field can be developed (see, e.g., Dürr et al. (2004)); other non-collapse interpretations can similarly be extended. The condition that some of the empirically equivalent theories fit better with more general theories than others is therefore not fulfilled—rather, the original question about which theory to prefer is repeated on the more general level, without an unambiguous objective resolution.

The story does not end here, for one may also consider the compatibility of the different interpretations of quantum mechanics with general space-time features, as specified by relativity theory. Here there is at least a *prima facie* problem for the Bohmindex theory, because this theory requires a notion of simultaneity (to make sense of *instantaneous* interactions) that is at odds with the standard interpretation of special relativity. However, the manoeuvre of constructing different empirically equivalent interpretations at a more general theoretical level can be repeated: it is possible to introduce a minimum of additional structure in relativity and to accommodate the Bohmian scheme in the resulting enriched relativistic spacetime. Importantly, this can be done without changing the empirical content of relativity theory (actually, various ways of doing this have been proposed, see e.g. Goldstein (2016)). So again the problem of choice can be elevated to the level of the more general theory, where there is once more empirical equivalence. The situation with respect to the Bohm interpretation remains ambiguous in this case: on the one hand, it is conventional wisdom in special relativity theory that a notion of global simultaneity is superfluous, which militates against Bohm; on the other hand this common wisdom derives from non-quantum considerations that might well be in need of revision in the face of quantum non-locality. As the Bell inequalities have shown, there is an element of non-locality in quantum mechanics regardless of its interpretation, and one might therefore argue (and many in fact argue this way!) that the combination of relativity theory and quantum theory requires an adaptation in the structure of relativistic spacetime anyway.

Summing up, it seems safe to say that no argument has been presented that singles out one of the various interpretations of quantum mechanics as objectively better than the others. There is no direct experimental evidence in the offing that might be able to enforce a decision between the alternatives; and the various more theoretical arguments—Laudan and Leplin style—about superfluity, testability and support do not convincingly identify one option as superior either.

7 Conclusion: Underdetermination in Quantum Mechanics

We are thus facing a real-life situation of empirical equivalence that is robust enough to make hopes that the decision procedures defended in the philosophical literature will before too long lead to an objectively best choice are utterly unrealistic. This diagnosis turns out to be in conformity with the attitude of the scientific community. On the one hand, physicists have no qualms whatsoever to speak—in general terms—about the unobservable parts of physical reality. Indeed, it is hard even to imagine what the practice of quantum physics without continuous reference to elementary particles, fields, strings, membranes, etc., would look like. These terms are not meant as mere mental tools without reference, but as denoting actual parts of reality. However, on the other hand, there is also a wide-spread conviction that very different views can be upheld about the precise details of the associated physical ontology. Are quantum objects living in one unique world, the only one in existence, or do they have doppelgängers in other worlds? Difficult to be sure! Are quantum particles entities à la Bohr, with contextual properties obeying complementarity and uncertainty relations? There is certainly something to be said for this option. But then again, is it not possible that the Bohm picture is right, so that particles possess definite locations and velocities, while interacting non-locally? Probably there are less enthusiasts for this alternative, but its scientific *tenability* is generally accepted. It is true that many find the non-local Bohmian interactions strange, but then, quantum mechanics displays strange features of non-locality anyway, and there seems to be a consensus that it is a matter of personal preference rather than of an objectively justified scientific stand which side one chooses.[13]

A number of diverse viewpoints about the nature of the quantum world thus coexist, each one with its own limited circle of convinced devotees. Many more physicists know about at least a number of these interpretations, without committing themselves to any of them—they refer in realist fashion to quantum objects, but refrain from becoming specific about their exact nature. Not infrequently, physicists make use of one picture or another depending on the kind of question that is being discussed. For example, sometimes it is easy to think of localized particles following trajectories while influencing each other superluminally, then again it seems to provide more insight to think of a system of "particles" as one undivided whole.

This situation should not be mistaken for one of "anything goes". In fact, the uncontested and shared core of empirical consequences of quantum mechanics imposes conditions on possible interpretations that make each of them necessarily distinctly nonclassical. Although the predictions of quantum mechanics do not fix a unique description of the physical world, they do give us a delineated field of possibilities. Indeed, the mathematical formalism and its empirical predictions exclude interpretations that, e.g., do not in some way reflect non-locality and contextuality. This has been shown in a long tradition of so-called "no-go theorems" harking back to the work of von Neumann in the early nineteen thirties (Von Neumann (1932), see also Dieks (2017)), in which the results of Bell (1964) and Kochen and Specker (1967) are some of the highlights. These theorems identify a number of characteristic quantum features that set the theory apart from its classical counterparts, and are differently represented in different interpretations. For example, in Bohm's theory Bell's theorem—excluding local realism—has the repercussion that interactions must be non-local and instantaneous, but in other

[13]In view of the perceived impossibility of an objective choice between the alternative interpretations, the proposal has recently gained popularity to be non-committal, by speaking only about the "information" that is present in the basic structure of the world; see, e.g., Bub (2016) and references contained therein. Opponents challenge this information-based approach in the foundations of quantum mechanics by asking what the information is information *about*—an implicit acknowledgment of realism.

interpretations (Copenhagen, modal interpretations) the theorem's conclusion is reflected in a feature of holism according to which properties of a composite system cannot be reduced to the properties of its components. In this way the various interpretations open up different perspectives, within different conceptual frameworks, of what it means to be a quantum theory.

This finally brings us back to the questions of rationality and objectivity of theory choice. As we have seen in Sect. 2, Laudan and Leplin warned against the dangers of empirical equivalence and underdetermination, which in their eyes are odious doctrines that have "wreaked havoc" in the philosophy of science and have encouraged deplorable forms of pluralism of belief and practice. How can science be rational if no objective evidence-based choices can be made between different descriptions of physical reality? The practice of modern physics supplies an answer to this rhetorical question.

This practice shows us that there is no harm in the coexistence of a plurality of world pictures that all do equal justice to what is established in experiments. Different world views of this kind may all in their own way bring out salient features of what physical reality is like, each making use of its own conceptual framework. There is no need for an objective decision between such interpretative options and it is no problem that individual scientists may feel drawn more to one of them than to another, on subjective grounds. Indeed, congenial ideas facilitate reasoning and make it easier to fathom the consequences of a theory and to achieve understanding—it is an an unnecessary impediment for scientific understanding and rationality to require one unique conceptual framework Regt and Dieks (2005). Rather than hampering rational discourse, the competition of different points of view furthers it.

References

Acuña, P. and Dieks, D. (2014). Another look at empirical equivalence and underdetermination of theory choice. European Journal for the Philosophy of Science, 4, 153–180. doi:10.1007/s13194-013-0080-3

Bell, J. S. (1964). On the Einstein Podolsky Rosen Paradox. *Physics*, 1, 195–200.

Beller, M. (1999). *Quantum Dialogue: The Making of a Revolution*. Chicago: University of Chicago Press.

Bohm, D. (1952). A Suggested Interpretation of the Quantum Theory in Terms of 'Hidden' Variables, I, II. *Physical Review*, 85, 166–179; 180–193.

Bub, J. (1997). *Interpreting the Quantum World*. Cambridge: Cambridge University Press.

Bub, J. (2016). *Bananaworld: Quantum Mechanics for Primates*. Oxford: Oxford University Press.

Curd, C, Cover J.A. and Pincock C. (eds.) (2013). *Philosophy of Science—The Central Issues*. Second edition. New York: W.W. Norton & Company.

Cushing, J.T. (1994). *Quantum Mechanics: Historical Contingency and the Copenhagen Hegemony*. Chicago: University of Chicago Press.

De Regt, H.W. and Dieks, D. (2005). A Contextual Approach to Scientific Understanding. *Synthese* 144, 137–170.

Dieks, D. and Vermaas P. (eds.) (1998). *The Modal Interpretation of Quantum Mechanics*. Dordrecht: Kluwer Academic Publishers.

Dieks, D., van Dongen, J. and de Haro, S. (2015). Emergence in holographic scenarios for gravity. *Studies in History and Philosophy of Modern Physics*, 52, 203–216.

Dieks, D. (2017). Niels Bohr and the Formalism of Quantum Mechanics. In *Niels Bohr and Philosophy of Physics: Twenty-First Century Perspectives*, edited by H. Folse and J. Faye. London: Bloomsburg Publishing.

Dieks, D. (2017). Von Neumann's Impossibility Proof: Mathematics in the Service of Rhetorics. *Studies in the History and Philosophy of Modern Physics*, to appear.

Dürr, D., Goldstein, S., Tumulka, R., and Zanghi, N. (2004). Bohmian Mechanics and Quantum Field Theory. *Physical Review Letters*, 93: 090402: 1–4.

Faye, J. (2014). Copenhagen Interpretation of Quantum Mechanics. *The Stanford Encyclopedia of Philosophy* (Fall 2014 Edition), Edward N. Zalta (ed.), URL = http://plato.stanford.edu/archives/fall2014/entries/qm-copenhagen/

Fine, A. (1984). The Natural Ontological Attitude. In J. Leplin (ed.), *Scientific Realism*. Berkeley: University of California Press, 83–107. Reprinted in Curd et al., (2013), 1191–1214.

Goldstein, S. (2016). Bohmian Mechanics. *The Stanford Encyclopedia of Philosophy* (Fall 2016 Edition), Edward N. Zalta (ed.), URL = http://plato.stanford.edu/archives/fall2016/entries/qm-bohm/

Griffiths, R.B. (2014). The Consistent Histories Approach to Quantum Mechanics. *The Stanford Encyclopedia of Philosophy* (Fall 2014 Edition), Edward N. Zalta (ed.), URL = http://plato.stanford.edu/archives/fall2014/entries/qm-consistent-histories/

Kochen, S. and Specker, E.P. (1967). The Problem of Hidden Variables in Quantum Mechanics. *Journal of Mathematics and Mechanics* 17, 59–87.

Kukla, A. (1993). Laudan, Leplin, and Underdetermination. *Analysis*, 53, 1–7.

Kukla, A. (1996). Does Every Theory Have Empirically Equivalent Rivals? *Erkenntnis*, 44, 137–166.

Laudan, L. (1990). Demystifying Underdetermination, in C. Wade Savage (ed.), *Scientific Theories*, vol. 14, *Minnesota Studies in the Philosophy of Science*. Minneapolis: University of Minnesota Press, 267–297. Reprinted in Curd et al., (2013), 288–321.

Laudan, L. and Leplin, J. (1991). Empirical Equivalence and Underdetermination. *The Journal of Philosophy*, 88, 449–472.

Leplin, J. and Laudan, L. (1993). Determination Underdeterred: reply to Kukla, *Analysis*, 53, 8–16.

Lombardi, O. and Dieks, D. (2016). Modal Interpretations of Quantum Mechanics. *The Stanford Encyclopedia of Philosophy* (Spring 2016 Edition), Edward N. Zalta (ed.), URL = http://plato.stanford.edu/archives/spr2016/entries/qm-modal/

Maxwell, G. (1962). The Ontological Status of Theoretical Entities. In H. Feigl and G. Maxwell (eds.), *Scientific Explanation, Space, and Time*, vol. 3, *Minnesota Studies in the Philosophy of Science*. Minneapolis: University of Minnesota Press, 3–15.

Musgrave, A. (1985). Realism versus Constructive Empiricism, in P.M. Churchland and C.A. Hooker (eds.), *Images of Science*. Chicago: Chicago University Press, 197–221. Reprinted in Curd et al., (2013), 1083–1108.

Vaidman, L. (2016). Many-Worlds Interpretation of Quantum Mechanics. *The Stanford Encyclopedia of Philosophy* (Fall 2016 Edition), Edward N. Zalta (ed.), URL = http://plato.stanford.edu/archives/fall2016/entries/qm-manyworlds/

Van Fraassen, B. (1980). *The Scientific Image*. Oxford: Clarendon Press.

Von Neumann, J. (1932). *Mathematische Grundlagen der Quantenmechanik*. Berlin: Springer. English translation by Robert T. Beyer: Von Neumann, John (1955), *Mathematical Foundations of Quantum Mechanics*. Princeton: Princeton University Press.

Wallace, D. (2012). *The Emergent Multiverse*. Oxford: Oxford University Press.

Is Uniqueness of Reality Predicted by the Quantum Laws?

Roland Omnès

Abstract The standard interpretation of quantum mechanics takes for granted an impossibility of deriving "wave function collapse" from the Schrödinger equation. One raises an opposite possibility, which would make collapse one of the major predictions of this equation. This proposal is based on properties of "local entanglement", and existence of a strong incoherence in the quantum states of a macroscopic system, from local entanglement with fluctuations in its environment. These two consequences of Schrödinger's equation, inexpressible by quantum observables, are shown able to imply collapse. An exceptional significance of the collapse problem, together with a novelty wanting thorough justification in these results, make them proposed only as conjectures, with could however bring more harmony into this essential part of the philosophy of physics.

1 Introduction

Eighty year ago, Schrödinger (1935) and Einstein (1935) expressed deep worries about the foundation of quantum theory. The main difficulty was a contradiction, between a universal evolution of quantum systems under the Schrödinger equation and the uniqueness of data in measurements, often expressed as the problem of "wave function collapse". This question, which became widely known as the "Schrödinger's cat problem", is still present today and the literature on the subject is so vast that one can quote only an anthology on its history (Wheeler and Zurek 1983) and a recent review of its present state (Laloë 2012).

Many answers were attempted. Some essays tried to modify in depth the interpretation of quantum mechanics (de Broglie 1956; Bohm 1952; Everett 1957) or revised its foundations (Bell 1964; Adler 2004). Attempts were made also to complete its physical content by extraneous phenomena (Ghirardi et al. 1986), or questioned its exactness (Pearle 1976; Ghirardi et al. 1990). New conceptions of

R. Omnès (✉)
Laboratoire de Physique Théorique, Université de Paris XI, Orsay, France
e-mail: roomnes@wanadoo.fr

E. Agazzi (ed.), *Varieties of Scientific Realism*,
DOI 10.1007/978-3-319-51608-0_17

understanding were also proposed and the problem is now a topic of much philosophical research regarding foundations of knowledge and the nature of Reality (d'Espagnat 1995; Caves et al. 2002; Epperson and Zafiris 2013).

Many physicists are more pragmatic. They do not pay an extreme attention to questions regarding the nature of Reality. They consider often that these problems are only waiting for answer: Time will tell. The best argument for this attitude stands in the prodigious advances of quantum physics, since it carries this burden of ignorance: There is no doubt that the quantum laws are perfectly valid for all practical purposes, and this is enough for a science.

A few advances regarding parallel questions have occurred however and one can mention two of them: One knows why classical determinism is valid with high precision at macroscopic scales (Omnès 1994). The compatibility of quantum theory with standard logic has also ben shown by "consistent histories", which discard many past paradoxes in interpretation (Griffiths 2002). One may also notice that these results required in both cases an enlargement of the standard interpretation of quantum mechanics, somewhat encouraging the prospect that the same procedure could apply to the collapse problem.

The research to be reviewed here is concerned with this problem. It will be convenient for a beginning to recall the status of this question, according to the "standard" interpretation of quantum mechanics. This interpretation, which can be found in most textbooks, was arrested particularly in two great books, by Dirac (1930) and by Von Neumann (1932). Von Neumann, especially, made an axiom of the condition that every statement regarding quantum physics should be expressed in terms of observables. (One recalls that, in the mathematical framework of quantum theory, an "observable" is associated with a "self-adjoint" operator in an abstract Hilbert space, which represents all the possible states of a physical system. But one will try to avoid as much as possible such technicalities).

Bohr did not accept completely this standard interpretation (Bohr 2010). He stressed that another concept, usually called "wave function collapse", was unavoidable. One may describe it as a hidden physical phenomenon, through which a unique result emerges at macroscopic scales when an actual measurement is performed. There are variants (Laloë 2012; d'Espagnat 1995), but one may consider, from a philosophical standpoint, that this mysterious collapse holds the key for the place where macroscopic Reality reaches its obvious uniqueness, which is the foundation of every science.

The present work reports on recently discovered new openings in this question. Briefly expressed, it aims at showing that collapse exists as a physical property, obeys a specific mechanism and this mechanism is contained in the Schrödinger equation, predicted by it and predicting itself unique data in observations (for instance a unique position of a pointer on a dial). An essential point will be also that this mechanism of collapse cannot be expressed in any way by means of "observables", in the formal sense of this word. Nevertheless, one will find that this mechanism results from other kinds of properties, resulting themselves from the Schrödinger equation. The price, which one will have to pay, will be to give away the standard interpretation of quantum mechanics. This venerable and very useful interpretation will not be forsaken, fortunately, but will have to be left aside sometimes, and particularly when collapse is concerned. Some remarkable aspects of collapse, until then ignored, will also come out from this change in perspective.

One will try to describe this conjecture in general terms, avoiding technical details as much as possible. A truly rigorous theory would need of course to go much deeper, but one acknowledges that such a level is not yet reached, which makes one speak of conjectures. The results are however so internally consistent, when suggesting from their own structure answers to the problems that they raise, that one dares to entertain the hope that they contain at least a significant part of a future right answer, which many of us are pursuing.

Even so, these ideas cannot hold into a unique brilliant vision, as some previous proposals did (like Bohm's idea of a unique reality at a macroscopic level (Bohm 1952), or Everett's of a wave function of the whole universe (Everett 1957), for instance). They show on the contrary a complex self-organization, through which the uniqueness of macroscopic Reality comes out only as a synthesis.

The basic elements of this process and their mutual relations will be introduced here in the following order: (i) Description of a notion of local entanglement between two quantum systems. (ii) Derivation of the existence, in the quantum state of an open macroscopic system, of a specific kind of incoherence, due to its local entanglement with fluctuations in its environment. (iii) Identification of an elementary phenomenon of "slips in coherence", which are the microscopic causes of collapse. (iv) A resulting explicit mechanism of collapse.

I must warn readers that several of these results, which open or seem to open new domains, encounter technical or mathematical, problems, often themselves quite new. They were discovered recently and sometimes approached only by means of simplified models. Much more work would be necessary to assert them with full confidence (or perhaps to discard them). One mentions at the present place these limitations as a preliminary warning, but without intending to come back to them. The aim of the next sections of this paper will be on the contrary to get as much clarity as its author could achieve.

2 The Notion of Local Entanglement

One first recalls the concept of entanglement between two physical systems, which is essential in quantum physics. Briefly expressed, it shows that, when two physical systems interact, their two states can never return to their initial independence, and their wave functions remain mingled together by algebraic knots, which nothing can unravel.

Schrödinger said that this property is the one by which quantum mechanics differs most from classical mechanics. The essence of Schrödinger's famous "cat problem", was that the state of a cat and the state of a radioactive source, together enclosed in a box, came out in an entangled state, out of which nothing could take them.

Mathematically, entanglement means that a wave function involving the cat and the radioactive source is a sum of two distinct wave functions, the first one being a product of a wave function of the cat, alive, with a wave function of the radioactive source, still intact. The second function is also a product of two wave functions involving a dead cat and a decayed source.

The algebraic character of entanglement is obvious, since its expression holds into a sum of two products (of wave functions). This character is hard to "see" intuitively, but it stands as the fortress in which one will have to find a weak point.

A noticeable event occurred in 1972, when Eliot Lieb and Derek Robinson discovered new properties of entanglement (Lieb and Robinson 1972). These were local properties, contrary to all known other ones. They did not involve all the atoms in a system, whereas everything was involved together in the box where Schrödinger's cat was locked up. Lieb-Robinson's discovery occurred however in the field of spins lattices, which is of importance in statistical physics but rather far from the domain of research regarding collapse, where it went practically unnoticed. This is why, ignoring it, I rediscovered it much later (Omnès 2012).

Perhaps this indifference of the community of research on collapse, was due to a belief, which John Bell expressed clearly: He wrote, essentially what most people believed more or less, which was that the deep problem of uniqueness of Reality was certainly not accessible by means of the standard methods of physics, which are only valid "for all practical purposes" (Bell 1987).

Since then, a remarkable book by Serge Haroche and Jean-Michel Raimond reported many experiments, mostly in quantum optics, which made sure that collapse never occurs at a microscopic scale, but only at a macroscopic one (Haroche and Raimond 2006). But since macroscopic physics is the domain of "fapp" methods (valid only for practical purposes), what does that mean?

There is certainly a lesson there: Collapse is entirely restricted to macroscopic conditions, and this is also true of the uniqueness of physical Reality, which can be seen as an emergence of the quantum world into the classical one.

One main point of the present work is that local entanglement holds the keys for collapse (I use the plural because one will see that there are two keys and the corresponding doors will have to be opened one after the other). The first one is concerned with local entanglement between a measuring system and a measured system.

One can introduce this notion of local entanglement in a case far away from quantum mechanics. This example involves a macroscopic gas, which is made of atoms: for instance an argon gas acting as the detecting part of a Geiger counter. One denotes it by B. It can express by a signal that an energetic charged particle has crossed it. One denotes this "measured microscopic system" by A.

One knows since Boltzmann how to describe the atoms in a gas by the equations of mechanics. One adopts for a moment this viewpoint, by assuming that the position and velocity of all atoms is given at some initial time 0, before the arrival of the external particle A. A computer follows the motion of all atoms thereafter.

As a preliminary, and an alternative to the notion of local entanglement, one will think of an "influence" of the particle A on the atoms, in the following sense: When the particle collides with one atom, the motion of this atom is strongly modified and its later motion conserves ever after an influence of this event. When comparing the difference, or lack of difference, between the state of motion of every atom in the case when A interacted with the gas and when it did not, one can distinguish whether a definite atom was "influenced", of not influenced by A. If the two data differ, one says that the atom was influenced by the particle A.

One can illustrate this phenomenon by imagining that all atoms in B carry initially a white color, whereas A is colored red. Influence is expressed then by the idea that, when the red particle A collides with a white atom a, it makes this atom become red ("influenced"). This red color is carried then to other atoms, either under direct collisions of atoms with A, or under collisions of an already red atom with a white one. When two white atoms ("non influenced") collide, they remain white.

One can conceive easily that the red color expands then, starting from the trajectory of the particle A, but a mathematical formulation shows much more. When seen from the standpoint of classical statistical mechanics, the collisions between atoms bring an expansion of the red color. The mathematical formulation of this expansion is governed by a diffusion equation (like the one, which describes the diffusion of heat, for instance).

If one denotes by $f_1(x,t)$ the probability for the atoms to be red, near a point x in the gas, and by $f_0(x,t)$ their probability for being white, the transfer of redness (which looks much like a contagion of measles) is due to collisions involving one red atom and one white one. Its probability is therefore proportional to the product $f_1(x)f_0(x)$ and yields a nonlinear process (since $f_0 = 1 - f_1$). Contagion is therefore governed by a nonlinear partial differential equation, which is nontrivial but can studied, and solved numerically.

This type of equation is known in the theory of nonlinear waves, where it is often characterized by a moving front, which separates in the present case a red region from a white one. More precisely, the gas is white in an outside region, which the front has not yet reached; behind the front, it shows shades of pink until it reaches full red at a distance of order of one mean free path of atoms. The front moves at a fixed velocity, which coincides with the velocity of sound in the case of a dilute gas.

Quite remarkably, these qualitative and semi-quantitative properties remain valid in the quantum case. It is more proper then to speak of "local entanglement" between A and B, rather than of an influence by A on B. One will use sometimes the abbreviation "*LE*" for "local entanglement", or for "locally entangled" when used in place of an adjective.

As a matter of fact, the existence and propagation of *LE are predictions of the Schrödinger equation*!

This prediction can look surprising (at least), since it is completely at variance with the standard interpretation of quantum mechanics. The reason is that *LE* cannot be associated with quantum observables, whereas the standard interpretation claims that everything possessing a physical meaning should necessarily be expressible by observables. There is no inconsistency in the difference, but there is certainly novelty.

There is something more: The previous probabilities of influence and of no influence, $f_1(x)$ and $f_0(x)$, can be still defined (by using "locally entangled quantum fields") and they make sense in quantum mechanics for a wave function ψ. They are then positive, smaller than 1 and sometimes equal to 1, but their sum is generally not equal to 1: It can be larger or smaller. In spite of their formal existence, these quantities cannot be therefore considered as having a meaning of probabilities.

But there is still more: When one considers a many-body system (like the gas, which one takes as an example), one must deal no more with an individual wave function, but with a "mixed state" (or "density matrix"), which involves an extremely high number of wave functions ("eigenfunctions" or "eigenvectors"). There are then compensations, between the contributions of different eigenfunctions and also of pairs of them, of the deviations from 1 of $f_1(x) + f_0(x)$. This sum becomes 1, with very high precision, when referred to the "real" mixed state of a many-body system. One can therefore consider them, legitimately, as being respectively local probabilities of *LE* and of no *LE*.

There is something startling in this extremely simple result, not because of its restriction to conditions of high complexity, but because it is valid in these realistic conditions. It forces one to get out of the domain of the standard interpretation and, at the same time, to discover that there is new physics in the domain of *LE*. One began to explore it some years ago, and found it much richer than could be expected at first sight.

3 On the Environment of a Quantum System and an Associated Form of Incoherence

One turns now to one of these openings: Long ago, Heisenberg called briefly attention on a possible determining action of environment on the state of a quantum system (Heisenberg 1958). Recently, WojciechZurek considered, with many interesting developments, the possibility that this action could be responsible for collapse (Zurek 2003). One considers it now in the light of local entanglement.

An environment can perturb the wave functions of a system and these perturbations grow with the number of particles, which are in contact with environment. The quantum state of a macroscopic system is therefore very sensitive to instability, except under very special conditions. In the present case of a Geiger detector, one can think of the environment as involving the atmosphere around it, and eventually a table below.

One advocates the existence of a strong influence of the nearby environment on the quantum state of a macroscopic measurement device. More than a direct action on the measuring system itself, this influence appears rather as a conditioning of its quantum state. More explicitly, one wants to show that the action of environment on a well-defined macroscopic object (namely one for which the notion of a specific quantum state makes sense), brings incoherence into the wave functions of that state. Still more explicitly, one will show that this incoherence is not associated with the bulk (or average) action of the environment, but with fluctuations in this action. The nature of this incoherence is closely linked with local entanglement, which is why it appears new, and why one looks at it now.

One will deal again with the example of a Geiger counter, denoted by *B*. The measurement is concerned with a particle, denoted by *A*. The counter is initially standing alone, with nothing to measure. One can think of it as consisting essentially of an atomic gas, enclosed in a solid box. One pays no attention to its support and considers that its environment is a surrounding atmosphere, under standard conditions of temperature and pressure.

Nothing could look apparently simpler that this object in that situation, but one will see that local environment maintains a permanent invisible turmoil in the gas, and also in the box. This unrest is much less obvious than the thermal motion of atoms, because it is contained in wave functions and one needs several hints for getting at it, as follow:

The direct action of environment consists in a multitude of collisions by atmospheric molecules on the box. One looks first at a unique collision, in a mathematical model where the environment would be an empty universe containing a unique molecule *M*, which hits the box at some time. The theory of local entanglement, which one sketched previously, implies that a wave of *LE* starts from the place where the molecule hits the box (solid state physics shows that this *LE* is carried through the box by phonons, before it reaches the gas). This is the kind of local entanglement, which one described before, now meant as *LE* with the outgoing state of the external molecule.

One found earlier that the associated *LE* wave moves at the velocity of sound. It takes some time Δt before it fills up the whole gas with its local entanglement, and thus realizes a state of complete algebraic entanglement. In a counter with size 10 cm, this time Δt is of order 10^{-4} s, which is quite significant.

One looks next at average effects of environment. One knows well enough how to deal with them and a famous formula, which extends one by Boltzmann to the quantum domain, gives an explicit formula for the average state of the gas, $Z^{-1}\exp(-H/k_{\mathrm{B}}T)$, where H is the total energy, T the temperature, k_B Boltzmann's constant and Z a normalization coefficient. No local entanglement is associated with this average so that the effects, in which one is presently interested, must be associated with fluctuations.

When considering an effect of fluctuations on the quantum state, one would rather speak of an "influence" rather than of an action, as one did earlier when introducing *LE*. Several features must be considered then and one begins with the most formal one, which is mathematical and as such can orientate towards the concepts that must be used.

One already wrote down the average state of the system, which one now denotes by $<\rho>$. Its main character is simplicity, when compared with the actual "mixed" state ρ of B (its density matrix). One may think of ρ as involving an extremely large number of wave functions expressing the thermal disorder in the gas.

This ρ is a matrix, with a tremendous number of lines and columns, but which is nevertheless a square matrix. The difference $\Delta\rho = \rho - <\rho>$ is also a matrix, but one will refine a little more on its algebraic aspects by splitting it into a part, ρ_+, which involves its positive eigenvalues, and a part—ρ_- involving the negative ones.

I make acknowledgements to readers considering that these mathematical manipulations are puzzling. Physicists would probably prefer hearing of positive fluctuations in the external flux of molecules as being above average, and of negative fluctuations below. The two present matrices ρ_+ and ρ_- are not exactly in one-to one correspondence with these positive and negative fluctuations however, although they are closely linked with them. To cut short a story, which could become long, one will say only that these two matrices *express* the occurrence of fluctuations in the present case, and that their size (which mathematicians call their trace) is determined by the strength of fluctuations.

One found that every external collision by a molecule leaves during a time Δt a mark of its occurence in the quantum state of the gas. One can find such marks also in the matrices ρ_+ and ρ_-, as a presence of so many *LE* waves in every one of their eigenfunctions. One denotes by N_f the number of these wavės (actually, if N_t is the total number of molecules, which collide on *B* during the time Δt, N_t is very large and N_f, which is the square root of N_t, is still quite large).

One pays then more attention to the random character of the fluctuations. One cannot identify individually external collisions, as belonging to excesses above average or lacks below. One can only know their number and assert that their properties must be random. This randomness has crucial consequences, requiring considerations of quantum mechanics and which one makes now.

Randomness of the collisions by molecules belonging to fluctuations implies that the place x where the molecule hits the box, its momentum p and its time of arrival are random. Quantum mechanics tell us then that a phase $\alpha = \Delta p.x/\hbar$ is present in the wave functions of atoms in the gas (Δp is the momentum transfer in the collision). An essential consequence of this effect is that every wave of local entanglement, associated with a molecule belonging to fluctuations, carries a definite phase, which is random. But in optics and in quantum mechanics, an occurrence of random phases is synonymous with incoherence! One can then sum up this first part of the analysis by the following.

Proposition 1 *Fluctuations in the influence of environment on the quantum state of the macroscopic system, generate incoherence in that state.*

More precisely, one may say that this incoherence is present in the matrices ρ_+ *and* ρ_-*, which one can use when testing the influence of environment.*

Proposition 1 is new and plays a central part in the present theory. One may consider it as the exact opposite of the main assumption in Everett's interpretation of quantum mechanics, which is an infinitely exact coherence of a wave function of the universe.

One will leave out this aspect and rather turn to an exercise in geometry. When closing eyes and concentrating, one can see in the state of the matrix ρ_+ *(for* *instance) a complex pattern of LE waves, which move everywhere along all directions, showing fronts with small width λ, the fronts crossing each other under their motion and their expansion.*

A substantial value of the number N_x of fronts, which overlap over a point x in the gas, implies an action of as many random phases of the previous type α, in the wave functions at that place. But every subset of these phases can add up together and yield another random phase β as their sum. These other various phases are also random and also mutually independent, so that their local number is about as large as the factorial $N_x!$. It shows that Proposition 1 does not only mean that there is some amount of randomness in the state of the system, but that there is a very strong incoherence.

A further comment is suggested by a cogent question, which is: If there is such a high amount of incoherence in most macroscopic systems, if not all of them, how could it be that nothing of that kind has never been observed?

The answer is twofold. It says first one that this incoherence is associated with local entanglement and, therefore, is inaccessible to observables. As a consequence, an observation, whatever it can be, needs absolutely projection operators for expressing its outcomes, so that it cannot be sensitive to this very special kind of incoherence. A second answer, which enforces the first one, is that the amounts of incoherence and its distribution are exactly the same in the matrices ρ_+ and ρ_-, so that their observable effects, if they had some, would cancel each other anyway.

Coming back then to the matrix ρ_+, for instance, one can give a look at one of its instantaneous eigenfunctions ψ, which is defined mathematically as representing one eigenvectors of the time-dependent matrix $\rho_+(t)$ at a sharp time t. Because of the very high number of composite phases β everywhere, and of the non-negligible width (of order λ) of the front where a phase α is active, one finds that the disorder in wave functions is huge and their correlations are of short range (of order λ). This situation can be expressed by another proposition, in which one denotes by N_c the number of atoms in a cell of space with size λ and which is:

Proposition 2 *Every eigenvector (or wave function) ψ of ρ_{B+} (or ρ_{B-}) splits under incoherence into a sum of independent components*

$$y = \sum_n \psi_n, \tag{1}$$

where every component ψ_n carries a distinct random phase, extends in space over a distance of order λ and involves a limited number of atoms, of order N_c, with

$$N_c = n_e \lambda^3, Z \tag{2}$$

(n_e denoting the number density of atoms in the macroscopic system).

4 The Notion of Slip in Coherence

In 1976, Philip Pearle proposed that collapse could be the outcome of a Brownian process, rather than being sudden and total, (Pearle 1976). The random character of collapse, as well as its agreement with Born's fundamental probability law, would come out then almost automatically. In 1990, Ghirardi, Rimini and Weber suggested another alternative as a freezing of wave functions, which would not obey quantum dynamics but would add to it (Ghirardi et al. 1990). When joined together, the two trends led to so-called continuous spontaneous localization (CSL theories) theories, which draw much attention in spite of their fundamental incompatibility with exactness of the quantum laws (Laloë 2012).

It turns out that the explicit mechanism of collapse, which one describes here, shares some features with the CSL model and proceeds also in a Brownian way, like in Pearle's approach. An essential difference of CSL theories with the present approach is however that although this approach stands out of the domain of the standard interpretation (because of its reliance on local entanglement), it relies anyway on standard quantum dynamics and takes as its unique basis the Schrödinger equation.

Like CSL theories (Pearle 1976; Ghirardi et al. 1990), the present theory relies on elementary events, which have only little individual effects, but which can bring out collapse by their accumulation in very large numbers. An essential difference between the two approaches stands however in the origin of these phenomena: This origin is unexplained in CSL theories, and their nature remains mysterious with only one identifiable character, which is that they break down the Schrödinger equation. The present theory relies on the contrary on explicit elementary events, which draw their existence from this equation, and it escapes a verdict of impossibility by the origin of these events in local entanglement.

One will call one of these elementary events a "slip in coherence", to mean that its effect is a little slip-up in the evolution of a wave function, like a little mishap in the standard interpretation. Its main effect is to transfer small amounts of quantum probabilities between different measurement channels.

Some formalism is necessary for defining such an event and showing how it works. As usual in measurement theory, one considers two quantum systems and one denotes them by A and B. The system A is the measured one and B the measuring one. One considers necessary that the system B be macroscopic. The physical quantity, which must be measured, is an observable X belonging to the system A. One denotes by $|k\rangle$ the state vectors of the system A, which are associated with different values of this observable, and one writes down the initial state of the system A as a superposition of these state vectors,

$$|A\rangle = \sum_k c_k |k\rangle. \quad (3)$$

A quantum probability p_k, equal to $|c_k|^2$, is associated with every channel in this expression.

When discussing a measurement of this quantum system, one can use the model of a Geiger counter B containing an atomic gas. It turns out then that the phenomena acting as slips in coherence, along the way to collapse, are simply collisions of two atoms in the gas, say a and b, which satisfy the three following conditions:

(i) The collision is incoherent.
(ii) The initial wave function of Atom a is locally entangled with some state $|j\rangle$ of A.
(iii) The wave function of Atom b is not locally entangled with the system A.

Condition (i) regarding incoherence is the most significant one and must be specified with special care.

As usual, incoherence is meant as a random phase. One encountered already this kind of phases when describing the action of environment, but it must be now analyzed more carefully.

Every collision by an atmospheric molecule on the box containing the gas is associated with the previous random quantum phase $\alpha = \Delta p.x/\hbar$. One already mentioned that fluctuations in the flux of external molecules produce incoherence in the quantum state of the gas. The average flux does not contribute. The fluctuations can be positive, when associated with excess in the flux, or negative when associated with a relative shortage.

The main effect of one collision is to produce an associated wave of local entanglement (with the state of the molecule, which bounced away after collision). The wave is carried by phonons when it crosses the solid box and then by atoms in the gas. One saw that the geometry of this wave (i.e. the position and the orientation of its wave front, at some time) is governed by the place x where the wave originated and with the time when it started. The associated phase α is an independent random parameter.

It might seem that one is emphasizing thus rather minute details, but this necessity is characteristic of quantum phases: One never sees them, but they can turn out essential: Quantum mechanics is so subtle.

One already mentioned another aspect of the waves of local entanglement with fluctuations in environment, which is the long time Δt during which they take to cross the system B, while continuing to affect the state of B. Without coming to needless details, one may mention an obvious consequence of this information, which is that one knows the number N_t of molecules, which have hit the box during that time. One knows also the number of them, which contribute either to positive or to negative fluctuations, namely $N_f = (N_t)^{1/2}$. Knowing that, and also considering the random character of the phase, linked with the uncontrollable parameter Δp, one knows much as a matter of fact, about this vast hubbub.

One knows even still more than meets the eye: This is a convenient place for calling attention to a property, which could be called pompously a "*principle of indeterminacy in fluctuations*". As a matter of fact, it consists only in a trivial comment on the notion of fluctuation, but one that will turn out so essential that it is worth attention.

What calls attention is the natural occurrence in the present formalism of two matrices, ρ_{AB+} and ρ_{AB-}, which are associated with excesses and with shortages in the action of the external flux, during the time Δt. The obvious property, or principle, which seems worth attention, is that absolutely no criterion could allow deciding: "This event *is* a positive fluctuation" or "Some events, among those that were predicted by averages and did not happen, can be identified as the missing ones." That would of course be awkward, but it teaches a lesson, which is that, since one can predict the numbers of excesses and of lacks here or there, the expected values of positive and of negative fluctuations make sense and one can compute their effects, as if their two sets were distributed randomly.

One uses this property of indeterminacy in the present study and one considers it not as a principle but as a character, which is inherent to the concept of fluctuations.

The main tool allowing conclusions from this notion of incoherence can be then derived. It consists in the splitting of every wave functions ψ of the macroscopic system B into a sum of mutually incoherent components ψ_n, as in Eq. (1) in Proposition 2.

One can use it by looking at the combination of two effects, which both rely on local entanglement. On one hand, there is a property of incoherence in the sate of the measuring system, which is due to its local entanglement with fluctuations in the environment. On the other hand, there is a direct local entanglement between the systems A and B. When taken together, they imply an expression of a wave function (or eigenvector) of the density matrix ρ_{AB+}, which is given by

$$|\psi_{AB}\rangle = \sum_{nk} c_k |\psi_{Bkn}\rangle |k\rangle. \tag{4}$$

This relation has essentially the same meaning as Eq. (1) regarding random phases, except that some subcomponents ψ_{Bkn} occur within the previous component wave functions ψ_{Bn}. The same random phase occurs in two wave functions, such as ψ_{Bjn} and $\psi_{Bj'n}$ (with $j \neq j'$), where the same index n characterizes this common phase. Different random phases occur on the contrary in wave functions, such as ψ_{Bkn} and $\psi_{Bk'n'}$, with different phase indices $n \neq n'$. (It may be useful to recall, to avoid misunderstanding, that these phases are not the elementary phases of the previous type α, but sums of several of them, and usually of many of them).

One can then easily derive the consequences of a collision between two atoms, when it satisfies the conditions (ii) and (iii) for a slip. One may call attention on the fact that Condition (iii), which requires that Atom b is not locally entangled, implies that its state is equally present in all the component wave functions ψ_{Bkn} sharing the same index n, and share accordingly the same random phase although they are algebraically entangled with different states $|k\rangle$ of A.

Two cases can happen then. In the first one, the two atoms belong to component wave functions with the same random phase (i.e. with the same index n). The collision is then coherent. Condition (i) for a slip does not hold and the (a, b) collision yields only an increase of local entanglement with the state of A with index j, by addition of Atom b to the set of locally entangled ones, as discussed at beginning of this paper. There is of course no transfer of quantum probabilities between different channels in that case.

The other case, which is the interesting one, occurs when all the conditions of a slip are satisfied, namely when the collision is incoherent in addition to conditions (ii) and (iii). The state of Atom a belongs then to a component wave function ψ_{Bjn} (notice the index j, which is the one occurring in Condition (ii)). As for Condition (iii), it requires that the state of Atom b be not locally entangled. It is therefore equally present in all the wave functions ψ_{Bkm}, where the index of phase m is different of n whereas the index k for measurement channels can take any value.

One can write down finally two simple formulas, which are necessary for a few last comments. They express the variations δp_k in the various quantum probabilities p_k, when the slips under consideration obey the conditions (ii-iii), with a definite index j in Condition (ii), they occur during a short time interval δt in a small region with volume δV, centered at a position x and containing a number δN of atoms. One gets then

$$\delta p_j = \ +Wp_j(\delta p_j = \ +Wp_j\big(1-p_j\big)(\delta t/2\tau)f_j(x)f_0(x)(\delta N/N_c), \tag{5a}$$

$$\delta p_{j'} = -\,Wp_jp_{j'}(\delta t/2\tau)f_j(x)f_0(x)(\delta N/N_c), \mathrm{for}\, j \neq j'. \tag{5b}$$

The number N_c was shown in Eq. (2). The + sign in (5a) means that there has been an increase in the probability p_j for the channel j, with which Atom a was locally entangled. The minus sign in (5b) means that all other channels suffered then a decrease in their probabilities.

The existence of these variations in quantum probabilities goes along with a significant first step towards collapse, by showing that these probabilities can vary. One sees that the sum of the right-hand side in (5a) and of the right-hand sides of (5b), for all values of j', vanishes, which means that the sum of all probabilities remains equal to 1.

The factor W in Eqs. (5a and 5b) is the probability for incoherence, which one can take as the trace of the matrix ρ_{AB+}. Its exact value is unknown, but one can establish an upper bound for it, which is: $W \leq 4/3\pi = 0.4\ldots$. This bound provides also presumably a sensible order of magnitude for W. Finally, the factors $f_j(x)$ and $f_0(x)$ are respectively the probabilities for local entanglement with Channel j and for non-local entanglement.

When the matrix—ρ_{AB-} governs this type of slip, in place of ρ_{AB+}, the plus sign in (5a) is replaced by a minus sign and the minus sign in (5b) by a plus sign.

5 The Mechanism of Wave Function Collapse

From the variations of quantum probabilities under a slip according to formulas (5a and 5b), an explicit mechanism of wave function collapse results. One will not describe it in detail and one mentions only a few significant points:

Equations (5a and 5b) dealt with slips towards a channel j from other channels j', under govern of the matrix ρ_{AB+}, but one must take also account of slips going from the channel j towards other channels j'. One must also consider the contributions by the matrix $-\rho_{AB-}$. Their combined effect cancels all the average values $< \delta p_j >$ but they let survive the correlation coefficients $< \delta p_j \delta p_{j'} >$ and the squares of standard deviations $< (\delta p)^2 >$.

An interesting remark is brought by opposite effects of the matrices $(\rho_{AB+}, -\rho_{AB-})$ and also of opposite transitions, $j \leftrightarrow j'$, from two channels. Since the two matrices ρ_{AB+} and ρ_{AB-} represented already fluctuations in the external flux and the final outcome, which consists only in correlations and squares of standard deviations, makes theses two matrices fight against each other, this means that the final results (which will include collapse), is due to fluctuations in the mutual compensation of positive and negative fluctuations.

This is a remarkable property, with no analog in physics, as far as I know (at least not in quantum mechanics). It came as a surprise when its character was realized and, presumably, would not have been anticipated under other approaches.

One may go then rapidly to the conclusion of these conclusions. It turns out that the quantities $< \delta p_j \delta p_{j'} >$ and $< (\delta p_j)^2 >$ are proportional to the time interval δt, which one considers. This behavior is the signature of a Brownian process. A famous theorem by Pearle (1976) implies then that the random variations in the quantum probabilities must end up necessarily with a situation where one quantum probability, say p_j, has become equal to 1 and the other probability vanished.

This conclusion is a prediction of collapse. Pearle's theorem asserts moreover that this Brownian probability is equal to the initial value of p_j in the state (3), and this is in perfect agreement with Born's fundamental probability rule. (The proof of Pearle's theorem must be slightly enlarged in the present case, but this is only a matter of mathematics owing to the finiteness of an individual slip). More details, including quantitative matters and the case of separate detectors, can be found elsewhere (arXiv:1601.01214v2).

6 Some Philosophical Comments

The place of this paper in this book, which is devoted to the philosophy of science, calls for a few philosophical comments.

Its motivation could not be else than philosophical. The problem of uniqueness of Reality is a major topic in the philosophy of science, among the most intriguing ones. It raises the kind of enigma, which can keep a man at the evening of his life under the charm of its mystery, even without an unreasonable expectation of finding a fissure in its carapace. Maybe one appears here, and I wish to comment its outcome as if it were true.

Its main feature is simplicity. Not a technical one, as I may say from the difficulty of putting order into its glimpses, but a philosophical simplicity. Reality appears just like many physicists have perceived it since more than half a century: An objective classical datum, set upon an objective science of a quantum world (space-time being another matter). Only a keystone, which would have insured a mutual equilibrium of their representations, was missing. These comments assume that a part of the answer could look like the one, which is proposed in this paper.

As it appears here, the uniqueness of macroscopic Reality would emerge from a quantum sea. Can one then say that there is a veil on this reality, to borrow d'Espagnat's striking expression? Yes, there would be a veil, but one, which would cover the two faces of reality, when seen with the eyes of philosophy.

I see personally no other basic veil than the one of mathematics. Quantum reality and macroscopic reality are joined together, as our science can see, by a mathematical unity and a mutual consistency of their laws. I believe that the greatest problem in the philosophy of science holds in a unique question: What is Mathematics? (Wigner 1967; Omnès 2004). To which one may add: What is the status of mathematics in a science, which aims at unity of its vision?

I do not propose an answer, but I may recall a remark, often heard and drawn from the history of mathematics: During more than two millenaries, mathematics looked as if inspired by (macroscopic) reality. Since about two centuries, it became its own field, much as if mathematics were discovering itself, more than finding its source in empirical reality or schematizing it. The irreplaceable place (or action?) of mathematics in the foundations of physical science, where it appears not only as a mode but also as a mold of thought, opens a view on an inner harmony of Being, in itself by itself, as Plato or Spinoza saw it.

References

Adler, S.; Quantum Theory as an Emerging Phenomenon, Cambridge U. P. (2004).
Bell, J.S.; Physics, **I**, (1964).
Bell, J.S.: Speakable and unspeakable in quantum physics, Cambridge U. P. (1987).
Bohm, D.; Phys. Rev. **85**, 166–180 (1952).
Bohr, N: Atomic Physics and Human Knowledge, Dover (2010).
Caves, C., Fuchs, A., Schack, R.; Phys. Rev. **A38**, 022305 (2002).
d'Espagnat, B.; Veiled Reality, Addison Wesley (1995), On physics and philosophy, Princeton University Press (2006).
de Broglie, L.; Tentative d'interprétation causale et non-linéaire de la mécanique ondulatoire, Gauthier-Villars, Paris (1956).
Dirac, P.A.M.: The Principles of Quantum Mechanics, Cambridge (1930).
Einstein, A., Podolsky, B. Rosen, N.; **47**, 777 (1935).
Epperson, M., Zafiris, E.: Foundations of Relational Realism, Lexington, N.Y. (2013).
Everett, H III, Rev. Mod. Phys., **29**, 454–462 (1957).
Ghirardi, G. C, Rimini, A, Weber, T.; Phys. Rev. **D 34**, 470 (1986).
Ghirardi, G. C, Pearle, P., Rimini, A.; Phys. Rev. **A 42**, 78 (1990).
Griffiths, R.B., Consistent Quantum Theory, Cambridge (2002).
Haroche, S., Raimond, J. M.; Exploring the Quantum, Oxford University Press (2006).
Heisenberg, W., Physics and Philosophy, Harper and Row (1958).
Laloë, F.; Do we really Understand Quantum Mechanics, Cambridge U. P. (2012).

Lieb, E.M., Robinson, D.W.: Commun. Math. Phys. **265**, 781 (1972).
Omnès, Converging Realities: Toward a common philosophy of physics and mathematics, Princeton U.P. (2004).
Omnès, R.: in Turing's Legacy, Proc. 2012 Meeting of the International Academy of Philosophy of Science, E. Agazzi, ed., FrancoAngeli, Milano (2013).
Omnès, R.: Scheme of a derivation of collapse from quantum dynamics, arXiv:1601.01214v2 (quant-ph) (This work is presently under careful revision of its proofs, arguments, and intelligibility).
Omnès, R.: The Interpretation of Quantum Mechanics, Princeton (1994).
Pearle, P.; Phys. Rev. **D 13**, 857. (1976).
Schrödinger, E.; Die gegenwärtige Situation in der Quantenmechanik, Naturwissensschaften **23**, 807, 823, 844 (1935), reprinted with English translation in [4].
Von Neumann, J.: Mathematische Grundlagen der Quantenmechanik, Springer, Berlin (1932). English translation by R. T. Beyer, Mathematical Foundations of Quantum Mechanics, Princeton University Press (1955).
Wheeler, Zurek, W.H; Quantum mechanics and Measurement, Princeton University Press (1983).
Wigner, E.P.: The unreasonable effectiveness of mathematics in the natural sciences, in: Symmetry and Reflections, M.I.T. Press, Cambridge (1967).
Zurek, W.H.: Nature Physics, **5**, 191 (2003).

Theories and Models: Realism and Objectivity in Cognitive Science

Jean-Guy Meunier

Abstract Scientific realism is often analyzed in the context of natural sciences theory. How does it behave in cognitive science theories? Some philosophers of science have proposed a pragmatic approach to the concept of scientific theory where models form an essential part of its construction. They play a role of mediation. Three distinct classes of models play such a role in cognitive science: (a) formal models, (b) physical models, and (c) conceptual models. Each of these classes challenges the realist thesis in specific ways.

1 Scientific Realism, Theories and Models

In philosophy, scientific realism aims at identifying the conditions under which a scientific practice can deliver truly objective knowledge of the world. And Thagard (2012) reformulated this classical this thesis: science is a specific type of *cognitive endeavour* that builds knowledge of reality. And one of the main ways this cognition is achieved is by constructing what is commonly called a *theory*. As Wright (1991) expresses directly: Scientific realism is a thesis on the truthfulness of a scientific *theory* about the world. And this realist thesis has generated a huge literature on various understandings of the conditions under which a *theory* "really" refers to the world it is about. But many anti-realists, for instance Giere (1988) think that the realist thesis cannot meet its own expectations. In this paper, we shall explore this realist thesis but in the context of cognitive science theories rather than in the classical "natural" sciences ones: What happens to the realist thesis when confronted with cognitive science theories? Do these type of scientific theories refer objectively to the real world?

J.-G. Meunier (✉)
Université du Québec à Montréal, Quebec, Canada
e-mail: meunier.jean-guy@uqam.ca

E. Agazzi (ed.), *Varieties of Scientific Realism*,
DOI 10.1007/978-3-319-51608-0_18

2 Theories and Models in Sciences

2.1 Theories

The philosophy of science has proposed many theories regarding what is a *theory*. A general one has been offered by Morgan and Morrison (1999:13). For these authors, a theory is "a discourse used by scientists to express "principles that govern a group of phenomena." Winther ads that "Effective scientific theories magnify understanding, help supply legitimate explanations, and assist in formulating predictions" (Winther 2016). But there are different visions regarding how these principles and explanations are realized. Today, the synthesis of these perspectives is inspired by the long-standing American[1] pragmatist tradition of Peirce, James, Dewey, and reformulated by Carnap (1937),[2] and Morris (1971) who propose three views regarding the nature of a scientific theory.

A first one, proposed by the logical positivists (Carnap 1937; Reichenbach 1938; Hempel 1965; Nagel 1961) and traditionally called the "syntactic" view, sees a scientific theory as a collection of axiomatic statements and theorems.

> Scientific theory is thus taken to be a syntactically formulated set of theoretical sentences (axioms, theorems, and laws) together with their interpretation via correspondence sentence (Winther 2016).

The second view of a theory is the "semantic view." Here, the language of a theory must be mathematical (Suppes 1967). Or, to be more precise, as formulated by van Fraassen (1970:327), a theory is "always a mathematical structure," but in which some interpretation, such as a Tarskian "model" for example, may define its semantics.

> A model is called a model of a theory exactly if the theory is entirely true if considered with respect to this model alone. (Figuratively: the theory would be true if this model was the whole world) (van Fraassen 1989:218).

Though such views regarding theories have been widespread, they have been highly criticized (Winther 2016) for their limits and constraints. And although they are formally acceptable, examples of such types of theories are rare and not easily applicable to many types of sciences (Papineau 2010; Leplin 1981). So a third view has been proposed: the pragmatic view.

---

[1]Many of the ideas were quite present with Duhem (1906) and Bachelard (1979).

[2]"If in an investigation explicit reference is made to the speaker, or, to put it in more general terms, to the user of the language, then we assign it to the field of *pragmatics*. (Whether in this case reference to designata is made or not makes no difference for this classification.) If we abstract from the user of the language and analyze only the expressions and their designata, we are in the field of *semantics*. And if, finally, we abstract from the designata and analyze only the relations between the expressions, we are in (logical) *syntax*. The whole science of language, consisting of the three parts mentioned, is called *semiotic*" (Carnap 1937, 3–5, 16).

This third view is grounded in the American pragmatism of Peirce, James, and Sellars and it has been influenced by Feyerabend, Hanson, Kuhn, Toulmin, Laudan, and Hacking. It is quite in opposition to the first two views. It has lately been renewed and reformulated by Rheinberger (1997), Giere (1999), Cartwright (1983), Morgan and Morrison (1999), as well as by many others.

According to this pragmatic view, most scientific theories are not formal and their explanation cannot be solely deductive or inductive. In fact, a theory is the result of complex cognitive processes that are *points of view*, *perspectives, enquiry strategies* regarding a phenomenon to be understood (Frigg and Hartmann 2012). And these perspectives are not just contextual (Scriven 1962). Such an understanding builds specific theory structures:

> To explain a phenomenon is to find a model that fits it into the basic framework of the theory and that thus allows us to derive analogues for the messy and complicated phenomenological laws which are true of it (Cartwright 1983:52).

In other words, as Winther says, a theory is constituted by a plurality of formal and informal components. For Cartwright (1983) and for Morgan and Morrison (1999), a theory of "a theory" must be more inclusive than what is proposed by the syntactic and semantic views, but without rejecting them. One way to achieve this is to approach the concept of *theory* through the concept of *model*. And models will become a dominant concept of the pragmatic view regarding theories.

2.2 Models in Science

The pragmatic view of scientific theories has underlined the role of models in science. For Cartwright et al. (1995), models should constitute the appropriate level of investigation to understand science. They play an important cognitive role in building theories. Knowing the world is not just an act of perception of specific and individual information signals. It is also the result of a complex process of categorization (Harnad 2005) and of structuring of these categories. And scientific theories are not reports on perceived bits of the world. Rather, they are more aptly described as reports on regularities in phenomena. In the words of Kant, science, as any other type of complex processes of cognition, requires synthesis and categorization. As Morgan and Morrison formulate it intuitively, their role "is to fit together… bits which come from disparate sources" (Morgan and Morrison 1999:15). They are mediators between theories and the world. But as synthesis and categorization are not given, they must be discovered and constructed, it is then that modelling intervenes.

Models are different from theories in that they focus more on the cognitive engagement of the scientific process than on the ontological commitment they may have with reality. In other words, models are not mini-theories in themselves. They are tools for the building of a theory. For Rheinberger (1997), models do not mainly serve to express what is known but mostly to "*explore what is unknown.*"

Unfortunately, as Giere (1999) recalls, the relation between a theory and a model is not a well understood relation. There does not exist necessary nor sufficient conditions to identify which model satisfies a theory.

> There are many senses of the term "model" in science, and these different kinds of modes serve distinct scientific ends (Giere 1999:1).

Naturally, as there are a multitude and a variety of possible points of view in a scientific enquiry, there will be a mosaic of models, as Baetu formulates so well:

> Models anchor the diverse pieces of the mosaic of knowledge to a description of a phenomenon, on the one side, and to the methods and tools, experimental or theoretical, used to obtain each piece of the mosaic, on the other (Baetu 2013:2).

Each model being an approximation, a point of view, a perspective, etc., a theory appears as the background architecture that gives them their coherence. Points of view become comparable, and from this differentiability in models, coherence judgments emerge. This multiplicity and variety of models play an important epistemic role in the construction and in the value of a theory. Models are not just representations of observational data, they are constructions that, in their interactions, point to some "reality." An anti-realist will focus on this constructivity feature of the models and will have to cope with the problem of explaining how, interactively, each partial model points to such reality. The realist will probably focus on models closest to observational practices and empirical verifications, but will, in turn, have to explain how other more conceptual if not physical models unveil reality.

3 Theories and Models in Cognitive Science

As we may recall, cognitive science brings together various scientific disciplines: psychology, philosophy, anthropology, linguistics, artificial intelligence, neurosciences, etc. Besides identifying the field's common objective as the science of cognition, it has not been an easy task to find some methodological unity among these sciences. But at least naming the field *science of cognition* was important. This signed a difference with the traditional names of the field: *knowledge theories* in English, *gnoseologia* in Latin, *théorie de la connaissance* in French, and *wissenschaften* in German. This name allowed the field to take a naturalistic turn, although a specific one: Indeed, the science of cognition could import concepts and methods from more formal and experimental sciences. For instance, Chomsky (1957) saw linguistic theory as modelling a grammar of a competence as sorts of automata. Newell and Simon (1994) identified a theory of intelligent cognition as a computable manipulation of symbols. For Bechtel (2005), a cognitive theory had to explain the mechanisms of the brain. These formulations gave the discipline a shared naturalist paradigm: A theory of cognition was to be formalized in a physical computational paradigm.

Dennett (1978) and Harnad (1994) criticized the exclusivity given to the "computational" view of cognition. It was a too limited and reductive approach. With Marr (1975), Pylyshyn (1984), and Newell (1994), Dennett proposed and a more integrated view of the science of cognition. For him, the cognitive sciences are a type reverse-engineering. A cognitive science enquiry must build at least three distinct explanatory stances for this reverse-engineering. And although these stances were named differently by these authors, they can be recategorized as (a) representational, (b) functional, and (c) physical types of explanations. The definition of these three types of explanations is quite specific although many variations are to be found in each cognitive science.

The classical understanding of these "stances" is that they are *points of view* regarding cognition. All three have different explanatory purposes. The first stance sees cognition as goal-oriented representations that are expressed in a non-formal language. The second one sees cognition as a set of operations and states and expresses them in a formal mathematical and computational language. And the third one sees cognition through its physical implementation. For Dennett, all three stances together contribute to the construction of a theory of cognition.

In this view, these stances correspond perfectly with the general definition given to the concept of models. And here, we shall also understand them as models. They are mediators in building a *theory* of cognition. A cognitive theory is not a single autonomous model working in one particular point of view. It is rather a set of models that interact in some manner so as to offer an overall theory of cognition.

We illustrate this multi-model approach of cognition by means of a case study: Jean Petitot's (2009) analysis of the cognitive perceptual Kanizsa triangle illusion. His approach contains three main models. The first one is built out of natural language sentences expressing the phenomenological representation one has of the illusion, such as: *I see 3 circles and angles creating black and white triangles, among which a black lined incomplete triangle and a full white lineless triangle,* etc.

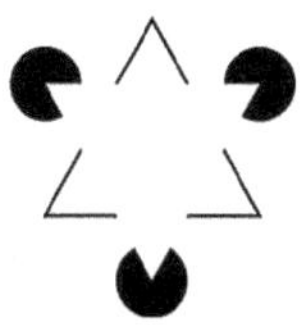

The second model expresses this perception of the triangle into algebraic, geometrical, and dynamical structures (equations on real numbers, force, energy filters, etc.), as represented by the following graph:

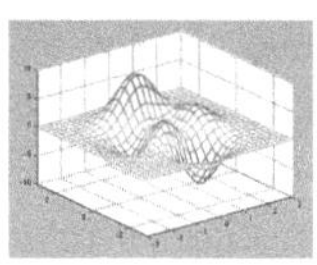

Finally, the third model shows the neural implementation underlying the cognitive operation. For instance, the model identifies the optical and neuronal components of the visual pathway that are activated or inhibited; it describes the path of signals in many brain areas, their neural interactions, the motor control centres, etc.

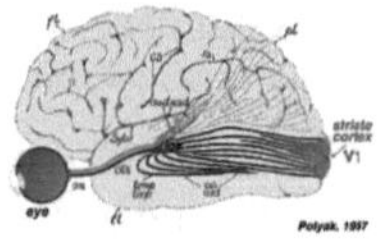

There have been many critiques of these distinctions regarding stances or models. We agree with many of them. The concepts are very general and fuzzy. But whatever the type and number of possible models, what appears clearly from an epistemological point of view is that the cognitive sciences build theories that are not just a single and controlled set of related formal and axiomatic languages expressing mathematical structures. These theories are in fact a collection of models where each one has a specific semiotic form, be it a natural language, a mathematical language or even an iconic form, for instance. The models interact with one another and they contribute, each in their own sphere, to the building of a specific theory. As McClelland regularly reaffirms: "Models are research tools that have their strengths and weaknesses, like other tools we use as scientists" (McClelland 2009:12).

4 Three Classes of Models

Science is a complex collection of cognitive processes, and these pertain namely to perception, conceptualization, categorization, discovery, explanation, justification, and action. Scientific theories are built from a complex interplay of models participating to these processes. They are mediators in the emergence of the knowledge of reality.

And in science, one model is usually not enough (Green 2013). They are multiple, and their classification becomes necessary. One of the main classes of models has been the mathematical one. Recently, Weisberg (2015), in his taxonomy, has added a class of computational models and a class of physical models. Inspired by the taxonomy put forth by Dennett, Marr, and Pylyshyn, and by the analysis of many concrete scientific practices in cognitive science, we shall propose a tripartite and more inclusive classification in which we will include many sub-models. Our three main types of models are: (a) *formal,* (b) *physical,* and (c) *conceptual* models. We see these classes as more inclusive and applicable to many other sciences. Cognitive science practices, compared to natural science, usually call upon one or other sub-models of these three main classes of models in building their theories. These three classes reveal finer properties of the internal dynamics of theories of cognition.

In this taxonomy, the functional stance is seen as a sub-type of *formal* models. The physical stance, often understood as a physical *implementation*, is here seen as one particular sub-type of *physical* models. The knowledge or representational stance is one particular type in the class of *conceptual* models. We shall explore these three classes of models but always in view of questioning the realist and objectivist thesis in the cognitive science context.

4.1 Formal Models in Cognitive Science

Practically all epistemologists suggest that the main if not the ideal classical model to be used in a theory is a formal model. It is seen as the prototypical way of doing science. No formal model, no serious science.

In the Hilbertian and Carnapian traditions, a formal system is first and overall an axiomatized system of symbols. Its syntax is highly controlled. Strictly speaking, such a formal system of symbols does not need a semantics for its lexicon and formulas. But normally, as it is applied in some particular domain, it will require interpretation. The various formal systems will then be differentiated by the importance given to the syntax or to the semantics. We shall here distinguish two main but related sub-classes of formal models: *mathematical* models and *computational* models.

The first class of models is that of *mathematical* ones. They are used mainly for expressing functional relations or regular dependencies among the entities identified by predicates describing them in the chosen domain of enquiry. For instance, in physics, the algebraic equation $F = ma$ expresses the functional relations between sets of numerical values pertaining to predicates: FORCE, MASS, ACCELERATION. And many variants of these mathematical models exist. For instance, in chemistry, they are mainly algebraic, probabilistic, geometric, and diagrammatic; in physics, the models are mostly algebraic, statistical, etc. In artificial intelligence and linguistics, they are mainly logical, algebraic, and algorithmic.

In cognitive science, many sets of mathematical languages are to be found. A first and most classical type of mathematical model have been inspired by the axiomatic and logical systems. In artificial intelligence (Newell and Simon 1976), *knowledge* is formalized through a set of physical symbols that become for instance predicates, propositions, or arguments, and they are expressed in a variety of notational systems such as *frames* (Minsky 1974), *conceptual* graphs (Sowa 2000), micro-worlds (Genesereth and Nilsson 1987), ontologies (Gruber 1993), and are used in different sorts of reasoning: common sense (McCarthy and Hayes 1969), non-monotonic logic (McDermott and Doyle 1980), description logic, semantic networks (Levesque and Brachman 1987), and causal reasoning (Pearl 2000).

Cognitive empirical psychology working on learning processes has explored probabilistic models namely for describing and explaining the evolution and stabilization of learning. We may consider for instance Hebb's law which is expressed as a simple linear algebraic equation.

$$w_{ij} = \frac{1}{p} \sum_{k=1}^{p} x_i^k x_j^k,$$

Connectionism (Rumelhart and McClelland 1987) has explored linear and non-linear algebra, while Smolensky (1986) has proposed tensor algebra. Hohwy (2014) used Bayesian probability for predictive reasoning. Cognitive linguistics (Langacker 1987; Talmy 2000; Evans 2009) offer topological structures with homomorphisms for the mental lexicon and statements from mental spaces, among other things. Neurosciences also described and explained brain processes with linear and non-linear, dynamic, chaotic, geometrical, and Bayesian systems. Finally, robotics includes a hybrid set of such models. Each choice depends on the functions of their modules and their interaction.

The second class of formal models is that of the *computable* models. These models are a subset of formal models (not to be conflated with mathematical models) in that they use *calculable* symbolic systems. A calculable system is one where the manipulation of symbolic expressions (irrespective of meaning) can decide whether a specific expression (equation, formula) belongs or not to the system. Turing (1936) demonstrated that calculability is equivalent to the "computation" performed by a physical and mechanical machine called a *Turing machine*. All calculable systems are computable systems. Later on, they were defined in algorithmic terms (Markov 1960), in productive terms (Post 1936), and in combinatorial terms (Curry and Feys 1958). Von Neumann (1945) has in turn demonstrated that certain physical architectures of computers are a specific sort of concrete Turing machine.

Thus, in a more general formulation, a functional expression of a mathematical system is calculable or computable, algorithmic, etc., if there exists an *effective procedure* to process inputs and systematically produce results. In other words, computation guarantees that the procedure will stop.

In cognitive science, computational models are of the utmost importance. Some claimed (Fodor 2008; Pylyshyn 1984; among others) that cognitive processes can only be described and explained by computational models. But not all cognitivists accepted this thesis. Some connectionists, neuroscientists, and philosophers preferred algebraic, dynamic system models, which they saw as external to the computational paradigm of cognitive science for it allowed a reduction of the theory to the physical brain (van Gelder 1997; Brooks 1991).

> We wish to replace the 'computer metaphor' as a model of mind with the brain metaphor as models of the mind (Rumelhart and McClelland 1987:I:75).

Still, following more technical examination[3] (Davis 1982; Utal 2003) these models remain computational. They even have been developed into the "neuro-computing" field of research (Sejnowski and Churchland 1992).

[3]Davis and many others have proven that computational machines that are parallel exist. And computational dynamicity will depend on the nature of the equation itself.

> Computation might be understood as a collection of active recursive functions operating on symbolic list structures. Alternatively, it might be understood as parallel-operating "knowledge sources" reading from, transforming, and writing complex symbolic expressions on a "black board" (Nilsson 2007).

As Mundici and Sieg (1993:30) have argued, computational modelling in cognitive science modifies the understanding of a scientific law. A law can be defined in a formal way as classically proposed, but computation helps to detail the proof. However, more importantly, when the function is not known, but a sample of its behaviour is observable (i.e. a sample of its extension), learning algorithms can approximate the function. The law then is not necessarily expressed by means of a general expression (i.e. intensionally), but can nevertheless be approximated by means of an algorithmic model like a perceptron, a random forest, or the like.

What happens then to the realist and objectivist thesis in the context of cognitive formal models? The answer to this question is quite direct. Strictly speaking, and as said above, a formal model does not have a semantics. It is cannot ask the realist question. The role of a formal syntax in such a system is precisely to allow formulas or sentences to be generated by the axioms and rules of the system. This is related to their decidability and computability properties. But these properties have limits. Indeed, by the first Gödel theorem, no formal system guarantees that all its sentences are decidable and that the system is complete. This has lately been strongly reaffirmed but in a different form by Chaitin:

> Everywhere, from mathematics to computer science, to physics, to mathematically-formulated portions of chemistry, biology, ecology, and economics (Chaitin et al. 2012).

The non-computability problem indirectly affects the realist and objectivist thesis in cognitive formal models. A solution to this may be to propose adjustments to the system, such as Turing's solution of Oracles or the adding of rules or axioms in the systems. But in doing so, realism or objectivity is not better guaranteed. This only adds more complexity: "Just add new axioms, increase the complexity of your theory" (Chaitin et al. 2012:36). In this perspective, the only solution would be to adjust the system by controlling its semantic and pragmatic interpretation which, however and by definition, is exterior to a formal system. And this addition is subject to all the problems of interpretation and adjustment that are classical in the semantic view of theories.

Still, even staying in the inner structure of the formalisms themselves, it has been shown that they are often riddled with hidden technical formal problems that affect the result's objectivity and truth. For instance, many abbreviation symbols for example: the integral symbol) have presented many problems regarding the manipulation of mathematical symbols (Woodhouse 1803; Koppelman 1971). Another one is the manipulation of variables (Desclés 2006). So much that Curry and Feys (1958) have proposed a language without variables: combinatorial logic. Gelfert (2011) demonstrated that many formal models are often manipulated so as to adjust and adapt them to the situation to be modelled.

Formal cognitive models do not escape these problems. The multiple conditions for a strict computational model are not that easily met. A typical example is

Fodor's (1983) modular proposition for modelling cognitive operations. This was explored systematically in AI projects such as expert systems, for instance. These types of programs contained specific computable modules. But because they were well-formed, they were seen as interchangeable with other modules in other programs. And it was hoped that they could operate just as well in this new environment and preserve truthfulness. But this ideal was problematic. Even though the modules, when exported into a new environment, were still formally identical, their semantics may have changed. Adjustments require heuristics or *oracles*, and this increased the complexity and the risk of non-decidability. It was opening the door for 'bugs' (Meunier 2003).

Other problems touch upon the multiplicity and complexity of functions, their sensibility to evolution, their variety among subjects, etc. In other words, the formal models, even if very mathematical and computational, are not exempt of adaptive interventions and their semantics may include biases. This challenges the realist and objectivist ideal in a particular way.

4.2 Physical Models in Cognitive Science

One important problem of formal models is their epistemic distance with causal explanation (Salmon 1984). For instance, the physics equation $F = ma$ does say that F is *caused* by the multiplication of m by a! Equations do not deliver a complete explanation for the human understanding of a phenomenon. According to the Humean tradition in the philosophy of science (Thagard 2000), a causal explanation is one that allows predictability.[4] So scientific theory will often include a second class of models: the physical models. They contribute to such causal explanations.

These types of physical models are used regularly in natural science. They are so integrated that often they become transparent as models. But many researchers (Baetu 2013; Rheinberger 1997; Leonelli 2007) have studied examples of these types of models. A typical one is Newton's laws of motion. It has among its formal models the equation or law $F = ma$. A concrete physical model for this equation takes the form of an implementation in specific and selected physical entities linked by observable causal relations. And this physical model concretely contributes to the semantics of the symbols "F," "m," "a," "=," and "x" (hidden multiplication symbol) by linking them to a numerical value associated to a concrete *mass*, a concrete *acceleration* for an effective *force*[5] and operations on these values.

[4]There are significant formal problems with this position, for not all causal explanations allow predictions. For instance, a chaotic explanation is not necessarily predictive and all predictive explanations are not necessarily causal. Correlation is predictive, but not necessarily causal.

[5]This could be reformulated in set-theoretical terms and constitute a formal Tarskian model.

When a many-model implementation is well controlled and systematized, it becomes, in fact, a many-individual physical experiment which can be considered as concrete proof of the formal model. And various statistical analyses will allow the generalization of the variations of a set of experimental numerical values. In other words, we can see experimental implementations of formal equations as physical models distributed over time.

These types of physical models are omnipresent in cognitive science. They define a first class of physical models which Bechtel (2008) and Craver (2016) call causal mechanical models:

> Mechanists insist explanation is a matter of elucidating the causal structures that produce, underlie, or maintain the phenomenon of interest (Craver et al. 2016).

> The term mechanism is as ubiquitous in psychology, cognitive science, and cognitive neuroscience as it is in the domains of biology to which philosophers have appealed in articulating the account offered above of what a mechanism is (Bechtel 2008:22).

The brain for instance is such a mechanism:

> Brain mechanisms, accordingly, are far more likely to be structured and organized in ways particularly suitable to the tasks they must perform (Bechtel 2008:3).

The brain is seen as a causal mechanism constituted by parts that are cells with various levels of organization that form a complex neural structure. Their interactions are made of electrochemical synaptic connections. Even if the brain is not a *machine* mechanism, it constitutes a *biological* mechanism. Because of this argument, the brain is modelled as a sort of computing machine. But it cannot be seen as an electronic computer. It is a neurocomputing machine. In neural theory, it can thus be seen as a physical mechanical model that instantiates/implements the computable functions given in the formal models.

In this same perspective, it follows that connectionist models are not physical models implementing formal equations. They are formal models: a set of linear and non-linear dynamic algebraic equations. They have effectively been called artificial neural net models and they are used to describe the dynamics of physical neurons.

More examples of physical models are found in other cognitive sciences. A cognitive psychologist may be confronted with sets of data originating from causal experiments where physical neurons of monkeys are actually stimulated these data in turn, can be submitted to learning algorithms (Michalsky 1983; Mitchell 1997) that can approximate the functions describing the causal relations among the data. And a formal model could express this function in an algebraic equation such as for example Hebb's law. In neuroscience, Eliasmith (2003) identifies such regularities in the activation and pathways of physical neurons and offers mathematical dynamic system models to explain these behaviour patterns.

The second class of physical models are strictly speaking *computer* models. As we have said above, computational models are mathematical or virtual models. Turing (1936) did in fact distinguish a virtual computing machine *A* and an actual physical machine *B*. The *A* machine was a diagrammatic formal computable function. The *B* machine, a mechanism made namely of a motor, some paper, and ink (often called the "the Turing machine"), *instantiated* physically the virtual machine *A*. Its purpose was to concretely and effectively implement the computation of a formal function. A contemporary computer is a *B* Turing machine with a Von Neumann architecture realized physically in an electronic circuit. It has *effective procedures* (Copeland 2000) that compute a "computable" function. In this sense, each *effective* computation by a computer is a physical model implementing formal functions.

Artificial intelligence programs are typically "implemented" in physical causal computer models. Often, these types of physical modelling of "intelligent behaviors" are called computer *simulations*. For instance, in simulating vision, a computer may receive through its own captor's input signals similar to the ones received by the cones, rods, and iris of the eyes. It then applies to these input signals some computer effective electronic process so as to produce a behaviour simulating an eye seeing something.

A variant of visual simulation is computer-based visual *representation* often called "visualization." The "causal" state transitions of the computing operations are projected on a monitor, where they are "illustrated" by colouring pixels in such a way as to be recognized by humans as images or as iconic figures, for instance. In many AI and cognitive sciences, visualization is a part of the proof methodology: For example, an IMR activation visualization uses colour intensity and tone to represent the degree of activation of computational functions translating the impulses in a specific brain area.

In more sophisticated visualization simulations, metaphorical names are given to these simulations. In information retrieval, paths of research may be called *maps*, or nets (Gentner 1983). Many complex cognitive behaviours (analysis, decision-making, strategies, narration, etc.) are simulated by means of "games." Complex "electronic machines," simulating very complex cognitive processes, are called "robots." They are complex computer instantiations of formal computational models of cognitive behaviours. Technically speaking, robots do not "think," "decide," "desire," or whatever. No more than there is a real "trash bin" or a real "desk" on the monitor screen.

How do these physical models relate to our realism and objectivity problems? The following argument by Craver (2006) explicitly reveals it. As he well stated:

> Constitutive explanations go beyond merely describing the phenomenon. They describe the mechanism responsible for the phenomenon, that is, the mechanism that explains its diverse features[6] (Craver 2006:153).

As explained when addressing the case of formal models, the physical model contributes to the semantics of formal models. Models are mediators between the formula of the formal models and the physical phenomenon itself. They are proxies (replicas, simulations, implementations) for the phenomenon. As physical entities, they possess observable properties or features. But not all of their properties and

[6]The devil's details for the realist and objectivist thesis applied to the physical models hide under the hood of the words *phenomenon* and *features* of the phenomenon.

features are relevant for a scientific theory. For example, if pushing a billiard ball on a carpet is taken as a physical model, not all the properties and features of this model are relevant for the Newtonian law $F = ma$. For instance, the colour and the type of marble of the ball are not relevant. A choice must be made, and this choice has criteria. A first one requires that the features chosen must be representable in the formal model. For instance, the mass is representable by "m" in $F = ma$. But another one is preliminary to this first one. For example, what, technically, is the physical property to be put into correspondence or to be associated with the mass? Is it its *weight?* Its *molecular structure*? Is it its *electromagnetism*? In natural science, instruments and controlled protocols are often the means by which these features are identified as pertinent and to which a numerical value can be associated. And strict observation language is used to express these features. Still, it has been shown that these means imply theoretical commitments on the part of the researchers.

In cognitive science, these problems are also to be found in the physical models. Choices of relevant features must be made. But here, the instruments and observation language cannot be controlled as easily when the physical models require complex interpretative acts. For instance, a computerized physical model may associate to its variables and operations certain physical entities and apply to them effective electronic computing processes. The dynamics of the processes and their results can be associated with some visualizations that are recognizable by scientists. For example, a brain scan will "dynamically" represent states of neural activation via the colouring of pixels of a computer monitor. These visualizations are then interpreted as "areas", "paths", etc. But these interpretations are committed to complex physical theories and are associated to concepts pertaining to phenomenal and introspective if not cultural accounts, as in the example in the Kanizsa triangle illusion referred to above. I see a *circle*, *triangle*, etc. A nice illustrative example of this is the brain scans of nuns in relation to their *verbal accounts* of their "spiritual experiences". The activated areas are interpreted as "spiritual areas". The authors of the study prudently said that these areas could not be identified as "God spots" (Beauregard 2006).

In other words, the semantic and pragmatic role that physical models play by their properties and features will often be influenced in their construction by some paradigmatic theories. In this sense, cognitive science physical models often require many implicit theoretical commitments if not even some tacit personal and cultural influences. This directly challenges the realist and objectivist ideal for the physical models.

4.3 Conceptual Models in Cognitive Science

Scientific research usually starts by an interrogation about a phenomenon. For example: *Why do apples fall?* Or: *Why is there a tide? Why do humans have emotions? How does the brain memorize places? Do neurons learn?* The answers to these questions cannot, for example, be given only in the form of equations, physical replica, or simulations of a phenomenon. A scientist usually adds some natural language sentences that for himself or for his epistemic community expresses a conceptualization of the problem and the way he would cope with it. In other words, he presents a conceptual model of the research problem.

Conceptual models share similar purposes with the other types of models. They are also mediators but they are used mainly for understanding, discovering, justifying, and communicating the research problem and solution explored in the scientific enquiry. They are not always expressed in natural language for they may take various other semiotic forms such as pictures, graphs, or films, all of which have their own ways of expressing conceptual structures. Still, these conceptual models have some specificity.

First of all, conceptual models determine the conceptual framework, the conceptual "space" (Gardenfors 2000), conceptual system (Brown 2007), and mental models (Johnson-Laird 1986) that are part of human explicit or tacit knowledge (Polanyi 1967), and through which humans ultimately understand the explanations given in a scientific theory. Secondly, they are heuristic in that they express various formulations of intuitions, hypotheses, and the methodologies upon which the scientific theory rests. Thirdly, they have a communicative role and they are mostly expressed in natural language, meaning the various contents will be shared with different epistemic communities.

These conceptual models are omnipresent in science, so much that they become transparent to the user. And even if they are not as rigorous as other models, they are still models in the strict sense because, as Cartwright says, models are an idealized and simplified representation:

> A model is by nature a simplified and therefore fictional or idealized representation, often taking quite a rough-and-ready form: hence the term "tinker toy" model from physics, accurately suggesting play, relative crudity, and heuristic purpose (Cartwright 1983:158).

A classical example explored in natural science is the Copernican theory, by Kuhn (1957). One of the important debates around this theory was not only about the mathematical equations describing and predicting the movement of planets. It was also about the conceptual space in which the formal model was understood. It clashed with the existing religious conceptual spaces.

In cognitive science, a nice example of a transparent conceptual model is to be found in M. Graziano's introduction to his neuropsychology book *Consciousness and the Social Brain*. Here is a sample of this introduction:

> The brain is composed of neurons that pass information among each other. Information is more efficiently linked from one neuron to another, and more efficiently maintained over short periods of time, if the electrical signals of neurons oscillate in synchrony. Therefore, consciousness might be caused by electrical activity of many neurons oscillating together (Graziano 2013:6).

Here, we are not reading equations or seeing some physical instance or replica of neurons. We are reading sentences that express some concepts specific to a conceptual framework built out of past and contemporary cognitive science and other theoretical influences, such as concepts from neuroscience, information theory, control theories, philosophical theories of causes, consciousness, attention, etc. They participate in Graziano's conceptual models of his theory of consciousness and attention. In fact, they take up some 80% of the book.

Such conceptual modelling is omnipresent in all sciences. We shall explore here three main types that are usually found in cognitive science: (a) the intentional, (b) the observational, and (c) the rhetorical models.

The first type, surely the most classical one, is the *representational* (Fodor 1981; Pylyshyn 1984) or *intentional* model. Such a model, says Dennett, explains the systems' behaviour by ascribing goals to it.

> One predicts behaviour in such a case by ascribing to the system the possession of certain information and supposing it to be directed by certain goals (Dennett 1971:224).

This type of model explores and explains cognitive phenomena by grounding them in a principle of rationality, a principle that ultimately allows inferentiality and normativity in the manipulation of representations (Sellars 1948; Brandom 1994).

> The practical uptake of specifically representational purport must include normative assessment of states, performances, and expressions—assessment of their specifically representational correctness (Brandom 1994:78).

Finally, these models, even formulated simply as in folk psychology, are the anchors for understanding. For humans, they serve as mediator in the interpretation and the understanding of cognitive phenomena.

A second type of conceptual models is the *observational models*. They are popular with the empiricists if not the experimentalist scientists:

> [E]xperiential data might be conceived of as being sensations, perceptions, and similar phenomena of immediate experience (Hempel 1952:740, 829).

Their aim is to report cognitive experiences as perceived by a cognitive agent. And if used in science, they have to be controlled, instrumentalized, and formalized. But whatever they are used for, they are usually expressed in natural language terms and sentences such as the following examples of first-person experience accounts: *I see, I fell, I hear, I listen, I perceive, I touch, I taste*, etc.

The sentences built out of these words refer to subjective data and become translated into variables and operations. Often, in some cognitive sciences, the observation may belong to inner experiences as described by an experimental subject. In his brain exploration by direct stimulation, Penfield (1959) could only describe the inner reaction of the subject by the phenomenological sentences and words given by the patient: *I see myself with my son, I recall my mother*, etc. This type of observational model has been shown to contain complex epistemological and epistemic problems specific to observation languages or even to other types of conceptually explicit semiotic forms such as iconic languages (maps, graphs, etc.).

A third type of conceptual model is a *rhetorical* conceptual model. In explaining a phenomenon, it explores various types of rhetorical strategies so as to mediate the understanding of complex phenomena. Many philosophers of science (Hartmann

1995; Knuuttila 2011) have underlined the presence of such types of models in science. As Morgan and Morrison say, one main task of models is "fitting together [...] bits which come from disparate sources [...] [including] stories" (Morgan and Morrison 1999:15). To express these problems, conceptual models will take rhetorical forms such as analogies, fictions, and metaphors, if not narration. Lakoff and Johnson (2003) are among those who regard metaphors and analogies as the main underlying process of thinking. They are omnipresent in natural and cognitive sciences.

As Hesse has shown, analogies and metaphors are often used in science to introduce and manipulate various theoretical entities: "most physicists do not regard models as literal descriptions of nature, but as standing in a relation of analogy to nature" (Hesse 1963:2011). A classical example of this is the famous illustrations of the Brownian movement through billiard balls or the atom as a planetary system (Rutherford and Bohr).

Artificial intelligence constantly uses metaphors in presenting its research programs. For instance, the behaviour of a robot cannot be explained only by a list of program lines. Azimo, for example, is presented as "thinking", "communicating", "meeting people", answering questions, etc. Such words referring to human cognition are analogically projected onto robots. Metaphors are found to be more "hard" and formal in connectionist models. Many of their concepts are explained in a conceptual model. They are translated in natural language by words such as *stimuli, impulses, point of gravity, attractors, stabilization, chaos*, or *equilibrium*, or by sentences such as: *Neurons communicate with other neurons. They talk to each other, exchange information.*

In cognitive science, one important form that a conceptual model often takes is called pop psychology. Many cognitive behaviours are then modelled through a variety and a mixture of general concepts, analogies, metaphors, and fictions. Concepts such as *motivation, beliefs, desires, attention, decision consciousness, will*, or *temptations* are of the sort. They belong to a Judeo-Christian folk philosophy of mental processes. For naturalists (Smart 1959; Armstrong 1968; Churchland 1988), such discourse is useless, and it should be discredited and abandoned. It does not belong to a scientific theory or model. They should be *reduced* to empirical "causal and physical models concepts".

Dennett (1978) and Fodor (2008) however recognize the role of these folk discourses in science. Analogies, fiction, and metaphors are heuristic means in thinking. We find them at the beginning of research where the *explanandum* is presented. They reappear at the middle of the research where the *explanans* are elaborated and inferences are made. And at the end, results of the research are anchored in the conceptual space of the individual or the epistemic community. Some complex mathematical equations are sometimes only understandable by humans if translated into folk-understandable terms such as *chaos, bifurcation, emergence, attractors, oracles*, if not *motivation, will*, and *communication*.

There probably exist other variants of the conceptual models. We have presented only three of them: the intentional models, the observational models, and the rhetorical models. Their difference lies in the content of the conceptual models.

They offer different points of view, emphasize different dimensions of the research problem and therefore determine different conceptual spaces. Theories are introduced, illustrated, and ultimately understood by means of conceptual models. They are a necessary part of scientific theory.

How does the conceptual model deal with realism and objectivity? As we may guess, the various conceptual models directly challenge once more the realist and objectivist thesis. They are often taken as proofs by anti-realists of the non-validity of the realist thesis. They are prototypes of the constructiveness inherent to scientific theories. Here are a few of the problems these models raise.

A first one pertains to the predicates the conceptual model chose to express the features and properties of the cognitive processes. Where do they come from? Most of them do not originate from the experimental schema or the rigorously controlled instruments. For instance, some of these predicate are in the narrative of personal introspection or cultural analysis. For example, the predicate "mourning" names a specific cognitive experience belonging to a personal or cultural experience. And from a purely lexical semantic point of view this predicat has its own social or cultural connotation. Even with standardized observation criteria, a scientist "observing" the Kanizsa triangle will rely on his introspection of the illusion. Other concepts are part of general culture. For example, the results of Rizzolatti's (1999) experiments on the brain of macaque monkeys have been translated into conceptual terms such as *mirror* and *neurons*, and some of their functions are translated as *empathy*. If one does not situate the concept of *mirror*, *neurons*, or of *empathy* in their own conceptual space, understanding will be difficult. Hence, because the predicates used in conceptual models are often grounded in subjective first-person experiences (introspection) or shared with third-person reports (culture), pure realism and objectivity in the conceptual model is challenged.

The second problem of the conceptual model pertains to the overall language chosen to express the concept. As said before, the language usually chosen is a natural language and not a formal one. Hence, the model will suffer of the many defects of a natural language. For instance, at the level of syntax, not all sentences or lexemes will be strictly controlled. And at the level of semantic and pragmatics, the various epistemic conventions for building the conceptual models are not always stable and explicit. Ambiguity will be omnipresent, formal rules of inference will not be strictly followed. Generalizations will be fuzzy and modality will be implicit. For instance, when describing the cognitive behaviour of an Alzheimer's patient, what does it mean to say *he is disoriented*? Does it mean: losing one's bearings or not knowing where one is in one's house? Does it not also imply a normative judgment?

As one may guess, such language directly affects the rigour of science and therefore contravenes to the realist and objectivist thesis.

Let's conclude by saying that maybe some user may think that the conceptual model is objective and expresses reality, but the language or semiotic forms chosen will often use predicates whose semantics and pragmatics are grounded in subjective and cultural conceptual spaces. These predicates suffer from the main defects (but also richness) of all natural semiotic forms: ambiguity, fuzziness, incorrectness in reasoning, etc. The realist thesis is therefore directly questioned in the conceptual models.

5 Conclusion

The epistemological notion of realism and objectivity, when applied to formal sciences, poses many epistemological problems. We have analyzed the problems that the realist and objectivist encounter in the theories of cognitive science. We have argued that it is not in regard to the concept of *theory*, be it understood in the syntactic or semantic manner, that the realist and objectivist positions can be best discussed. We preferred the pragmatic approach for it understands a theory as a set of interacting models.

We have also argued that there exist at least three different classes of models that interact constantly in cognitive science. As Braillard and Malaterre (2015) remind us, this interaction has to be better understood. It lacks precision. It seems interesting to see that it is this interaction of models that allows the emergence of an objective apprehension of reality.

The realist and objectivist problem may pose the question of reality. But as Schiffrin expresses it, it is not the role of models to pose ontological questions. Their main function is cognitive, they "enable progress in our understanding of this vastly complex system" (Schiffrin 2009:736). Still, we have argued that the realist and objectivist thesis are impaired in each type of model. Each one is theory laden and allows many epistemic biases to interfere. Formal models are robust only regarding decidability. Their semantic association with physical models is fragile. Experiments have to be repeated in order to ensure some stability of the apprehend of reality. Most of all, conceptual models, expressing conceptual spaces, are dependent upon viewpoints, on culture, if not on pure subjectivity and situations.

Still, we believe, the question of realism and objectivity in cognitive science invites us to see these terms as not only predicable about theory, but also about models. This seems an important change in the exploration of the realist and objectivist questions.

References

Armstrong, D. M., (1968). *A Materialist theory of Mind*. Routledge and Kegan Paul, London.

Bachelard, R., (1979). Quelques aspects historiques des notions de modèle et de justification des modèles. In P. Delattre & M. Thellier, (Éd.), *Elaboration et justification des modèles*. Paris: Maloine.

Baetu, T. M., (2013). When Is a Mechanistic Explanation Satisfactory? Reductionism and Antireductionism in the Context of Mechanistic Explanations. *Romanian Studies in Philosophy*

of Science. Volume 313 of the series *Boston Studies in the Philosophy and History of Science*, 255–268.
Bechtel, W., (1998). Representations and Cognitive Explanations: Assessing the Dynamicist's Challenge in Cognitive Science, *Cognitive Science*, 22, (3), pp. 295–318.
Bechtel, W., (2008). *Mental Mechanisms: Philosophical Perspectives on Cognitive Neuroscience*, London: Routledge.
Brachman, R., Levesque, H., eds., (1985). *Readings in Knowledge Representation*, Los Altos: Morgan Kaufmann.
Braillard, P., Malaterre, C., (2015). *Explanation in Biology*, Springer.
Brandom, R. B., (1994). *Making It Explicit: Reasoning, Representing, and Discursive Commitment*, Cambridge, MA, Harvard University Press.
Brooks, R. A., (1991). Intelligence without representation. In Haugeland, J., (ed.) *Mind Design II*, Routledge, (1997), pp. 395–420.
Brown, H. L., (2007). *Conceptual Systems,* Routledge Studies in the Philosophy of Science, Routledge.
Burch, T. K., (1999). Computer Modelling of Theory: Explanation for the 21st Century. *PSC Discussion Papers Series* 3-1.
Carnap, R., (1937). *The Logical Syntax of Language*, London: Kegan Paul, Trench, Trübner & Co.
Cartwright, N., Shomar, T., Suárez, M., (1995). "The Tool Box of Science: Tools for the Building of Models with a Superconductivity Example," in *Theories and Models in Scientific Processes*, (Poznan Studies in the Philosophy of the Sciences and the Humanities, Volume 44), Amsterdam: Rodopi, pp. 137–149.
Cartwright, N., (1983). *How the Laws of Physics Lie.* Oxford University Press.
Chaitin G. J., Doria, F. A., da Costa N. C. A., (2012). *Gödel's Way: Exploits Into an Undecidable World*, Boca Raton, CRC Press.
Chang, H., (2011). "The Philosophical Grammar of Scientific Practice" in *International Studies in the Philosophy of Science*, 25, (3): 205–221.
Chomsky, N., (1957). *Syntactic Structures.* The Hague: Mouton.
Churchland, P. M., (1988). *Matter and Consciousness.* Cambridge, (Mass.): MIT Press.
Churchland, P. S. & Sejnowski, T. J., (1992). *The Computational Brain*, The MIT Press.
Copeland, B. J., (2000). "Narrow Versus Wide Mechanism Including a re-examination of Turing's views on the mind-machine issue," *Journal of Philosophy 96,* pp. 5–32.
Craver, C. F., (2006). When mechanistic models explain, *Synthese*, 153(3), pp. 355–376.
Curry, H. B., Feys, R., (1958). *Combinatory logic*, Vol.1, Amsterdam, North-Holland.
Davis M., (1982). *Computability and Unsolvability, Dover.*
Duhem, P., (1906). *La théorie physique*: *Son objet et sa structure*, Paris: Chevalier et Rivière; transl. by P.W. Wiener, *The Aim and Structure of Physical Theory*, Princeton, NJ: Princeton University Press (1954).
Dennett, D., (1978) *Brains Storms.* Cambridge: MIT Press.
Desclés, J. P., & Cheong, K. S., (2006). Analyse critique de la notion de variable, *Math. & Sci. hum. Mathematics and Social Sciences*, 44e année, n 173, 2006, (1), p. 43–.
Eliasmith, C., Anderson, C. H., (2003). *Neural engineering: Representation, Computation, and Dynamics in Neurobiological systems.* Cambridge, Mass: MIT Press.
Evans, V., (2009). *How words means Lexical concepts, cognitive models, and meaning construction*, Oxford University Press.
Feyerabend. P., (1975) *Against Method: Outline of an Anarchistic Theory of Knowledge*, Verso Books, New York.
Fodor J. A., Pylyshyn Z. W., (1988). Connectionism and Cognitive Architecture: A Critical Analysis. *Cognition.* Mar; 28 (1–2):3–71.
Fodor J. A., (2008). *LOT2: The Language of Thought Revisited*, Oxford, Oxford University Press.
Fodor, J. A., (1981). *Representations.* Cambridge: MIT Press.
Fodor, J. A., (1983). *Modularity of Mind*: *An essay on faculty psychology* Cambridge: MIT Press.

Frigg, R., Hartmann, S., (2012). "Models in Science", *The Stanford Encyclopedia of Philosophy,* (Fall 2012 Edition), Edward N. Zalta, (ed.), URL = <http://plato.Stanford.edu/archives/fall2012/entries/models-science/>.

Gardenfors, P., (2000). *Conceptual Spaces.* Cambridge, (Mass.): MIT Press.

Gelfert, A., (2011). Mathematical formalisms in scientific practice: From denotation to model-based representation, *History and Philosophy of Science*, 42(2), pp. 272–286.

Genesereth, M., Nilsson, N. J., (1987). *Logical Foundations of Artificial Intelligence*, San Mateo, California: Morgan Kaufmann.

Gentner, D., (1983). Structure mapping: A theoretical framework for analogy. *Cognitive Science 7 (2)*, pp. 155–170.

Giere, R. N., (1988). *Explaining Science: A Cognitive Approach*, Chicago: University of Chicago Press.

Giere, R. N., (1999). *Using Models to Represent Reality*, in L. Magnani, N. Nersessian and P. Thagard, (eds.), *Model-Based Reasoning in Scientific Discovery.* Plenum Publishers: New York, pp. 41–57.

Givón, T. (Ed.), (1979). *Syntax and Semantics 12: Discourse and Syntax*, New York: Academic Press.

Godfrey-Smith, P., (2009). Models and Fictions in Science. *Philosophical Studies*, 143(1), pp. 101–116.

Graziano, M. S. A., (2013). *Consciousness and the Social Brain*, Oxford University Press.

Green, S., (2013). *When one model is not enough: Combining epistemic tools in systems biology. Studies in History and Philosophy of Science Part C: Studies in History and Philosophy of Biological and Biomedical Sciences*, 44(2), 170–180.

Gruber, T. R., (1993). "A translation approach to portable ontology specifications", (PDF), *Knowledge Acquisition*, 5(2): pp. 199–220.

Harnad, S., (1994). Computation is Just Interpretable Symbol Manipulation; Cognition Isn't, *Minds and Machines*, 4(4), pp. 379–390.

Harnad, S., (2005). "To cognize is to categorize: Cognition is categorization", *Handbook of categorization in cognitive science*, p. 20–45.

Hartmann, S., (1995). *"Models and Stories in Hadron Physics", in Morgan and Morrison 1999*, pp. 326–346.

Hempel, C. G., (1965). *Aspects of Scientific Explanation and other Essays in the Philosophy of Science.* New York: Free Press.

Hesse, M. B., (1963). *Models and Analogies in Science.* London: Sheed and Ward.

Hohwy, J., (2014). *The Predictive Mind*, Oxford University, Press.

Johnson-Laird, P. N., (1986). *Mental models: Towards a cognitive science of language, inference, and consciousness*, Harvard University Press.

Knuuttila, T., (2011). Modelling and representing: An artefactual approach to model-based representation. *Studies in History and Philosophy of Science*, 42(2), pp. 262–271.

Koppelman, E., (1971). The calculus of Operation and the Rise of abstract algebra. *Arch. Hist. Exact Sci.*, 8(3), pp. 155–242.

Langacker, R.W., (1987). *Foundations of Cognitive Grammar.* Stanford University Press.

Lakoff, G., Johnson, M., (2003). *Metaphors we Live by.* The University of Chicago Press.

Leonelli, S., (2007). What is in a Model? Combining Theoretical and Material Models to Develop Intelligible Theories. In Manfred Dietrich Laubichler, Gerd B. Mülle (eds.) *Modeling Biology: Structures, Behaviors, Evolution*, MIT Press.

Leplin, J., (ed.), (1981). *Scientific Realism.* Berkeley: University of California Press.

Levesque, H. J., Brachman, R. J., (1987), "Expressiveness and tractability in knowledge representation and reasoning". *Computational Intelligence*, 3(1), pp. 78–93.

Markov, A., (1960). The Theory of Algorithms. *American Mathematical Society Translations*, series 2, n 15, pp. 1–14.

Marr, D., (1975). *Computer vision*, P. Winston, (Ed.), McGraw Hill.

McClelland, J. L., (1998). *Connectionist Models and Bayesian Inference.* In *Rational Models of Cognition*, M. Oaksford & N. Chater (Eds.), Oxford UP.

McDermott, D., Doyle, J., (1980). Non-monotonic logic I. *Artificial intelligence*, 13(1-2), pp. 41–72.
Meunier J. G., (2013). Computers as Models of Computers and Vice Versa. *Epistemologia*, 2013:18–37.
Michalski, R. S., (1983). *A theory and methodology of inductive learning*. Springer.
Minsky, M., (1974). "A framework for representing knowledge", Tech. Rep. 306, *Artificial Intelligence Laboratory*, MIT.
Mitchell, T., (1997). *Machine Learning*, McGraw Hill.
Morgan, M. S., & Morrison, M. (1999). *Models as mediators: Perspectives on natural and social science* (Vol. 52). Cambridge University Press.
Morris, C. W., (1971). *Writings on the General Theory of Signs*. The Hague: Mouton.
McCarthy, J., and Hayes, P. J., (1969). Some philosophical problems from the standpoint of artificial intelligence. In Meltzer, B., and Michie, D., eds., Machine Intelligence volume = 4. Edinburgh University Press. 463–502.
Mundici, D., & Sieg, W., (1995). "Paper Machines", *Philosophia Mathematica*, 3(1), pp. 5–30.
Nagel, E., (1961). *The Structure of Science: Problems in the Logic of Scientific Explanation*, New York: Harcourt, Brace & World.
Newell, A., Simon, H., (1976). *Symbol Manipulation*. In Ralston A., Meek C. L., (eds.), (1976), pp. 1384–1389.
Newell, A., (1994). *Unified Theories of Cognition*. Cambridge, MA: Harvard University Press.
Neumann J. V., (1945). First Draft of a Report on the EDVAC Published in IEEE Annals of the History of Computing, Volume 15 Issue 4, October 1993, pp 27–75.
Nilsson, N. J., (2007). The Physical Symbol System Hypothesis: Status and Prospects in M. Lungarella et al., (Eds.): *50 Years of AI, Festschrift, LNAI 4, 85, 9–17.*
Papineau, D., (2010). "Realism, Ramsey Sentences and the Pessimistic Meta-Induction", *Studies in History and Philosophy of Science*, 41(4), pp. 375–385.
Pearl, J., (2000). *Causality: Models, Reasoning, and Inference*, Cambridge, England: Cambridge University Press, ISBN 0–521-77362-8.
Peirce, C. S., (1931-58). *Collected Papers* C. Hortshorne, P. Weiss, and A. Burks, Cambridge, MA: Harvard University Press.
Penfield. W., (1959). "The interpretive cortex", in: *Science*, 129(3365), pp. 1719–1725.
Petitot J., (2009). *Morphogéométrie de la vision*, Paris, École Polytechnique.
Piccinini, G., (2007). "Computing Mechanisms", *Philosophy of Science*, 74(4), pp. 501–526.
Polanyi, M., (1967). *The Tacit Dimension*, New York: Anchor Books.
Post, E. L., (1936). Finite combinatory processes—formulation. *The Journal of Symbolic Logic*, 1 (03), pp. 103–105.
Pylyshyn Z. W., (1984). *Computation and Cognition: Toward a Foundation for Cognitive Science*. Cambridge, MA, MIT Press.
Reichenbach, H., (1938). *Experience and Prediction: An Analysis of the Foundations and the Structure of Knowledge*, Chicago: The University of Chicago Press.
Rheinberger, H. J., (1997). *Toward a history of epistemic things: Synthesizing proteins in the test tube*, Stanford: Stanford University Press.
Rizzolatti, G., Fadiga, L., (1999). "Resonance Behaviors and Mirror Neurons". *Archives Italiennes de Biologie*. 137: 85–100.
Rumelhart, D. E. & J. L. McClelland, (Eds.), (1987). *Parallel distributed processing: Explorations in the microstructure of cognition*, Vol. 1, (Chap. 7, pp. 282–317), Cambridge, MA: MIT Press.
Salmon, W. C., (1984). *Scientific Explanation and the Causal Structure of the World*, Princeton, NJ: Princeton University Press.
Scott, D. S., & Strachey, C., (1971). *Toward a mathematical semantics for computer languages*. Oxford University Computing Laboratory Programming Research Group-Library 8–11 Keble Road, Oxford OX, 3QD Oxford, (0865) 54, 141.

Scriven, M., (1962). "Explanations, Predictions, and Laws", in *Scientific Explanation, Space, and Time* (Minnesota Studies in the Philosophy of Science: Vol. 3), H. Feigl and G. Maxwell (eds), 170–230. Minneapolis: University of Minnesota Press.

Sejnowski, T. J., & Churchland, P. S., (1992). *The Computational Brain*. Cambridge, Mass: MIT Press.

Sellars, W., (1948). *Concepts as Involving Laws and Inconceivable Without Them. Philosophy of Science*, 15(4), pp. 287–315.

Schiffrin, R. M., (2009). Perspectives on Modeling in Cognitive Science 2, *Topics in Cognitive Science*, 2(4), pp. 736–750.

Smart, J. J. C., (1959). "Sensations and Brain Processes", *Philosophical Review*, 68(2), pp. 141–56.

Smolensky P., (1986). Information Processing in Dynamical Systems: Foundations of Harmony Theory. In Rumelhart D. E., McClelland J. L., (eds.), pp. 194–281.

Sowa, J. F., (2000). *Knowledge Representation: Logical, Philosophical, and Computational Foundations*, Brooks Cole Publishing Co., Pacific Grove, CA.

Suárez, M., (2003). "Scientific Fictions as Rules of Inference" In: Mauricio Suárez, (ed.): *Fictions in Science. Philosophical Essays on Modelling and Idealisation*, Routledge: London, pp. 158–178.

Suppes, P., (1967). "What is a Scientific Theory?", in S. Morgenbesser, (ed.), *Philosophy of Science Today*. New York: Basic Books, pp. 55–67.

Thagard, P., (19xx). *The Cognitive Science of Science*: *Explanation, Discovery, and Conceptual Change*, MIT Press.

Talmy L., (2000) *Toward a Cognitive Semantics* Cambridge, MA: MIT Press 2000.

Turing A. M., (1936). On Computable Numbers, with an Application to the Entscheidungs problem, *Proceedings of the London Mathematical Society*, 42(2), pp. 230–265.

Uttal, W. R., (2011). *Mind and brain: A critical appraisal of cognitive neuroscience*. Cambridge, MA: MIT Press.

van Fraassen, B. C., (1970). "On the Extension of Beth's Semantics of Physical Theories," *Philosophy of Science*, 37(3), pp. 325–339.

van Fraassen, B. C., (1989). *Laws and Symmetry*, New York: Oxford University Press.

van Gelder, T., (1997). Dynamics of Cognition. In J. Haugeland (ed), *Mind design II*, MIT Press.

Weisberg, M., (2015) *Simulation and Similarity: Using Models to Understand the World* Oxford: Oxford University Press.

Winther, R. G., (2015). "The Structure of Scientific Theories", *The Stanford Encyclopedia of Philosophy*, (Spring 2016 Edition), Edward N. Zalta, (ed.), URL = <http://plato.stanford.edu/archives/spr2016/entries/structure-scientific-theories>.

Woodhouse, R., (1803). *The Principles of Analytical Calculation*. Cambridge 1803: 212.

Wright, C., (1993). *Realism, Meaning and Truth*, Oxford: Blackwell. Aaron Stump, Programming Language Foundations. Wiley, 2014.

Realism as the Methodological Strategy in the Cognitive Science

V.A. Lektorski

Abstract The author discusses philosophical and methodological presuppositions of investigations in Cognitive Science. In this context a traditional philosophical and psychological idea of consciousness as the only certainty is critically analyzed, and the understanding of mental content in psychology and earlier cognitive science is studied. A special attention is given to a popular idea of situated, embodied and inacted Cognitive Science and to discussions about its philosophical interpretation. The idea of "the methodological solipsism as a strategy of Cognitive Science" by J. Fodor is criticized, as well as the interpretation of the situated Cognitive Science as going beyond the dichotomy of Realism-Idealism (A. Varela and others). The thesis about Activity Realism as an adequate interpretation of the idea of situated Cognition is argued. In this context the author analyzes the notion of "affordances" by J. Gibson, the idea of an interconnection between the Real World and an epistemic agent (the idea of "many worlds"), the role of actions and operations of a cognitive being in forming contacts with real objects. The problem of illusion and reality is analyzed, as well as relations between Naïve and Scientific Realism. A special attention is given to artificial objects as a special kind of existence.

1 Introduction

During the last 20 years anti-realism in different forms has become popular in Russian research in the sphere of humanities, human and cognitive sciences (philosophy, historical studies, psychology, neurosciences, etc.). At present some scholars and scientists in Russia think that philosophical realism in understanding science (scientific realism) and ordinary experience (naïve realism) is an anachronism, that philosophy, science and social life have shown its failure and that its defense is impossible.

V.A. Lektorski (✉)
Institute of Philosophy of the Russian Academy of Sciences, Moscow, Russia
e-mail: v.a.lektorski@gmail.com

E. Agazzi (ed.), *Varieties of Scientific Realism*,
DOI 10.1007/978-3-319-51608-0_19

Nevertheless, I make such a defense and argue the following thesis: a certain conception of realism is a philosophical position that not only adequately interprets facts of cognition and consciousness, but also formulates the best methodology for contemporary research in this field.

1.1 Consciousness as the Only Certainty?

Doubts about a possibility to know the world as it really is, that is, independent of its cognition, arose already in ancient times. They are expressed in particular in the famous relativist words of Protagoras: "Something is as it appears to anybody". But such ideas were successfully refuted, So philosophy in Antiquity and Middle Ages had ontological orientation—the world determines the possibilities of its cognition. Certainly, beliefs and knowledge, illusion and reality were sharply distinguished and even opposed to each other. Philosophy arose as a critique of those things that are considered as indubitable from an ordinary point of view. But ordinary opinions and even illusions were considered as having ontological correlates.

Subjectivism in understanding knowledge and cognition appeared in the framework of the "epistemological turn" in philosophy in 17th century. It is connected with Descartes's "discovery of consciousness". The existence of consciousness and Ego as its bearer was recognized as self-evident and undoubted in distinction from the world out of consciousness: things, other people and even a body of the Ego itself. It is possible not only to be mistaken about what is happening in the world, it is possible to have doubts concerning the existence of the latter. But it is impossible to be in doubt about the existence of the Ego and states of consciousness. In the sphere of consciousness there are no distinctions between illusion and reality: if I am thinking, I am really thinking, if it seems that I am sorry, so I am really sorry, if I decide to do something, so I really have decided to do it, and it is not something illusory.

For three centuries European philosophy and human sciences (although there were a number of schools and trends) proceeded from the idea that only the subjective world, but not the objective one, is immediately present to a human being. Here are words of a contemporary Russian specialist in cognitive science V.Allakhverdov about "the most perplexed and the most grand philosophical problem": "The content of consciousness is the only that is known and about which we can be certain. We can know about the existence of things only owing to that content. But how can we know what things really are, if we know only the content of consciousness?" (Allakhverdov 2003, p. 37). If we are closed in the subjective world, we cannot compare the content of consciousness with the objective world. And it is logical to conclude that my consciousness is the only existent. It is solipsism as a result of such understanding. According to Kant, it is a scandal in philosophy, although he himself shared the idea about the impossibility to compare contents of consciousness with what exists out of it. So it is not simply a strange fantastic idea, but a necessary conclusion from a thesis which seems evident: the content of consciousness is the only about which we know anything with certainty.

It is clear that in ordinary life, out of the sphere of philosophizing, one cannot seriously be a solipsist without self-contradiction (B. Russell wrote about a philosopher who had written to him a letter with an astonishment: why other philosophers don't share such a convincing idea as solipsism?).

The majority of philosophers did not share solipsist ideas, although in the 20th century some thinkers were not afraid to call themselves solipsists, and the problem of solipsism is being discussed now (I will write about it later). But in any case the idea of consciousness as a specific closed world was considered obvious for a long time.

It was shared by empiricists and rationalists, by psychologists and anti-psychologists. For Berkeley existence is a collection of sensations, "simple ideas": esse est percipi. For Kant the world of objects exists only in the framework of experience, which is constructed by the Transcendental Subject from sensations and a priori forms of sensitivity and a priori categories of understanding. The distinction of illusory and real worlds is within consciousness: the illusory is that which is not consistent with the interconnection of the components of experience.

The phenomenology of Husserl, influential in the 20th century, made some modifications in transcendental philosophy: it stressed a distinction between acts and intentional objects of consciousness. But according to phenomenology intentional objects are in the sphere of consciousness. A question about relations between intentional objects and the outer world is "in brackets", phenomenology abstains from answering it. From this point of view realism in epistemology and philosophy of science is an uncritical and naïve naturalistic position, which can be shared in ordinary life and even in scientific investigations, but which cannot be admitted in sophisticated philosophical reflection. According to Husserl phenomenology must be understood as an "egology"—a study of Transcendental Ego. So Husserl was interpreted as close to solipsism.

2 The Psychological Justification of Anti-realism

It is interesting, how such an understanding of consciousness—as something principally distinct from all natural phenomena—could be adopted by science. In the second half of the 19th century a new science appeared, which tried to investigate mental phenomena with experimental methods, copying natural sciences: physics, chemistry, biology. It was experimental psychology. But how is it possible to study experimentally an object that is so unlike all natural phenomena and from a philosophical point of view (which was shared by a lot of psychologists) even constitutes other objects? An answer to this question was formulated, and it was considered as an orientation of the psychological practice.

A scientist creates specific conditions in a laboratory, which afford to establish results of experimental impacts on an object of research. It is a common feature of all experimental researches in all sciences. But a psychological experiment has a specific character. A psychologist studies another person, who has consciousness.

Results of experimental impacts in psychology are some changes in another person: they can be expressed in outer phenomena (actions of another person, words), and in such a case an experimenter can fix them. But these results are necessarily also changes in the states of consciousness of the person under research. The latter changes are present only to the other person, but not to the psychologist. The latter can know about these changes only with the help of words by another person who is aware of them. Consciousness is aware of itself. Introspection (self-observation) is the specific method of psychological investigations.

So scientific psychology tried to combine two positions. The first proceeds from the idea that cognition and consciousness deals with the real world, existing independently from them. It is the position of common sense. It is as well the position of science. It is impossible to carry out an experiment or create a theory, if one doesn't presuppose that there is something real, which resists to outer impacts, something about which one can get new knowledge. Such a realist attitude is "built in" the structure of every cognitive practice: ordinary and scientific ones. It is possible to say, using Kant's phraseology, that it is the necessary condition of the possibility of knowledge. When a psychologist carries out experiments with another person, she/he deals with a specific object, really existing in space and time and having consciousness. The consciousness is understood as connected with a body and, first of all, with a nervous system and a brain (by the time of the beginning of experimental psychology a lot of facts about a connection between psychic states and activation of parts of cerebral cortex were gathered). It is a position of "the third person" in relation to another person and her/his consciousness. But there is also a position of "the first person". It is occupied by a person who is under investigation. It is awareness of the world and of oneself, and as the world is present to a person only through the states of her/his own consciousness (from the point of view of the philosophy and psychology of that time), this position is one of self-consciousness.[1]

But in such a case a paradox appears. On the one hand, one thinks that consciousness and brain exist in the world—otherwise it would be impossible to study them. On the other hand, a psychologist considers the world as a content of consciousness.

3 Other Unusual Problems Arosen in Scientific Psychology

The famous Russian psychologist, the founder of experimental psychology in Russia, G. Chelpanov wrote about them at the beginning of the 20th century. His views were very similar to the theoretical attitudes of other leading psychologists of that time.

[1]A psychologist, occupying a position of "the third person" in relation to a person under her/his study, at the same time occupies a position of "the first person" in relation to oneself.

Everyone perceives objects in the world as existing in space and time, as colored and often sounding. Meanwhile physics shows that objective correlates of subjective colors and sounds are very dissimilar to them: they are different lengths of light and sound waves. In addition physiological study of structures and functions of sense organs (by the beginning of the 20th century psychology worked in close alliance with physiology: the founder of experimental psychology V. Wundt wrote about "physiological psychology") discovered its limitations: humans don't perceive certain light and sound waves, which can be perceived by other living beings. The very idea about a difference between subjective feelings of colors and sounds and real processes in the world was formulated long before G. Chelpanov and V.Wundt. It is an old problem about "primary and secondary qualities", which was discussed in 17th and 18th centuries by Locke and other philosophers. But G. Chelpanov, using results of physiological psychology of the beginning of the 20th century, wrote that not only colors and sounds in their subjective appearance don't exist in the objective world, but that also so called "primary qualities" are not objective. Space forms which a human being perceives, G. Chelpanov wrote, are connected with specific features of human visual and tactile sensory systems. Other living beings perceive space in other ways. Only physics and mathematics can say what really space is, although they don't have a definite answer to this question till now: it is possible that space has not 3 dimensions, but a lot of them. The situation with time is similar: our perception of time is dependent on how many images can be in the field of consciousness simultaneously. It is determined by human physiology and psychology. Beings with another biology and way of life (for example, a mosquito) perceive time in other ways. So according to G. Chelpanov colors, sounds, space and time exist only in consciousness. Naive realism is incompatible with physics, physiology and psychology from this point of view. Only science can say what really exists, but not ordinary knowledge (Chelpanov 1912).

But in such a case one faces the following situation. Sciences study natural phenomena (including a neuronal system and a brain) using data of experience. The latter is presented in forms of naïve realism. For example, in our experience a brain is white, grey and blue colored. It has a space structure. But from the point of view of physics and physiology in the interpretation of G. Chelpanov colors and space exist only in consciousness. So science denies presuppositions from which it proceeds. But if G. Chelpanov had written only that, his position could be understood as a kind of scientific realism: science gives a genuine image of reality, ordinary knowledge is situated in a sphere of illusions. But G. Chelpanov made an addition to the idea about the opposition between science and naïve realism: he asserted that the very distinction between consciousness and objective reality is appearance: only the content of consciousness is certain, and scientific theories of objective reality are constructions of consciousness. (Chelpanov 1912, pp. 250–300). So his position is essentially subjectivism: psychophysiology of the human being prevents one from getting the objective reality, but the very scientific knowledge of a nervous system deals not with something real, but with phenomena of consciousness, which are only certain.

Psychology in its development refused some old ideas, in particular that of separate sensations: they were understood in old psychology as expressing not objective stimuli, but specific features of sensor systems. In the 20th century another idea was accepted in psychology: not sensations, but an integral image of a situation (Gestalt) is perceived. Separate sensations can be singled out of perception as a result of a special analysis. Laboratory conditions in which sensations were experimentally studied were artificial, not corresponding to the genuine nature of perception. But the new psychology didn't refuse the main idea of the old psychology: a human being deals not with the world itself, but with its representations, which cannot be compared with the objective reality. Some new experiments supposedly supported this idea.

For example, the same picture can be perceived either as an image of a rabbit or as an image of a duck. Transition from a certain mode of perception to another one ("gestalt-switch") doesn't depend on what is going on in an objective situation, but is connected, as gestalt-psychologists think, with processes within the nervous system of the perceiver. It was supposed that the latter determines existing types of gestalts. So it is senseless to say what gestalt corresponds to a real situation.

There is a later approach to understanding perception, which is shared now by a lot of psychologists and specialists in cognitive science. It has been elaborated by a famous psychologist R. Gregory on the base of investigating visual perception (Gregory 1970).

According to R. Gregory in this process perceptual etalons ("object hypotheses") are put on sensory information. This work of the brain can be interpreted as hypothetical attempts of human reason to answer the question "what is it?". Hypotheses that are suggested by reason (they are unconscious) can correspond or not correspond to a real situation. In the latter case an illusion arises: one perceives what doesn't exist. In contemporary psychology a number of such illusions have been demonstrated. For example, "Ames room": a person observing the interior of a room through a narrow hole perceives people in a room with dimensions which completely don't correspond to the real ones known to a perceiver. Another illusion: a photo of a mask, made from a rear side, is necessarily perceived as made from a front side. R. Gregory thinks that in such cases there is an incorrect interpretation of sensory information: the brain suggests a perceptual hypothesis which works well in ordinary conditions, but doesn't work in unusual situations. These perceptual hypotheses are partly inborn and partly appropriated during life in interactions with the world. The set of such hypotheses determines possibilities of perception: perception can happen only as a result of interpretation of sensory information. Nowadays the human being is facing a great danger: humans can create such things, which will act on visual sense organs, but which could not be interpreted, because corresponding perceptual hypotheses will be absent. In such a case a human being will not be able to see.

But if a cognizing human being is closed in the world of one's own representations, what are the reasons for saying something about the objective world?

In our time another conception appeared which, using some contemporary scientific ideas, is trying to justify philosophical solipsism. It was formulated in the framework of an inter-disciplinary movement, which spread primarily in Germany and Austria for the last 30 years. It was called "radical epistemological constructivism". Its founders are some specialists in biology, neurocybernetics, psychology, system theory: E. von Glasarsfeld, H. von Forster, G. Roth etc. Main ideas of this movement have been elaborated in a theory of autopoiesis by U. Maturana and F. Varela. According to their theory the specific feature of autopoetic systems is that their elements produce certain functions, and these functions produce elements, which produce functions etc. ad infinitum. Living systems are understood as self-producing and self-referential. Outer impacts on such systems play the role of an impulse for creating inner structural changes, which serve for the self-sustaining of the system. Autopoetic systems first of all deal with themselves, but not with the outer world. A kind of structural changes in such systems appears from the "inner" point of view as cognition of the world, but it is a construction of reality, not a relation to the world, but a self-relation. H. von Forster wrote about a principle of circularity: a human being in this conception is a human creation, which in principle doesn't differ from illusion. H. von Forster and E. von Glasarsfeld refer to G. Berkeley as their philosophical forerunner and think that they can be called epistemological solipsists (Glasarsfeld 2001, pp. 31–43).

One of the leading theoreticians of the contemporary inter-disciplinary movement, called cognitive science—the philosopher J. Fodor—formulated the idea of methodological solipsism as a strategy of cognitive research (Fodor 1980, pp. 63–73). According to him the only fruitful way of investigating cognitive processes (they were understood in cognitive science as determining all mental phenomena, including emotions and the will) must be based on understanding them as a computational processing of information by the brain. The character of this processing is determined by interrelations between inner mental states—syntactic features of structures, written on the inborn "language of thought"—and doesn't dependent on relations of these states to the outer reality, in other words doesn't depend on the semantics of these structures. So from this point of view a researcher of cognitive processes must not take into account whether these processes correspond to the world outer of a brain and what this world is.

4 Cognition and Action

Meanwhile the development of cognitive research has lead to the appearing of a principally new strategy. It is connected with a drastic revision of the main presupposition of studying cognition and knowledge, which philosophers, psychologists, specialists in cognitive science considered as indubitable: the idea that the content of consciousness is the only that is known and which is certain.

Already in the first half of the 20th century G.E. Moore suggested to refuse this idea. Consciousness refers not to oneself, but to the world, as it is "open" to the world, he maintained. Consciousness is "transparent", and any attempt to describe a mental content becomes a description of the world outer to consciousness. So it has no sense to invent "proofs of the existence of the outer world", which were called by Kant "a philosophical scandal". Because this problem doesn't exist. Ordinary intuition about the existence of the real world is "built in" the structure of consciousness and cognition and is more evident than theories of the world, which are constructed by science. G. Moore in his famous article "Proof of the External World" asserted that statements. "Here is my left hand", "Here is my right hand", accompanied by corresponding gestures, express genuine knowledge (Moore 1959).

At our time a conception appeared that refuses the traditional understanding of mental processes and states and formulates a principally new, revolutionary paradigm of cognitive research, proceeding from ideas of epistemological realism. It is called "ecological approach" to visual perception, elaborated by the famous psychologist D. Gibson. Researchers who develop the principal Gibson's ideas—gibsonians and neogibsonians—call their approach in cognitive science "embodied, situated and inactive" (Gibson 1979).

D. Gibson has based his theory on investigating visual perception. In the history of the studies in this field for several centuries perception was considered as a combination of elementary sensual entities: sensations, sense data. Empiricists thought that this combination happens spontaneously, without activity of a subject (associations of sensations). Intellectualists thought that a subject plays an active role, building, constructing perceptual experience from sensual information with the help of some rules, standards, etalons. In the 60-th and 70th of the 20th century, when cognitive science started, the interpretation of perception as a result of brain activity on a base of mental representations (perceptual object-hypotheses) became a common opinion.

Empirical sensualism and intellectualism in understanding perception are opposite to each other. But both of them share a common presupposition. It includes two points.

The first one. It is supposed that in the process of perception a subject deals with phenomena of consciousness and is closed in the field of the latter (although it is understood in the first case as a more or less passive register, and in the second case as an active constructor). The outer world is considered as a trigger. It acts on sense organs as a cause, creates certain "prints", and later is excluded from the game. Consciousness (and therefore the brain) deals only with these "prints". But if perceiving is such—and it seemed that it cannot be otherwise—it is impossible to understand how a cognizing subject can deal with the outer reality.

The second point. "Prints" were considered as ideal entities, on the base of which another ideal entity arises: "percept". The latter is "projected" to reality in some way which is incomprehensible.

D. Gibson refuses both these points. He proceeds from the idea that perception is not a manipulation of "prints", but an interaction of a perceiving agent with the outer world. Perception exists only in this interaction. It is not an ideal entity in the "inner world" of consciousness, not a thing, but a process, not a "percept", but perceiving. It is a process of extracting information from the real world. Perception is not given to a sensory system, and it is not constructed by a brain. It is possible owing to actions of a cognizing being. Something in the world can be perceived or not. It is possible to perceive better or worse. Actions of perceiving belong not to consciousness or brain (although it is impossible to perceive without a brain), they are real actions of an agent in relation to the outer world (perception is "inacted"). So a process of perceiving includes not only consciousness, and not only a sensory systems and a brain, but also the body of an agent (perception is "embodied") and a part of an environment, which participates in this process (perception is "situated"). Perceiving is not simply a phenomenon of consciousness. It is an event in the real world, a necessary part of life.

Extracted information—in distinction from sensory signals, which according to old conceptions of perception create separate sensations—says about the features of the real outer world, and it is those features that are correlated with demands of a cognizing agent and with possibilities of its actions (Noe 2004).

There is another important idea in Gibson's theory. For understanding, for example, visual perception one should use not those theories which are elaborated in physics and geometrical optics. Because in the case of real visual perception (not as it was studied in traditional psychology) a cognizing agent deals with objects compatible with the dimensions of a body and included in activity. It is not a physical world, but the immediate surroundings. For a human being they are what is recognized by naïve realism: trees, mountains, rivers, seas, buildings, other people, a colored and sounding world. It has its own ontology, distinct from the ontology of physics and even specific laws of the spreading of light: it is then so called "ecological optics".

Reality must be understood as multilayer. Different levels of reality are not reduced to each other, and at the same time there are dependencies between them. There is the micro world, but there is also the macro world. It would be strange to assert that a chair doesn't exist, that it is only a cloud of atoms and elementary particle in a certain part of space and time (although there are physicists who assert this). Modes of existence at each level don't exclude, but presuppose each other. So, for example, the "ecological optics" by D. Gibson doesn't refuse the physical conception of light spreading: the point is that under conditions, which exist on the surface of the Earth, light, reflected from different objects many times, spreads in accordance with "ecological optics".

Naïve realism and ecological realism are not in mutual opposition. Their objects are at different levels, which don't exclude, but presuppose each other (a similar idea concerning real referents of theoretical objects in the process of theory change has been elaborated by Agazzi (2014).

D. Gibson asserts that his conception can solve in particular the old problem of "primary" and "secondary" qualities. So called "secondary qualities" (for example, color) are not features of corresponding light waves and are not determined by the specific nature of sense organs, but features of surfaces and structures of objects of the surrounding world, about which a cognizing agent is informed through spreading of light stream, understood from the point of view of "ecological optics". (Gibson 1967, pp. 169–170).

D. Gibson stresses that each living being selects out in the world what affords its actions. These affordances exist objectively, in the surrounding world itself, but are selected by living and acting beings differently depending on their body dimensions, demands, and the specific features of their actions. So the reality not only has different levels. It is manifold, and a cognizing agent deals with only some of its features. For example, a person who is sitting at a table, a dog which is close to its host near a table, and a cucumber which is moving around a table's leg, perceive the same real object—a table. But each of them perceives it in a different mode. For a dog a table doesn't exist as something that can be used for food or writing texts. A cucumber cannot perceive a table as something whole. All these beings live in the world in which a table really exists, but they perceive it in accordance with affordances for their actions. If there were extraterrestrial intelligent beings they would perceive the world, in particular, objects on the Earth, in other modes than humans do.

But there is a problem in this connection. If living beings select in the world only those objects and features that correspond to the specific nature of their sensory systems and possibilities of actions, then it seems that different beings live in different worlds, which not only don't intersect, but can exclude each other (Chemero 2003, pp. 181–195). For example, colors and forms of things, which we see, don't exist for a bat, which doesn't have visual organs and orients itself in the world with the help of echolocation (Nagel 1974, pp. 435–450) . But if one thinks that different cognizing agents live in such worlds, this position is equal to acknowledging that there is no common real world, that it is better to assert that there is not a selection of different aspects of a common world, but a construction of different worlds according to the specificity of agents. It is the position of F. Varela and his co-authors in the famous book (Varela 1992), which was a starting point (together with works by D. Gibson) for the development of "embodied", "situated" and "inacted" approaches in contemporary cognitive science. F. Varela refuses the thesis of idealism that a cognizing being deals only with the content of its own consciousness. But he doesn't agree also with an idea of realism that features of the world don't depend on the process of interaction between it and a cognizing agent. F. Varela asserts that actions of a cognizing being determined by its bodily nature construct a world in which this being lives and which it cognizes. Cognition doesn't simply depend on action, he maintains, it is an action (Varela 1992, pp. 130–170).

But the genuine nature of the embodied approach to cognition can and must be understood in another way. Different cognizing beings live in the common world, although they select out its different aspects. But these ones are objectively connected with others, which are not fixed by some living beings. So an acting and cognizing being in the process of interaction with those parts of the surrounding which are accessible to it interacts at the same time with those aspects of the world which it doesn't perceive in a direct way. For example, a cucumber, which is moving around a table's leg, fixates only a small part of a leg's surface that is in its visual field. But features of this surface depend on the form of the leg as a whole and on how the leg is connected with the table. So a cucumber is really interacting with the same table at which I am sitting, although I perceive it in other way than it. As to the human being she/he goes out of the limitations of her/his own sensory systems and can perceive with the help of instruments what she/he cannot perceive directly, and can with the help of scientific theories understand how a bat or a cucumber cognize the world (Lektorski 2013).

5 Illusion and Reality

It is possible in the framework of embodied and inacted approach to cognition to suggest a new solution of some problems that have been discussed in philosophy and psychology for a lot of time.

One of them is the problem of the relations between illusion and reality. It seems that the content of perception is the same independently of the fact that one perceives what really exists or has an illusion. Because it was a common opinion that only this content is directly given to a perceiver, one could think that it is impossible to be certain whether perception refers to a real situation. It seemed that psychological experiments (for example, "a room of Ames") support this opinion. And in the case of a "gestal-switch" (a rabbit or a duck) it is senseless to speak about a distinction between illusion and reality.

But perception is actually not a static state of consciousness, but a process of extracting information, carried out in actions of a cognizing agent with the aim of inspecting a real situation. So it is impossible to understand, for example, the nature of visual perception by studying the perception of a drawing—as it was made by several generations of psychologists who proceeded from the idea that the brain interprets a picture on the eye's retina (the latter exists in 2 dimensions, so it seems similar to a picture) in the same way as a human being sees a drawing. But real perception can be understood if a researcher starts from studying the perception of real situations with which a human being or other cognizing being deal, and only after that studies perception of drawing, pictures, schemes, paintings and other artificial representations.

Illusions happen in real life. They are not inventions of consciousness, but are conditioned by real situations of perceiving. For example, a spoon in a glass of water seems to be broken. One knows that it is an illusion, as a spoon out of water is strait and even when it is put in a glass with water a spoon is perceived as strait with the help of tactile reception. But this illusion is not a mistake of a perceptive system. According to laws of light refraction a spoon which is partly under a surface of water must look objectively as broken. If the situation were another it would be an aberration of consciousness. Let's imagine such a situation. A little child sees a spoon as broken in a glass of water for the first time in life. She/he cannot take it out of water and touch it. In such a case she/he could not distinguish between illusion and reality: she/he would perceive a spoon as really broken.

In the case of "Ames room" a psychologist creates an artificial situation, when an observation of the interior of a room is made from a single point of view and through a narrow hole: the observer cannot see the interior from another point of view (because she/he is not permitted to move), cannot make the hole wider. In other words, an observer is under such conditions, when she/he can do only one thing: to stay in the same place and passively see a scene inside a room. As soon as the observer is allowed to move and to see the interior from another point of view, the illusion disappears.

Perceiving presupposes active examination of a situation with the help of actions. They are not chaotic, but are carried out with the help of a scheme, which determines a way of examining and is formed as result of former interactions of a cognizing agent with surroundings (Neisser 1976). A scheme is not such a mental representation that exists between a cognizing agent and realty and blocks getting reality and is the only present to an agent. A scheme really is a mode of an interaction of a cognizing being with the real world for extracting information (sometimes these schemes are called "activity oriented representations"). A scheme determines a certain "horizon of expectations". In such cases when this horizon doesn't correspond to a real situation owing to some objective causes an illusion arises. The latter can easily disappear when there is a possibility of examination of the situation. It can be stable if there are no such possibilities.

Here is an example. A person regularly goes from home to a subway station passing near a certain house. One day she/he walked near this house and recognized it. But when she/he went round it, realized that the house didn't exist any more: there was only one wall of the former house left. The wall was illusionary perceived by that person as a whole house. It turned out that at night the house was destroyed in order to build a new one at that place. So the first perception was an illusion. But as there was a possibility to continue moving and to go around the wall the illusion disappeared.

One cannot examine drawings and other pictures as it is possible in relation to real objects and situations. So one cannot distinguish illusion and reality in such situations, to decide, for example, what exists: a rabbit or a duck. One cannot get rid of an illusory perception of two sections on a paper as unequal, although she/he knows very well that they are really equal (the famous Muller-Lyer illusion). But in real life one never mixes a rabbit and a duck and never makes a mistake concerning dimensions of two real sticks which she/he is manipulating.

So the mental content of an adequate and of an illusory perception is not the same—in distinction from how it was understood in traditional philosophy and psychology. Because the content of perception is not a certain static ideal entity, but information that is extracted from a real situation by a process of active examination. If there are no obstacles for such examination, perception is adequate, if such obstacles exist or if such examination is impossible in principle (as it is, for example, in cases of drawings), then illusions are possible and sometimes can be very stable.

It is impossible also to examine those situations that are presented in dreaming and hallucinations. "A test for reality" doesn't work in such cases. Meanwhile a lot has been written about the supposed impossibility to distinguish a content of dreaming and hallucinations from a content of adequate perceptions (an idea of "a world as dreaming").

6 Reality and Artificial World

The human being creates a number of artificial representations of reality. They are drawings, paintings, sculptures, visual pictures on TV and computer screens, descriptions in oral speech, written texts, including literature, philosophy, science. Such representations are impossible without material bearers: stone, canvas, speech sounds, paper, screens etc. Activity of constructing and using such representations presupposes consciousness. But these representations, being constructed, form a special form of reality, as they exist independently of their cognition by a certain individual. The latter can know not all the ideal contents of such representations. But the function of these representations is not that they are a form of reality existing in itself, but that they represent the outer reality. A content of perception is determined by interactions of a cognizing being with a real situation: so it is possible to separate illusion and reality. But it is often not easy to distinguish between what are real referents of artificial representations and what belongs to them as such, because it is impossible to examine them as it is possible in relation to real situations. But the human being lives in a world of such representations, which form her/his nature in a certain sense.

Nowadays an opinion is popular according to which artificial representations don't represent anything, but create what seems to be an outer reality. This is constructivism in epistemology and human sciences. From this point of view unobservable theoretical objects in science don't have real referents, and a psychologist, studying a human being, constructs psychic states in the latter in a process of interacting with her/him.

It is true that there are such theoretical objects that have no real referents. These are the so called idealized objects (a point mass, an ideal solid body etc.). Their real existence is excluded by scientific theories. But there are other unobservable objects (atoms, electrons etc.) whose real existence is presupposed by contemporary science. What is a difference between the first and the second kinds of objects?

Some specialists in philosophy of science (Hacking 1983; Agazzi 2014) think that it is the activity of an experimenter, a possibility of manipulating unobservable objects (for example, measuring coordinates or impulses of elementary particles) that confirms the real existence of such objects.

A psychologist can create some psychological states in a person with whom she/he interacts, in particular, certain understanding of the past events, some features of self-image and so on. But a human being is not only an object of psychological study. She/he lives a real life, interacts with other people, is doing something. If she/he is engaged in life with representations about oneself which don't correspond to whom she/he really is, it is discovered very soon and compels such a person to change her/his own self-image.

Realism is the most fruitful strategy of cognitive research. I called earlier my understanding of epistemological realism a constructive realism (Lektorski 2013). Now I think that it can better be called Activity Realism. It is close to that conception of realism which has been elaborated by Agazzi (2014), (Lenk 2003)Lenk. In my opinion Activity Realism includes the embodied, situated, inacted approach of contemporary cognitive science.

References

Agazzi E. Scientific Objectivity and Its Contexts. Springer. Heidelberg, N-Y, Dordrecht, London. 2014.
Allakhverdov V.M. A methodological journey through the ocean of the unconsciousness to a mysterious island of consciousness. "Rech" publishers. SPB, 2003, p. 37 (In Russian).
Chelpanov G. Brain and Soul. Criticism of Materialism. Kushnerov I.N. and Ko publishers. M., 1912 (in Russian).
Chemero A. An Outline of a Theory of Affordances// Ecological psychology. 2003. 15: 181–195.
Fodor J. "Methodological solipsism Considered as a Research Strategy in Cognitive Science// Behavioral and Brain Sciences. Vol.3., 1980. pp. 63–73.
Gibson J. J. New Reasons for Realism. "Synthese". 17:2, 1967. Pp. 169–170.
Gibson J.J. The Ecological Approach to Visual perception. Houghton Mifflin. Boston, 1979.
Glasarsfeld E. von. The radical constructivist view of science. In: A. Rigler (Ed.) Foundation of Science. 2001. Vol. 6, No. 1–3, pp. 31-43.
Gregory R. L. The Intelligent Eye. Weidenfeld and Nocolson. L., 1970.
Hacking, I.: Representing and Intervening. Cambridge University Press. Cambridge - New York 1983.
Lektorski V.A. Le realism constructif dans l' epistemologie et les sciences cognitive// Revue philosophique. de la France et l'estranger 2013. 138 année – tome CCIII, 171–186.
Lenk H. Grasping Reality. World Scientific. Singapore. 2003.
Moore G.E. Proof of the External World. In: G.E.Moore. Philosophical papers. Allen and Unwin. L., 1959, pp. 127–148.
Nagel T. What Is It like to be a Bat?// Philosophical Review, 83, 1974, pp. 435–450.
Neisser U. Cognition and reality. Freeman. N-Y. 1976.
Noe A. Perception in Action. The MIT Press. Cambridge (Mass.), L., 2004.
Varela F., Thomson E., Rosh E. The Embodied Mind. The MIT Press. Cambridge (Mass.), 1992.

Causation and Scientific Realism: Mechanisms and Powers without Essentialism

Amparo Gómez

Abstract This paper is based on the assumption that the most plausible metaphysics behind the scientific image of the world is causal realism. A theory of causality is defended within the framework of the new mechanical philosophy, and therefore, in terms of mechanisms. This theory is substantiated on properties, dispositions and powers. In sum, the paper aims to show that causality operates through mechanisms formed by entities whose properties have the power or disposition to affect other properties (through interactions), thereby producing certain effects; and this is what *causation* means. The analysis is situated within new dispositionalism, but shies away from new essentialism, showing that a causal ontology of properties (causal powers) can be sustained without recourse to essentialism.

1 Introduction

The discussion presented in this paper is based on the assumption that the most plausible metaphysics behind the scientific image of the world is causal realism. In large part, as Samuel Alexander argued, 'the only reason for attributing reality to something is that it has causal powers' (Glennan 2010a, 373). Thus, the world is constituted by events that are causally related, it is not a Humean world of independent events; causality is discussed in Aristotelian terms, rather than linguistic ones as Humean empiricists do. The aim of this contribution is, first to lay down the basic elements of a theory of causality in terms of mechanisms (within the framework of the New Mechanical Philosophy); second, to explore the ontology that underlies this conception of causality presenting arguments in support of properties, dispositions and powers; and third, to argue that such an ontology can be sustained without relying on essentialism, contrary to what most New Dispositionalists have held thus far.

A. Gómez (✉)
University of La Laguna, Tenerife, Spain
e-mail: agomez@ull.es

E. Agazzi (ed.), *Varieties of Scientific Realism*,
DOI 10.1007/978-3-319-51608-0_20

2 Causality: Mechanisms Versus Laws

Classical mechanical philosophy approached causality in terms of fundamentally universal, deterministic laws.[1] The new mechanical philosophy understands causality in terms of mechanisms or, rephrasing Mumford's expression, 'passing around of mechanisms'.[2] Causality works through mechanisms, and precedes and is more basic than laws; in fact, causality by mechanisms brings laws themselves into question.[3]

There are at least two positions regarding laws among the new mechanists. Firstly, some argue that the abstract general laws of science are descriptions of covariations and relations generally mathematically expressed[4]; They are not causal laws since they do not account for causal mechanisms underlying the covariation or relationship they assert. Causality refers to mechanisms, which are an alternative to laws (Cartwright 1989, 2007a, b; Bogen 2005; Machamer 2004; Glennan 2005 and Andersen 2011). Mechanisms explain laws, and laws are descriptions of regularities.[5]

The second position argues that all laws, universal or statistical, are like black boxes which must be opened to reveal the mechanisms hidden inside in order to establish true causal explanations. As Elster (1998) notes, mechanisms are the antonym of laws, but they do not exclude them. Thus, although black box laws are accepted, causality must be investigated by looking at mechanisms. In this respect, Glennan (1996, 49) argues that 'all but the fundamental laws of physics can be explained by reference to mechanisms (…), a theory of causation according to which events are causally related when there is a mechanism that connects them'. The idea is to make the black boxes transparent; in other words, to identify the causal mechanisms involved. The explanatory strategy is to go from black boxes regularities to mechanisms because what underlies most causal connections are not laws but mechanisms (Glennan 1996, 1997, 2002; Machamer et al. 2000, further MCD, among others).

Regularity does not imply causation, even though it may be a powerful reason to expect it, as Little (1991) believes. The deductive-nomological (D-N) explanation is satisfied by assertions of regularities which leave causal mechanisms and processes unspecified. These regularities can be quite stable (Glennan 2002 refers to them as

---

[1]Causality is a notion intrinsically connected with the notion of time (Agazzi 1980, 299).

[2]I am referring to Mumford's paper (2009, 94–111), although he uses the term 'powers' there, not mechanisms.The expression appears in the title of the paper ('Passing powers around').

[3]Laws are merely covariations, and causal explanations are based on mechanisms.Therefore, as Andersen argued recently, mechanisms in fact replace laws. See Andersen (2011, 325–331). She argues this point in her response to Leuridan's work (2010), in which the author held that mechanisms depend on laws, rather than replacing them.

[4]In Mayntz'swords, 'the main difference between a mechanism approach and a covering law approach is that … "laws" are basically general statements about covariation' (Mayntz 2004, 240–241).

[5]Therefore, the 'search for mechanisms means that we are not satisfied with merely establishing systematic covariations between variables or events' (Hedström and Swedberg 1998, 7).

'robust sequences'), less stable or even unique ('unstable sequences' Glennan 2002).[6] But an adequate causal explanation requires the causal mechanism involved in the observed correlation to be specified. The universal laws of D-N models emphasise prediction; mechanisms emphasise causal explanation.

There is no need to postpone explaining the events of the world until we have knowledge of their laws: according to Cartwright (1983), there may be a covering law including the different factors which together produced an effect, but what is important is that 'our ability to give this humdrum explanation precedes our knowledge of that law. On the Day of Judgment, when all laws are known, these may suffice to explain all phenomena. But in the meantime we do give explanations; and it is the job of science to tell us what kinds of explanations are admissible' (Cartwright 1983, 51–52). We have a causal explanation identifying the mechanisms operating in a singular case (for example, the death of Cartwright's camellias) or in general cases (the tendency for women to be paid less than men for performing the same job).

Explaining an event, therefore, consists of showing where it lies in the causal structure of the world, which involves describing the process through which the mechanism (or set of mechanisms) in question produced it.[7] Thus, 'it is not surprising that much of the practice of science can be understood in terms of the discovery and description of mechanisms' (MCD 2000,1–2). Scientific explanation does not consist in logically connect, as in the covering law model, the explanans and the explanandum and thus inferring the latter from the former.

Both positions understand that causality operates through mechanisms; the difference lies in how laws are conceived. For the proponents of one position, laws are never formulated at the level of causal mechanisms: they are inevitably black boxes. For advocates of the other position, black box laws are temporary, existing as such only until the black boxes can be opened to reveal the nuts and bolts that will allow us to formulate the true causal laws of nature. In any case, laws do not constitute a basic ontological category, and 'when we treat them as such we get a distorted picture of the nature of causal relations' (Glennan 2010a, 380). Mechanisms can tell us more about causality than laws can.

Below causality in terms of mechanisms will be characterized, starting with a discussion of mechanisms themselves.

3 Mechanisms

In basic terms, causal mechanisms can be characterized as space-time events that produce results. Mechanisms can be stable or unstable, singular or general. The mechanisms characteristic of sciences are most often stable and general, but

[6]Mechanisms can explain both types of regularities (robust and unstable).

[7]As Salmon has pointed out (1984, 22 and 1998, 66).

unstable mechanisms are also scientific, as are singular ones. Glennan points out that stable mechanisms 'are certainly of great importance, but there are many events that do not occur as a consequence of the operation of stable mechanisms. Events of natural and human history are often the consequence of causal processes that are ephemeral and capricious' (Glennan 2010b, 251).

The definition of causal mechanisms as events that produce results highlights their active dimension. In Glennan (1996, 52)'s words, 'One cannot even identify a mechanism without saying what it is that the mechanism does'. Mechanisms are made up of entities 'doing something', in other words, 'a kettle boils, a bomb explodes, a hen lays an egg' (Glennan 2009, 327).[8] In this way, entities produce or cause effects. A solar eclipse is an event produced by the alignment of the Moon and the Sun, due to the paths of their orbits. And this type of alignment causes eclipses of this kind.

The new mechanists agree on the role of entities, but they disagree on the nature of this 'doing something'. There are two basic theories here, that of Glennan (1996, 2005, 2010a, 2011) and that of MDC (2000). Glennan holds that the activity of the entities is produced when they interact under certain conditions. This is a monist ontology: causal mechanisms produce results through the interaction of entities. The position put forth by MDC is dualist: the entities cause effects through the activity of these entities; therefore, mechanisms include entities and activities. The notion of *activities* refers to the behaviour of the parts of the mechanisms.[9] Basically the activities are the producers of effects, and therefore they are materialised causes. According to MDC (2000, 6) the notion of 'cause' is general, and more specific causal verbs make it more meaningful.

The concept of interaction is hold in this work. Therefore, it is defended a mechanistic monist position in which diverse kinds of effects—biological, physical, chemical, neurological, social, etc.—are produced by the interaction of entities under specific conditions.[10] Entities interacting under specific conditions are called *causal mechanism*. The steps involved in a complete description of a mechanism must be continuous; the description must be free of gaps that render specific steps unintelligible (Machamer 2004).

There are at least two ways to understand at causal mechanisms: as sequences of interconnected events that produce an effect, as Salmon (1984) argues, (the force of the wind produces a specific wave which sinks a boat), and as complex systems (which form part of a higher-level mechanism), the view held by Glennan (1996).[11] In the latter case, the parts of a mechanism are lower-level mechanisms. Thus, for

[8]Entities comprise a vast range of very different objects, from massive bodies such as stars and galaxies to fundamental particles such as quarks, photons, neutrons and neutrinos.

[9]The debate between the proponents of the two positions is quite interesting. See its reconstruction in Tabery (2004).

[10]Salmon (1984, 275) affirms that 'mechanisms are composed of processes and interactions'.

[11]In Glennan's words (1996, 51), 'analysis of causal connections in terms of mechanisms is only meaningful when there are ways (even if indirect) of acquiring knowledge of their parts and the interactions between them'. Salmon and Glennan are examples of authors who defend the two

Glennan, an interaction involves a change in a property of one of the parts of the mechanism producing a change in a property of another part of this mechanism, which is understood as a complex system. Salmon sees causality as causal nexuses, while Glennan sees it as causal systems.

According to Salmon, a sequence of interconnected events might occur once or once in a while (fragile sequences), but a sequence may also repeat itself systematically and invariably, as is the case, for example, with lunar eclipses (robust sequences).[12] This is due simply to the stability or robustness of the entities that interact and the regularity of these interactions. From my point of view, the robustness or stability of mechanisms do not require Glennan's systems approach; his approach also presents a problem of circularity (which he addresses by considering mechanisms hierarchically).

In this paper an eclectic position is maintained, admitting that there are cases which mechanisms form part of a complex system, but this is not always so, and many mechanisms are interconnected sequences. It seems a good idea, understanding, as Glennan does, that the parts of mechanisms are often mechanisms themselves. But we must also bear in mind that mechanisms are simple at some point. The dissolving of sugar in a liquid entails only the properties of the sugar crystals and those of the liquid. This is a case of a simple sequence of events made up of entities that interact to produce an effect. It does not represent an obstacle to establishing many different types of laws.

In both approaches, explaining phenomena involves providing a detailed exposition of how they were produced by interaction between entities. Causal analysis based on mechanisms involves knowing which entities are at work and the interactions between them. It is this knowledge which allows us to explain the occurrence of an event, as is the case of the death of the camellias Cartwright planted in her garden. On this occasion we have an explanation of a particular event, if we are able to establish the mechanism involved: the soil was poor in certain components, the temperature was too high, etc. The specific history leading from these entities' interactions to the death of the camellias explains the occurrence of the event.

On the other hand, identifying mechanisms of a certain type allows us to explain general events, or to account for certain types of events such as cognitive dissonance reduction and lunar eclipses. In this case, the causal chains that make up mechanisms are regularly activated given certain conditions, producing a certain type of effect. Thus, insofar as these mechanisms are neither unique nor unrepeatable, they can be identified as probabilistic, tendencies or even universal. A number of mechanisms as tendencies can be found in the social sciences: the

(Footnote 11 continued)

main positions on mechanisms, although there are other positions as well, for example that of Torres (2009).

[12]Glennan (2002) maintains this division in his systems approach to mechanisms.

'forbidden fruit effect', the 'spillover effect', the 'compensation effect', 'loss aversion', the 'consumption effect' and 'wishful thinking', among others.

An entity, a bomb that has blown up, for example, produces certain effects on another entity (a person's body) with which it interacts, producing specific effects (the person's death). In order to explain the effect produced, we must identify the mechanism that caused it (entities interacting). But, has the event been explained? Not quite, because in order to formulate a causal explanation we must determine how it was that those entities produced the effect they produced when they interacted. In other words, we must specify what makes the bomb capable of producing the effect it produced and what makes the person's body capable of reacting how it did to the explosion. To put it in general terms, what gives certain types of entities the capacity to produce the effects they produce when they interact? Not all interactions are causal. The answer to this question lies in the notion of 'properties'.

4 Properties

When entities interact, it is their properties that affect and/or are affected in the interaction. Certain properties of an entity interact with the properties of another entity, producing changes in that entity. Entities are material objects with properties, and causality operates around properties. Glennan (2002, 344) provides a clear explanation of the nature of the interaction between entities: 'an interaction is an occasion on which a change in a property of one part brings about a change in a property of another part'.[13] And this is what 'causation' means.

In an interaction, the properties of one entity act on the properties of another entity, producing changes (in other words, effects) in this property. If, for example, a particular *property*—postcapillary vasodilation—of a *chemical entity* (nitroglycerine), produces a *change* in a particular *property* of an *organic entity*—relaxation of the arterioles anda decrease in vascular resistance and arterial pressure—with which it has interacted, we are looking at a causal mechanism that has produced an effect. Vasodilation occurs because the properties of nitroglycerine interact with the properties of cardiovascular system and thus produces relaxation of the peripheral arteries and veins, decreasing cardiac output and reducing oxygen demand by the heart muscle. Not all properties are causally relevant; causal relevance distinguishes the properties that produce certain effects from the properties that do not have that power.[14]

---

[13]This notion of transferring a change in a property is borrowed from Wesley Salmon's concept of transmitting a mark or a conserved quantity, with the important switch to an invariant, change-relating capacity from Woodward (Salmon 1984, 1998, and Woodward 2000).

[14]As Salmon (1984, 143) illustrates by pointing out that there is a difference between a moving car and its shadow. According to Glennan (2010, 365) 'Causal relevance is essentially a counterfactual notion'.

Finally, dispositional properties are not independent. They are not dispositions simple, and separate, they are relational properties; in other words, their identity or essence is determined by their relations with other properties, rather than intrinsically. The relational nature of properties is accepted by non-essentialist new dispositionalists like Mumford (2009), but also by essentialists like Bird (2007), and by those who take one position or the other depending on the properties involved, like Chakravartty (2008). Thus Bird affirms'According to dispositional essentialism the essence of such a property is determined by its relations to other properties' rather than purely intrinsically (Bird 2007, 524, 527). For non-essentialists, it is the identity of the property that is established relationally. Mumford (2004, 95) emphasizes that, 'there is nothing at all more to' a property than 'its relations with other properties'. And as Bird writes (2007, 533), 'For the dispositional monist, identity of properties is dependent on something else, rather than being primitive (the latter view is quidditism). The something else is the pattern of manifestation relations'.

The relational nature of dispositional properties raises the regression or circularity problem. Insofar as the identity of each property is relational rather than fixed, regression is clearly present. New dispositionalists respond to this objection by arguing that more than regression, there is circularity, an acceptable circularity we must learn to work with. As Mumford (2009, 101) argues'The circle is not too tight. It is big enough for us to grasp an adequate-enough understanding of what our original property is'. Simply put, dispositionalism entails interrelated properties and its nature and identity is determined by that.

Properties are relational, and they cause or are causally affected. In order to account for the production of effects by properties, let's take our analysis one step further and look at another factor, that of 'dispositions' or 'powers'.

5 Powers

Causal properties are dispositional (rather than categorical) and dispositions are powers: 'a power is essentially a disposition to do something' (Mumford 2009, 96). Therefore, 'Dispositional monism is the view that natural properties and relations are "pure powers"' (Bird 2007, 513). In the right dosage, nitroglycerine has the property, that is, the disposition or power, of relaxing arterioles, thereby decreasing arterial pressure. Cyclic guanosine monophosphate has the power of relaxing the peripheral arteries and veins, reducing cardiac output and the oxygen demand of the heart. Water has the power to dissolve sugar crystals.

Among the new dispositionalists[15] we find two positions on the dispositional nature of properties. Authors such as Mumford hold that all properties are

[15]Authors such as Mumford (2005, 2009, 2010, 2011, 2013); Ellis (2001, 2005); Molnar (2003); Chakravatty (2008, 2011); Bird (2005, 2007); Shoemaker (1980, 2011), among others.

dispositional (a view which Molnar 2003 has called'pandispositionalism'). But other authors, such as Ellis, disagree arguing that there are some properties that are not dispositional. Specifically, they do not consider geometric and structural properties to be dispositional. Thus, they divide properties into two types: dispositional and non-dispositional.

The new dispositionalists do agree that the relations between properties and causal powers are real, not merely analytical. The natural properties that exist in the world include a series of causal powers which are dispositions or powers. Mumford (2011, 54) explains,'Causes and powers are closely connected (…) A vase breaks when struck, for instance, because of its fragility, while sugar dissolves in water because it is soluble'. The idea is that an ontology of powers offers a good explanation of causality, something that was explored by earlier authors such as Bhaskar (1975), Salmon (1982), Harre and Madden (1975), among others.

Mechanisms were defined above as entities 'doing something'. Now, we shall go one step further to say that the way in which entities 'do something' is through powers, which are activated under certain conditions (when the properties of the entities interact under these conditions). The sugar dissolved in my tea is no more than the manifestation of a liquid's power of dissolution when it interacts with sugar crystals, which are characterised by their power of solubility ('dissolvent' and 'soluble' are properties of liquids and sugar crystals, respectively). Both are dispositions or powers which materialise and cause the sugar in my tea to dissolve.

Powers are the basic building block of a dispositional metaphysics. They are activated and interact to produce results. *Transfer* notion has place here since dispositionalism can be understood as a transfer theory insofar "causation will be the passing of the powers" (Mumford and Anjum 2013, 108). Therefore, the transference would not be as in the case of Salmon (1984, 133) energy, information, marks, signs and propensities. The theory of special relativity requires us to distinguish between causal and pseudo causal processes. Causal processes are capable of transmitting energy, information, propensities, etc. pseudocausal processes are not. A car moving on the road is a causal process, it interacts with its surroundings in a number of ways, transmitting causal influence. Its shadow does not (Salmon 1984, 143).[16] For Salmon, the basic causal mechanism is a causal process that transmits *propensities*.

[16]Regarding causal processes, Salmon (1984) considers two cases to be particularly important: those in which there are shared causes for two or more effects (this goes back to Reichenbach's common cause principle) and those in which there are two causes for the same effect (both cases can also exist together). This entails complex causal interactions, and is an interesting notion to consider when explaining improbable events and seemingly random coincidences.

While Salmon frames causality mainly in terms of a theory of propensities, Cartwright (1989, 2007a, b) defends a theory of capacities. At the heart of her theory lie *causal capacities*: C causes E because C has the capacity to do so, and this capacity is something that Cs carry with them from one situation to another (Cartwright 1989, 146).[17] The property of being aspirin carries with it the capacity to relieve headaches. Capacities are closely related to powers, although Cartwright (2007a, 25) prefers the notion of *capacities*.[18]

The new dispositionalists (Mumford, Ellis, Molnar, Chakravartty, Bird and Shoemaker, among others) prefer the concept of powers. My own analysis adopts the notion of powers while acknowledging that the concepts of capacities and powers are very closely related. Thus, in both cases, for instance, one must distinguish between the exercising of a power or a capacity and its manifest results, as Cartwright and Mumford, among others, have demonstrated. This is an important point raised by the new dispositionalists, one which opens the door to indeterminist causality.[19]

As Mumford (2009, 102) reminds us, 'powers are the causes of their manifestations'. But the fact that a manifestation does not occur doesn't mean that the powers (or capacities) are not there. The properties of nitroglycerine, and thus its powers, are in the pills even if they do not manifest, because nobody took them, for example, or someone took them along with another medication that blocks their manifestations. Powers do not have to manifest in order to exist; they are dispositions.

We must distinguish between the existence of a power (or capacity), its exercise and its manifest results:'Dispositions can fail to manifest because of some interfering factors or because the conditions just happen not to be conducive' (Mumford 2009, 95). The exercise of a power can produce manifest results but it also can be interfered with, causing the manifest results not to be produced. The main reasons why the results of a power may be interfered with are interactions, either between powers themselves or with other factors of the situation.

Therefore, finks or masks prevent the manifestations of a power from occurring in the circumstances in which they would normally be activated (Vihvelin 2004, Smith 2003, Fara 2008, Clarke 2009, and many others). Thus, 'A poison's power to kill when ingested can be masked by an ingested antidote. A glass's fragility can be masked by internal packing that prevents breakage even if the glass is struck' (Clarke 2009, 325). The gravitational pull between two celestial bodies can be completely masked because the pull from a third body, a nearby planet, for instance, is much stronger. But, as Mumford (2009, 106) points out, 'The force

---

[17]Cartwright (1989, 9) likens her conception to those of Salmon, Eells and, in particular, Spohn. She argues explicitly for the centrality of capacities, and explicitly distances from Hume, and even more so from Hempel and Nagel, while she agrees with S. Mill.

[18]She remarks in her introduction that her capacities might well be called either 'propensities' or 'powers' (Cartwright 1989, 9).

[19]See Gómez (2015). DesAutels (2011, 914–925) believes that mechanisms should be characterized stochastically.

between *a* and *b* is still there. It could make a difference if it were not, for instance. But it is completely defeated and unable to act'.

In the interaction between powers, one power can neutralise another, as when an acid and a base interact (as Cartwright (1989, 163) notes in her discussion of capacities). For example, in social sciences, the interaction between the power to adapt preferences to conditions (as in the case of sour grapes discussed by Elster) and the power for *wishful thinking* (preferring that which is beyond one's possibilities) can cancel the effects of the former, thus neutralising it. The interaction (in an individual) between the *spillover effect* (if a person follows a pattern of behaviour P in X they will follow this same pattern of behaviour in Y) and the *compensation effect* (if a person does not follow a pattern of behaviour P in X they will not follow this same pattern of behaviour in Y) can be cancelled (both dispositions o powers).

The fact that the manifest results of an exercised power may not occur (when not interfered with) also depends on environmental circumstances. The exercise of the power (or set of powers) combined with other environmental factors is what explains this fact. The result may differ from one occasion on which the capacity is exercised to another.

All these facts strongly affect conditionality and pose important difficulties, even for the conditional reformulated by Lewis (1997) , since he admits that the intrinsic properties which constitute the causal bases of powers are altered by the working of finks and masks. The fact that some dispositions are indeterminist also poses difficulties for the conditional. In this case, even if there is no interference, the stimuli for the manifestation of a disposition do not guarantee that it will occur: 'the stimulus might be present, the causal basis retained and all masks absent, and still the manifestation might or might not occur' (Clarke 2009, 326). For Cartwright (1989, 55), 'the only way to state a true conditional (…) is roughly this: If the capacity is triggered properly and *is not interfered with,* then the canonical manifestation will result'. What new dispositionalists' analyses show is that the problem is the same for both powers and capacities: they can be affected in many ways, with the result that their manifest results do not occur. New dispositionalists have continued to refine the formulation of the conditional in order to respond to the objections posed.[20]

Therefore, even if a power (or disposition) has a universal type of manifestation, sometimes this manifestation may not occur. This means that causality, while universal, is not deterministic. Mumford (2009) holds that powers involve a*sui generis* modality, and he therefore places powers somewhere between necessity and contingency: 'The modal force of a power is neither entirely necessary nor entirely contingent but something in between' (Mumford 2009, 95). The existence of a power is not determined by its continuous explicitation in each of its manifestations.

[20]Nevertheless, too many problems remain unresolved for them to be considered to have achieved this (as Clarke shows in his critical analysis of the diverse reformulations of the conditional).

However, the manifestations must take place once the interference is gone. Therefore, we can explain why a power does not manifest itself and, if possible, eliminate the interfering factor in order to demonstrate that if it were not for this factor the manifestations would occur, which means that the power exists even if it does not manifest itself.

For now, we are dealing with powers whose manifestations are sometimes cancelled out or masked by interactions with other properties or other factors of the situation. To assert a property is simply to say that a certain entity has the power to behave in a certain way, producing certain effects or manifestations if nothing prevents it from doing so.

Besides being relational, another interesting aspect of powers is that they work together with additive effects (and sometimes subtractive effects, as Mumford (2009) points out).[21] Events are produced by powers working together, but also by powers acting against each other. This happens when they act jointly but in opposite ways: two horses pulling a barge from opposite banks of a canal (Mumford 2009, 103). The key idea is that the production of a result usually involves many powers which contribute to the production of a result individually, but this production or causation is the sum of all the powers involved (even though they sometimes have subtractive effects, as Mumford notes). This is where the idea of *contribution* to the production of an effect comes into play, and this notion also enables us to explain how it can be that, given the presence of *power X,* sometimes an effect is produced and sometimes it is not. Or to put it another way, this helps us understand how the same power can be involved in the production of different effects. Mumford (2011, 57) states that 'This last point also brings to light what will be a crucial factor for us. The causes of an effect are often very complex, involving many different powers of many different things'.

Therefore, the effect produced may vary, as it is the result of a set of powers, all of which contribute to its production. But this set of powers is not necessarily absolutely stable or fixed. One or more of the powers that make it up might change. Thus, *power X,* whose contribution has not changed, is present in the context of a set of powers in which there have been changes, and therefore the final effect produced by the set of powers is different from the one produced on other occasions. This is how a different event can be produced even if the contribution made by a particular power (but not the final result) always remains the same.

The notion of *contribution*, which emphasises both the individual and joint contributions of powers to the final result, was proposed by Mumford. In his words, 'This contribution is what we have to equate with the power's manifestation' (Mumford 2009, 104). Therefore, it is not that the same power has different manifestations and causes different effects, but rather that insofar as these effects are produced by a set of powers, any change in this set translates to different or variable

[21]See Mill (1843)'s work on the composition of causes. Chakravartty (2008, 170) introduces the notion of *sociability*, arguing that 'specific sets of properties are always found together'.

results. As in the case of distinguishing between the exercise of a power and its manifestation, to explain how the same power can be involved in the production of different results imply to account for why causality is not deterministic.

6 Natural Kinds Without Essentialism

Many new dispositionalists (Ellis, Shoemaker, Bird and Molnar, among others) adopt essentialist positions, accepting the existence of causal powers, natural kinds and essences. There is even a tendency to draw parallels between 'the new dispositionalism' and 'the new essentialism'.

Their main argument is that causal properties are essential, intrinsic, and that things are what they are, or of the kind they are, by virtue of these properties. Properties determine natural kinds, which are *constituted* and *characterised* by the essence of intrinsic properties. However, many authors recognise, as discussed above, that these essences are not determined intrinsically, but rather by their interactions with other properties.

The members of natural kinds are instances of properties and their essences. Kind membership is determined by essences: 'nothing could acquire any set of kind-identifying properties without becoming a thing of this kind' (Ellis 2001, 237–238). Particulars are instances of real essences of intrinsic properties, which constitute natural kinds.

New essentialism is opposite to the categorical approach, which holds that natural properties are categorical rather than dispositional. The same property can exist in another possible world (Armstrong 1997). New essentialism does not invoke the Kripke-Putnam theory, in which direct reference is extended to natural kinds. However, as Mumford (2005, 428) points out, 'Ellis's essentialism is of the Kripkean variety'. Mumford accepts N. Salmon's (1982) critique of this theory, namely, that it does not present a conclusive argument for essentialism about natural kinds.

The question to explore below is whether it is possible to accept the existence of dispositions, causal powers and natural kinds and reject essentialism. The central issue is whether we can accept the existence of natural kinds without accepting essentialism. The answer, I will argue, is positive: it is possible to acknowledge the existence of natural kinds (and certainly of dispositions and powers) without committing to the existence of essences. Dispositionalism entails accepting the thesis that causality involves dispositions, powers, and also natural kinds, but not necessarily essences. Properties are not necessarily constituted by dispositional essences.

What theory on natural kinds allows us to avoid essentialism? Mumford (2005) offers an interesting argument. He develops an account of natural kinds that considers anti-essentialist. However, some aspects of his approach, one could argue, contain echoes of essentialism. This would be related to his approach on universals. Mumford holds that universals are real, as are particulars: 'Universals and particulars are distinct and both real" (Mumford 2005, 433). However, immediately after this statement he affirms that a universal exists'only in its instances', not in some Platonic universe:'There is not some transcendent form that exists besides' (Mumford 2005, 433).

Universals exist immanently in the particulars that carry them. This is so in the case of properties, which he defines in relation to natural kinds as 'a universal but accidental characterising attribute [property]' (Mumford 2005, 434). Natural kinds are 'characterised by their attributes [properties]' (Mumford 2005, 434). They are neither intrinsic nor essential, but they are universal. Therefore, the universality of natural kinds, which is affirmed by Mumford, is based on properties, which exist immanently in the totality of the particulars, (and each of them), that make up that kind. It is their shared property which makes the particulars are similar each other and constitute a universal natural kind. However, argues Mumford (2005, 435) These kinds can be characterised by attributes [properties] but a further assumption that any of these characterising attributes is essential is neither required nor independently motivated'. But it is true that without this shared property (*redness*, for example, or *solubility*) there would be no universality, nor natural kinds. This is evident in Mumford's discussion of types and modes.

The types are universal, while modes are particular. Modes belong to objects, and 'The basic things that exist are objects-bearing-modes' (Mumford 2005, 433). But, modes, what are modes of? They are modes of types, of course, and types are simply properties considered by themselves, rather than in their instances, their modes. Mumford sums this up nicely when he says that 'Attributes (properties) are, therefore, *types* of modes-borne-by-objects' (Mumford 2005 433). Properties (or attributes), redness for example, are simply 'natural types of modes', those that are borne by objects. But, what is that, borne by objects, making them constitute a natural kind? The answer here is redness, *that* which makes the property to be of that type.

So what makes a property to be of a certain type? One response could be 'what the property is', *that* which exists immanently in the objects and constitutes their modes, -of which the property is the type:*'redness', 'solubility', 'fragility'*. But what is *redness*? What is *fragility*? What is *solubility*? Applying Mumford's argument, one could say that these are all 'natural types' existing beyond of, their modes (despite Mumford's insistence that they are 'types of modes'); or at least existing parallel to their modes. If this is the case, it raises the question of what extent it does not involve essentialist reminiscences. Mumford rejects this idea but one could contend that his arguments do seem to invoke it. We must not forget that types are universals real, existing, according to Mumford.[22]

At this point we can ask: Is it possible to accept natural kinds and reject even the smallest hints of essentialism? The answer I propose below is in line with Mumford's view for the most part, although without accepting the existence of real universals. Natural kinds are constituted by totalities of objects with certain

[22]See Lowe (2006, 41)—although he speaks of kinds rather than types.

properties, being red or soluble or explosive, for example. Natural kinds are sets of objects (entities) with a certain property (or properties) and therefore with certain dispositions or powers. Natural kinds are made up of objects with a certain property, disposition or power (for example, liquids that have the power to dissolve sugar crystals).

It could be said that properties are types, but this is merely a linguistic and conceptual resource that helps to classify and differentiate properties in order to define or characterise them. Therefore, I agree with Mumford (2005) that natural kinds are *constituted'* by the omnitemporal totality of their member objects or particulars' (Mumford 2005, 434) and that they can be *defined* or 'characterised by their attributes' (Mumford 2005, 434). But this is simply a perfectly valid linguistic and conceptual resource.

Thus, 'natural types' are not something of the world (real) but rather of language. What *is* of the world are the sets of objects with shared properties, dispositions and powers which constitute natural kinds; and to constitute a natural kind all that is needed is 'resemblance'.[23] Natural kinds are constituted by entities that share a property or properties (being soluble, fragile, irritable) and therefore certain dispositions or powers.

I agree with Chakravartty (2008) that one can expect members of a kind that share properties to behave similarly in similar circumstances. This allows us to make generalisations, including many different types of laws, although as Chakravartty reminds us: 'In many cases, however, causal generalisations are susceptible to exceptions and *ceteris paribus* qualifications' (Chakravartty 2008, 160). The behaviour of the members of a kind depends on their causal powers, not on any essence. What matters is the possession of a causally efficacious property—a power or property that confers a power.[24]

In my approach, natural kinds are constituted by sets of objects with certain shared properties and dispositions or powers; essences play no part here. This is a departure from Plato but not from Aristotle, because as Chakravartty (2008) argues, the concepts of powers and essences are not inextricably linked, contrary to what some Aristotelians believe. Therefore, causal generalisations are not determined by the essential properties of kinds of entities (objects). The behaviour of entities is determined by their causally efficacious dispositional properties.

[23]Mumford's notion (2010, 434).

[24]This is what Chakravartty (2008, 155) argues regarding *cluster kinds* (natural kinds of things without essences). However, he holds that there are kinds of things that appear to have essences, and therefore *essence kinds* with essential properties.

7 Conclusion

The new approach to causality that is being developed from the new mechanistic and new dispositional philosophy is proving very fertile ground and has been applied in a number of fields of the natural and social sciences. These philosophies represent a promising research area and offers innovative perspectives on the analysis of issues such as causation, laws and explanation.

This paper has aimed to clarify some foundational issues around causality in order to provide metaphysical support for the scientific image of the world. In it has been analyzed the concept of mechanism from the new dispositionalism (from a monist perspective) and has been demonstrated that it is possible to do so without adopting an essentialist position. In brief, it is understood that causality operates through mechanisms, which are made up of entities, and effects are produced through the interaction of the properties of these entities. Properties are dispositional, and they therefore have the power to affect other properties in a certain way, producing certain effects; and this is what it means *causation*. The metaphysics of this mechanistic and dispositionalist conception of causality leads us to natural kinds and a non-essentialist account of them.

There are still many unanswered questions regarding laws, the nature of properties, dispositions and powers, natural kinds and essences. Even, the very notion of causality is understood differently by different authors. But this approach to causality is very promising and is constantly providing new contributions to some of the most fundamental philosophical issues.

Acknowledgements This article has been supported by the Spanish Ministry of Economy and Competitiveness (Research Project FFI2012-33998 and Research Project (FEDER) FFI2015-64529-P).

References

Agazzi, E.: Time and Causality, in Dalla Chiara (1980), pp. 299–321 (1980).

Andersen, H. K.: "Mechanisms, Laws, and Regularities." *Philosophy of Science* 78 (2) (2011): 325–331.

Armstrong, D. M.: *A World of States of Affairs.* Cambridge: Cambridge University Press, 1997.

Bhaskar, R.: *A Realist Theory of Science,* Leeds Books Limited, Leeds (1975).

Bird, A.: "Laws and Essences." *Ratio (new series)* 4 (2005): 437–461.

Bird, A.: "The regress of Pure Powers?" *The Philosophical Quarterly* 71(1) (2007): 513–534.

Bogen, J.: "Regularities and Causality; Generalizations and Causal Explanations." *Studies in History and Philosophy of Biological and Biomedical Sciences* 36 (2005): 397–420.

Cartwright, N.: *How the Laws of Physics Lie,* Oxford University Press, Oxford, New York (1983).

Cartwright, N.: *Nature's Capacities and their Measurement*, Oxford University Press, Oxford, New York (1989).

Cartwright, N.: *Causal Powers. What are they? Why do we need them? What can and cannot be done with them*, London School of Economics and Political Science, Technical Report N0 04/07, London, (2007a).

Cartwright, N.: *Hunting Causes and Using Them: Is There No Bridge from Here to There?*, London School of Economics and Political Science, Technical Report N0 05/09, London, (2007b).
Chakravartty, A.: Inessential Aristotele. Powers without essences, in Groff (2008), pp. 152–162 (2008).
Chakravartty, A.:"Scientific Realism and Ontological Relativity." *The Monist,* 94 (2): 157–180, 2011.
Clarke, R.: "Dispositions, Abilities to Act, and Free Will: The New Dispositionalism." *Mind* 118 (2009): 323–351.
Dalla Chiara, M. L. (ed.): *Italian Studies in the Philosophy of Science*, Reidel Publishing Company, Dordrecht, Boston, London (1980).
DesAutels, L.: "Against Regular and Irregular Characterizations of Mechanisms." *Philosophy of Science* 78(5) (2011): 914–925.
Elster, J.: A plea for mechanisms, in Hedström and Richard (1998), pp. 45–73 (1998).
Ellis, B.: *Scientific Essentialism,* Cambridge University Press, Cambridge, (2001).
Ellis, B.: "Universals, the essential problem and categorical properties." *Ratio* (new series) 4 (2005): 462–472.
Fagan, M. B.: Explanatory Interdependence: the Case of Stem Cell Reprogramming, in Pierre-Alain Braillard and Malaterre (2015), pp. 387–412 (2015).
Fara, M.: "Masked Abilities and Compatibilis." *Mind* 117 (2008): 843–865.
Glennan, Stuart. "Mechanisms and the Nature of Causation."*Erkenntnis* 44(1) (1996): 49–71.
Glennan, S.: "Capacities, Universality, and Singularity." *Philosophy of Science* 64(4) (1997): 605–626.
Glennan, S.: "Rethinking Mechanistic Explanation." *Philosophy of Science* 69(3 Supplement) (2002): S342–S353.
Glennan, S.: "Modeling mechanisms." *Studies in the History and Philosophy of Biology and Biomedical Sciences* 36 (2005): 443–464.
Glennan, S.: "Productivity, relevance and natural selection." *Biology and Philosophy* 24 (2009): 325–339.
Glennan, S.:"Mechanisms, Causes, and the Layered Model of the World." *Philosophy and Phenomenological Research* 81(2) (2010a): 362–381.
Glennan, S.: "Ephemeral mechanisms and historical explanation." Erkenntnis 72 (2010b): 251–266.
Glennan, S.: Singular and general causal relations: A mechanist perspective, in Illari, Russo and Williamson, (2011), pp. 789–817 (2011).
Gómez, A.: "Are the Social Mechanisms the Antonym of Laws?" *Epistemology. Italian journal of philosophy of science* 40 (2015): 29–44.
Groff, R. (ed.): *Revitalizing Causality. Realism about causality in philosophy and social sciences*, Routledge, London, New York (2008).
Harre, R and Madden E. H.: *Causal Powers: Theory of Natural Necessity,* Blackwell, Oxford (1975).
Hedström, P. and Swedberg, R.: Social mechanisms: An introductory essay, in Hedströmand Swedberg (eds.) (1998), pp. 1–31 (1998a).
Hedström, P. and Swedberg, R. (eds.): *Social mechanisms: An analytical approach to social theory*, Hedströmand Swedberg (1998), pp. 1–31 (1998b).
Illari, Ph. Russo, F. and Williamson, J. (eds.): *Causality in the sciences,* OUP, Oxford (2011).
Little, D.: *Varieties of social explanation: an introduction to the philosophy of social science*, Westview Press, San Francisco (1991).
Leuridan, B.: "Can Mechanisms Really Replace Laws of Nature?" *Philosophy of Science* 77 (2010): 317–40.
Lewis, D.: "Finkish Dispositions." *Philosophical Quarterley* 47 (1997):143–158.
Lowe, J.: *The four-Category Ontology: A Metaphysical Foundation for Natural Science,* Clarendon Press, Oxford (2006).

Mayntz, R.: "Mechanisms in the Analysis of Social-Macro Phenomena." *Philosophy of the Social Science* 34 (2) (2004): 237–259.
Machamer, P., Darden L. and Craver C. F. "Thinking about Mechanisms." *Philosophy of Science* 67(1) (2000): 1–25.
Machamer, P.: "Activities and causation: the metaphysics and epistemology of mechanisms." *International Studies in the Philosophy of Science* 18(1) (2004): 27–39.
Marmodoro, A. (ed.): *The Metaphysics of power: Their Grounding and Their Manifestation, Routledge* (2010)
Mill, J. S.: *A System of Logic Ratiocinative and Inductive,* Routledge & Kegan, Londres (1843).
Molnar, G.: *Powers: a Study in Metaphysics,* Oxford University Press, Oxford, New York (2003).
Mumford, S.: *Laws in Nature,* Routledge, London, USA, Canada (2004).
Mumford, S.: "Kinds, Essences, Powers." *Ratio (new series)* 4 (2005): 420–436.
Mumford, S.: "Passing powers around." *The Monist* 92 (2009): 94–111.
Mumford, S. and Anjum R. L.: "A Power Theory of Causation." in Anna Marmodoro (ed.), pp. 145–159 (2010)
Mumford, S.: "Spoils to the Vector: How to Model Causes If You Are a Realist About Powers." *The Monist*, 94 (1) (2011): 54–80.
Mumford, S. and Anjum R. L.: C*ausation A very short Introduction,* Oxford University Press, Oxford, New York (2013).
Salmon, N. U.: *Reference and Essence,* Blackwell, Oxford (1982).
Salmon, W. C.: *Scientific Explanation and the Causal Structure of the World,* Princeton University Press, Princeton (1984).
Salmon, W. C.: *Causality and Explanation*, Oxford University Press, Oxford, New York (1998).
Shoemaker, S.: Causality and Properties, in van Inwagen, (ed.) (1980), pp. 109–135 (1980).
Shoemaker, S.: "Realization, Powers, and Property Identity." *The Monist* 94 (1) (2011): 3–18.
Smith, M.: Rational Capacities, or How to Distinguish Recklessness, Weakness and Compulsion, in Stroud and Tappolet (2003), pp. 17–38 (2003).
Stroud, S. and Tappolet Ch. (eds.): *Weakness of Will and Practical Irrationality,* Oxford University Press, Oxford (2003).
Tabery, J. G.: "Synthesizing Activities and Interactions in the Concept of a Mechanism." P*hilosophy of Science* 71(1) (2004): 1–15.
Torres, P. J.: "A Modified Conception of Mechanisms." *Erkenntnis* 71(2) (2009): 233–251.
vanInwagen, P. (ed.), *Time and Cause,* D. Reidel Publishing Company, Dordrecht, Netherlands (1980).
Vihvelin, K.: "Free Will Demystified: A Dispositional Account." *Philosophical Topics* 32 (2004): 427–450.
Woodward, J.: "Explanation and Invariance in the Special Sciences." *British Journal for the Philosophy of Science* 51(2) (2000): 197–254.

Objectivity in Mathematics: The Structuralist Roots of a Pragmatic Realism

Gerhard Heinzmann

Abstract This paper proposes a reconsideration of mathematical structuralism, inaugurated by Bourbaki, by adopting the "practical turn" that owes much to Henri Poincare. By reconstructing his group theoretic approach of geometry, it seems possible to explain the main difficulty of modern philosophical eliminative and non-eliminative structuralism: the unclear ontological status of 'structures' and 'places'. The formation of the group concept—a 'universal'—is suggested by a specific system of stipulated sensations and, read as a relational set, the general group concept constitutes a *model* of the group *axioms*, which are exemplified (in the Goodmanian sense) by the sensation system. In other words, the shape created in the mind leads to a particular type of platonistic universals, which is a model (in the model theoretical sense) of the mathematical axiom system of the displacement group. The elements of the displacement group are independent and complete entities with respect to the axiom system of the group. But, by analyzing the subgroups of the displacement group (common to geometries with constant curvature) one transforms the variables of the axiom system in 'places' whose 'objects' lack any ontological commitment except with respect to the specified axioms. In general, a structure R is interpreted as a second order relation, which is exemplified by a system of axioms according to the pragmatic maxim of Peirce.

1 From Objects to Structures

From antiquity to the nineteenth century and even up to now, two theses are being among the most debated subjects in philosophy of mathematics:

This paper draws on two talks given at the AIPS-conference at A Coruña and at the University of Toulouse. The work was carried out as part of the ANR/DFG project *Mathematics: Objectivity by Representation.*

G. Heinzmann (✉)
Université de Lorraine/CNRS & Archives Henri-Poincaré, Nancy, France
e-mail: gerhard.heinzmann@univ-lorraine.fr

E. Agazzi (ed.), *Varieties of Scientific Realism*,
DOI 10.1007/978-3-319-51608-0_21

(a) According to the Aristotelian tradition, mathematical objects such as numbers, quantities and figures, are thought of as entities belonging to different categories.
(b) Mathematical objects are extra-linguistic entities existing independently from our representations of them in an abstract world. They are conceived by analogy with the physical world and designated by singular terms of the mathematical language.

Although the Aristotelian thesis and the ontological 'Platonism' were countered by nominalism and early tendencies of algebraic formalization, they become more problematic when mathematicians like Niels Abel thought the relations before the *relata* or, like Hermann Hankel, maintained that mathematics is a pure theory of forms whose purpose is not that of treating of quantities or combinations of numbers (see: Bourbaki 1968, 317).

2 Bourbaki's Hypothetico-Deductive Structuralism

In the 1930s, Bourbaki finally defended the view that mathematics does not deal with the traditional mathematical objects at all, but that objectivity is solely based on the stipulation of structures. Bourbaki inaugurated an axiomatic-structural point of view that could seemingly work without the need of metamathematics in Hilbert's sense. Indeed, metamathematics being "finite" and contentual, it would be an exception to the slogan that mathematics is only about formal structures.

The hypothetico-deductive Bourbakist foundations were explicitly designed as neutral with respect to philosophical foundations and opposed to this effect to the practical turn in philosophy of mathematics. In fact, the practical turn is positioned in a field of tension between pragmatism and the working mathematician: Wittgenstein and Bourbaki. What unites both is that they refuse the hypostasis of mathematical objects either from a philosophical perspective or from the point of view of mathematical practice. The 'working mathematician'[1] Henri Cartan, one of the founders of Bourbaki, wrote in 1943:

> The mathematician does not need a metaphysical definition; he must only know the precise rules to which are subject the use he has in mind […] But who decides upon the rules? (Cartan 1943, 1b [transl. G.H.]).

This sounds Wittgensteinian but is not really so: according to Cartan, the first mathematical reasoning on a certain area intuitively obeys certain rules and if difficulties arise, the use is adapted, etc. Thus, a mathematical reality is created through practice. What is the criterion of practice and rules that result in that reality? In his historical notice on set theory, Bourbaki

---

[1]An expression used by Bourbaki (1949).

> "recognized that the 'nature' of mathematical objects is ultimately of secondary importance, and that it matters little, for example, whether a result is presented as a theorem of a 'pure' geometry or as a theorem of algebra via analytical geometry. In other words, the essence of mathematics […] appeared as the study of *relations* between objects which do not of themselves intrude on our consciousness, but are known to us by means of *some* of their properties, namely those which serve as the axioms at the basis of their theory" (Bourbaki 1968, 316–317).

Bourbaki considers the fact of addressing 'the problem of the nature of beings' or of 'mathematical objects' as a 'naive point of view", "half-philosophical, half-mathematical" (Bourbaki 1948, 40) and abandoned the philosophical problem of object-individuation in favor of structures.

3 Some Difficulties of Contemporary Structuralism

In general, the philosophical thesis expressing that mathematics is not concerned with individual objects, but with "systems" or properties of objects that share a common structure, is called *structuralism*. A "system" is a domain of objects provided with certain functions and relationships fulfilling certain conditions. Structuralism is called *non-eliminative*, if it does not presuppose any ontology for particular objects. According to this kind of structuralism, mathematical objects are incomplete, because they lack an internal structure. A structure appears to be an abstract form or entity, and what the constants in mathematical propositions denote are not individuals, but just 'positions' or 'places' in this structure; they do not have an identity outside of the structure. The objects 'denoted' in mathematical propositions are structure-dependent. Thus, the 'objectual' character of Mathematics is a way of speaking. Structuralism is called *eliminative*, if it questions the existence of abstract mathematical structures and maintains that the nature of mathematical objects is exhausted by their positions in these structures. An example is the set-theoretic structuralism, where mathematics is considered as the study of different structures built up from sets that were thought to exist (Bourbaki). Both versions of structuralism involve two main tasks for the philosopher:

(a) He must clarify the relationship between 'structure', 'system', 'theory', 'subject' and 'position' or 'place' and argue for or against their respective ontological significance.
(b) He should provide an explanation why mathematicians do not realize about what they actually speak, or if they realize it, that they do not expressly refer to it and continue to use singular terms.

Indeed, in eliminative structuralism, the ontological commitment of mathematical objects or at least sets, remains unexplained, while in non-eliminative structuralism, we are faced with two versions, each of which leads to its own problems: if we adopt a version in re, we consider that a 'structure' is all that can be instantiated by a system, but there is not an independent entity 'structure'.

'Positions' in the structure are treated in terms of a function, that is to say, '3', for example, refers to the object in the position '3' in all systems instantiating the structure. The structure depends on systems that instantiate it and presupposes an ontological background. This background is the crucial point that cannot be itself included in a structural approach. If we adopt a version ante rem, we consider that the structure exists independently of the systems that instantiate it, and its 'propositions' express a generalization with respect to all systems that instantiate the structure (realistic interpretation of the structure). The 'positions' in the structure are then treated as 'objects' in a grammatical sense lacking any non-structural property. For example, the object '3' is the successor of the object '2'. Do we have uniqueness of objects? In no way, unless we assume the uniqueness of the isomorphism between systems instantiating the structure. For example, let (E, $\leq$) be an ordered set such that there is an increasing bijection (i.e. an isomorphism for the order) of (Z, $\leq$) in (E, $\leq$)—where Z is the set of all integers and $\leq$ its usual order. Without the use of additional information, you cannot specify an individual element of E![2]

But there is not only the problem of the ontological status of places but also the problem of the identity of the structure itself: to define, for example, truth with regard to a structure requires a criterion of identity or equivalence for the structure; this can lead to circular reasoning or the resurrection of a Platonist attitude. For example, if one defines by abstraction a structure by fundamental properties that apply to all objects that instantiate the structure (see Linnebo and Pettigrew 2014): what is the status of these objects? But if there is no structure-identity, we cannot compare, for example, the 'natural numbers' as sub-structure of 'real numbers' and 'natural numbers' as 'objects' of the structure of 'natural numbers'.

Christopher von Bülow argues carefully and convincingly that neither structuralism *in re* nor structuralism *ante rem* can explain the function of 'places' in the respective version of structuralism. He then proposes to interpret "structures as universals or, to be more precise, as properties of systems. A universal (or one-over-many) [...] is not a set or a class, nor in any other way reducible to entities of a different kind". He sees structures "as simples in the sense of not being composed of entities of other kinds" Bülow 2009, 9).

In the following, we will also interpret structures as 'universals' but neither as simples nor as resorting to a supersensible world; they are archetypes, characterized in a threefold way:(a) they are "functions (and not inborn contents) of our intellect" (Agazzi 2014, 440), i.e. (b) tools for the creation of structured forms suggested by stipulated systems that (c) exemplify in Goodman's sense the structures. This position can be read as an emanation of Poincaré's philosophy of geometry.

[2]I am grateful to Alain Genestier for this precision.

4 Poincaré Reconsidered: Structures as "Mixed-Universals"

Poincaré holds the modernist sounding structural view that we have no pre-axiomatic understanding of geometric primitives, that rigor demands that we eliminate appeals to intuition with respect to metrics in geometry, and that pure (metric) geometry is neither true nor false: it is the result of conventions. He argues that what science

> "can attain is not the things themselves, as the naive dogmatists think, but only the relations between the things; apart from these relations, there is no knowable reality" (Poincaré 1902, 25).

But Poincaré is not in general, i.e. with respect to all abstract entities (relations), an epistemological Platonist[3]: this is even confirmed by his predicative attitude.

Two questions arise:

(a) How to conceive relations without *relata*? (= 'relationalism')
(b) What are the links between Poincaré's epistemological 'relationalism' and the geometrical conventions?

For the rest of this paper, I argue for the thesis that his 'relationalism' and his conventionalism are in fact two different aspects of his structural approach, which attenuates the problem of *ante rem* structuralism.

Whereas for Hilbert the expressions in geometric systems are by construction *schematic* axioms (which are neither true nor false), for Poincaré the expressions in geometric systems are by construction apparent hypotheses, which are neither true nor false, too. According to Hilbert, the mathematical formalism requires "finite" metamathematics in order to demonstrate the consistency of formal mathematical systems, according to Poincaré, it is necessary to explain certain hypotheses with respect to metamathematical standards and to decline the variants of these hypotheses in different scientific disciplines (see: Heinzmann and Stump 2013) . Both approaches have their own difficulties: very general and well known by Gödel's theorems for Hilbert, much less known for Poincaré[4]: for our general purpose, they can be neglected here.

According to Poincaré, the set of relations that hold between the geometric primitives constitute the form, not the matter, of geometric objects, and the form is what is studied in geometry:

[3]I call *epistemological Platonism* a doctrine for which we possess a cognitive faculty in mathematics that plays there a role similar to perception in physics. On the contrary, an *ontological Platonism* posits mathematical entities without necessarily giving the explanation of their cognitive accessibility.

[4]His construction of the dimensions of geometrical space is viciously circular because the choice of the Euclidean group was grounded on Lie's classification of transformation-groups operating on R^3 (see Heinzmann and Nabonnand 2008).

"What we call geometry is nothing but the study of formal properties of a certain continuous group; so we may say, space is a group" (Poincaré 1998, 41).

The link between Poincaré's ontological 'relationalism' and his geometric conventions becomes now visible by recalling the first level of the construction of geometric space. It is obtained by choosing the language of groups to serve as the tool of reasoning about representations of muscular sensations.[5] Similarly to Carnap's *Aufbau*, the starting point (guided by experience) is for Poincaré the definition of two two-place relations satisfying certain *minimal* empirical conditions: an *external change a* (with '*x a y*' for '*x changes in y* ***without*** *muscular sensation*') and an *internal change S* (with '*x S y*' for '*x changes in y* ***accompanied by*** *muscular sensations*'). Further, he proceeds to a *conventional* classification of external changes: among external changes some can be compensated by an internal change, others cannot. If they can, experience teaches us only "that the compensation is approximately produced"; it gives the mind only "the occasion to accomplish this operation", but "the classification is not a raw fact of experience" (Poincaré 1898, 16). If compensation is possible, the changes are called **changes of position,** if not **changes of state**. In this way, he obtains the following result: *modulo* an identity condition with respect to the compensation by internal changes, Poincaré defines the equivalence class of changes of position and calls it a *displacement*. Displacements form a group in the mathematical sense and it depends on the choice of its sub-groups whether the group corresponds to Euclidean or non Euclidean geometry.

At first glance, Poincaré's approach seems just to be an abstraction process leading to a form of invariance. Nevertheless, in reality, the faculty to create the general concept of a group is the expression of a form of our understanding "existing in our mind". The set of relations satisfying the group axioms resembles an *ante rem* structure, which is *exemplified* (in a Goodmanian sense) by the specific displacement-structure (= transformation group). In other words, the form in the mind leads to a special kind of epistemologically accessible universals without that one has "the possibility of deducing by purely logical means the particular form of the universal" (Agazzi 2014, 442). The formation of the group concept—a 'universal' or a second order form (concept) —

$$G\left(A^1, A^2, A^3\right)$$

is *suggested* by a specific sensations system

$$G'\left(S^1, S^2, S^3\right),$$

which is the material of the form (a vague part of the extension of the concept). The form is the conventional 'abstraction' of the various imagined laws of sensations

[5]Muscular sensations are themselves a stipulation suggested by, but not abstracted from the empirical world.

corresponding to groups of axioms. Read as relational set (i.e. extensionally), the general group G is a model of the group axioms and these axioms are 'only' exemplified by the sensation system G', which does not correspond *exactly* to the axioms. The elements of displacements groups are complete and independent entities with respect to the axioms of the group. But by analyzing the subgroups of the displacements groups (common to geometries with constant curvature), the variables of the axiom system are transformed into 'places' that lack ontological independence: they are depending on decisions that are taken concerning the property of distance, for which exists a choice between different possibilities — such a choice was not yet possible with respect to the general axioms of group!

Thus Poincaré defends a structural point of view without completely disengaging the structure from an "ostensional" aspect and this allows him to dispense with a consistency proof. What matters here is the general idea that the faculty of construction of the general concept of group pre-exists in our minds and led to a universal, and that this faculty is suggested by an imagined system of sensations. The concept obtained can be read as a model of an axiom system, in which originally the domain of quantification is composed by independently given elements, which lose progressively their independence and become incomplete. The genesis of the geometrical metric structure is neither seen as the creation of the concrete material from the universal (structure) nor as the creation of the universal from the concrete (sensations), but as the advent of relations linked to the concrete in a *semiotic analysis* of the universal. Contrary to *in re* structuralists, Poincaré's universal, i.e. the general group structure, is not ontologically but epistemically dependent on exemplifications. Nevertheless, Poincaré doesn't speak of this structure as such but uses it as a meta-mathematical tool for his psycho-physiological genesis of *real* actions with *imagined* sensations: the psycho-physiological model is the *ratio cognoscendi* of the existence of the faculty to build the general group structure in our mind.

The exemplification of the structure by big varieties of systems (imagined sensations/mathematical and physical facts/objects) is not a logical but a semiotic operation. It's only the structure that gives the common character to systems: the building up of the structure is a mastery occasioned by concrete systems (samples), which do instantiate *and* exemplify the structure.

The universal as mathematical structure of the first level is not underdetermined but, as relational property, indeterminate. If we stay within the framework of a Tarski-semantic, such vague 'objects' probably imply the move to a non-classical logic (Evans 1978, 208): it seems contradictory to say that there *are* objects and at the same time that it is indefinite whether they are identical or not. Indeed, if an object is undetermined, it differs from another with respect to the property of being identical. Therefore, by contraposition of Leibniz-identity, the two objects are different. Contradiction! If it is found that logic led to some sorites with regard to vague concepts, we have at least two possibilities:

(a) to solve the problem through a special logic appropriate to vague situations
(b) to find a better understanding of the relationship between a precise language and the linguistic practice in question.

We choose the second possibility. Based on the thought of Peirce, we explain vagueness as indeterminacy of meaning in terms of indeterminacy of an exemplification of a concept in a dialogue (Williamson 1994, 48, note 28). In this sense a vague 'object' or structure R is interpreted as a second order relation

$$R(P_1, \ldots, P_n) \approx \alpha'(P_1, \ldots, P_n) \wedge \beta'(P_1, \ldots, P_n) \wedge \ldots\ldots$$

where '$\approx$' signifies the exemplification of 'R' in a semiotic sense, by a system of axioms α', β'.... The meaning of 'R' is then to develop according to the pragmatic maxim of Peirce:

> "Consider what effects that might conceivably have practical bearing you conceive the object of your conception to have. Then your conception of those effects is the whole of your conception of the object" (C.P., 5.422; cf. Peirce 1878/79, p. 48 and C.P. 5.402).

The experimental perspective involved into the maxim is here performed by the exemplifications and limited by the constraints of definitions and conceptual formations of formalisms. The formalisms and proofs with respect to which the universal is a model in the usual model theoretic sense, take here a hypothetical form, 'as if' their variables referred to clearly identifiable objects instantiating the universal. The working out of universals as structures is still 'growing' and the extended structures still remain incomplete in principle, both with respect to their proper identity and the identity of their objects.

5 Conclusion

The here proposed pragmatic insight into the relation between structures and systems leads to an alternative interpretation of Quine's thesis of the incompleteness of mathematical objects and the ideas to which they belong: the incompleteness is neither an epistemic deficiency possessed finally by all objects according to Poincaré, nor a purely verbal accommodation which in fact hides an ontological commitment with respect to a set theoretic progression (Quine 1986, 401), but a functional peculiarity of a new sort: it is the result of a stipulation of a system of relations connected with the relatively concrete by a practice of semiotic analysis linked to our capacity of harmonization. To quote Albert Lautman, the structure is a scheme "incarnated in the very movement" of mathematical work (Lautman 2011, 83).

References

Agazzi, Evandro. 2014. *Scientific Objectivity and its Contexts*. Cham/Heidelberg/Ney York/Dordrecht/London: Springer.

Bourbaki, Nicolas. 1948, L'architecture des mathématiques, in: F. Le Lionnais (Hrsg.). *Les grands courants de la pensée mathématique*, Paris: Blanchard: 35–47; Engl. Translation: *The American Mathematical Monthly* 57, No. 4: 221-232.

Bourbaki, Nicolas. 1949. Foundations of Mathematics for the Working Mathematician. *The Journal of Symbolic Logic* 14, Number 1:1–8.

Bourbaki, Nicolas. 1968. *Elements of Mathematics. Theory of Sets*. Paris: Hermann.

Bülow von, Christopher. 2009. What Is a Place in a Structure? https://www.researchgate.net/publication/255883932_What_Is_a_Place_in_a_Structure.

Cartan, Henri. 1943. Sur le fondementlogique des mathématiques. Revue scientifique (1943/01): 1–11.

Evans, Gareth. 1978. Can there be Vague Objects?. Analysis 38 (4): 208.

Heinzmann, Gerhard and Nabonnand, Philippe. 2008. Poincaré, Intuitionism, Intuition and Convention. In One Hundred Years of Intuitionism (1907–2007), M. v. Atten, P. Boldini, M. Bordeau& G. Heinzmann (eds). Basel/Boston/Berlin: Birkhäuser, 163-177.

Heinzmann, Gerhard and Stump, David. 2013. Henri Poincaré. http://plato.stanford.edu/entries/poincare/.

Lautman, Albert. 2011. Mathematics, Ideas and the Physical Real. London/New York: Continuum International Publishing Group; English translation of Les Mathématiques, les idéeset le réelphysique. Paris: Vrin, 2006.

Linnebo, Øystein and Richard Pettigrew. 2014. Two Types of abstraction for Structuralism. The Philosophical *Quarterly* 64, No. 255: 267–283.

Poincaré, Henri. 1898. On the Foundations of Geometry. *The Monist* IX: 1–43.

Poincaré, Henri. 1902. *La Science etl'hypothèse*. Paris: Flammarion. English translation in: *The Value of Science. Essential Writings of Henri Poincaré*. New York: The Modern Library, 2001.

Quine, Willard Van Orman. 1986. Reply to Charles Parsons. In *The Philosophy of W.V. Quine*, ed. L. E. Hahn and P. A. Schilpp, 396–403. La Salle: Open Court.

Williamson, Timothy. 1994. *Vagueness*. London/New York: Routledge.

Mathematical Truth Revisited: Mathematics as a Toolbox

Reinhard Kahle

Abstract We discuss the notion of truth in Mathematics as relative to certain structures, very much in line with Bernays's conception of "*bezogene Existenz*". Looking to some concrete examples, we argue that even so-called non-standard structures may have their own rationale. As a result, and in accordance with Bourbaki, structures turn out to be *tools* and have to be judged with respect to their usefulness rather than with respect to a concept of mathematical truth simpliciter.

1 Introduction

If I were to show you a hammer and ask you whether this hammer was *true*, you would probably consider this as an ill-posed question. Furthermore, if I were to show you a hammer and a screwdriver, and ask you which of these two is *truer* than the other, you could reasonably consider the question as complete nonsense.

Hammers and screwdrivers are *tools*, and as such we don't ask for their "truth", let alone in a comparative manner. We would rather ask whether they are useful; and as such, for which purpose: a hammer is useful to drive a nail into the wall, a screwdriver to do the same with a screw (and better not to do it the other way around).

We will argue that it is similarly odd to ask for truth (simpliciter) in Mathematics. This is, in fact, in accordance with modern mathematical self-conception, resulting from the shift of understanding caused by the discovery of non-Euclidean Geometries. The aim of the paper is to show that there is, however, still a rationale for truth in Mathematics which, however, has to be relativized to certain structures—including the possibility of non-standard structures.

Work partially supported by the Portuguese Science Foundation, FCT, through the projects *The Notion of Mathematical Proof*, PTDC/MHC-FIL/5363/2012, *Hilbert's 24th Problem*, PTDC/MHC-FIL/2583/2014, and the *Centro de Matemática e Aplicações*, UID/MAT/00297/2013.

R. Kahle (✉)
CMA & DM, FCT, Universidade Nova de Lisboa, Lisbon, Portugal
e-mail: kahle@mat.uc.pt

E. Agazzi (ed.), *Varieties of Scientific Realism*,
DOI 10.1007/978-3-319-51608-0_22

2 Truth in Mathematics

Euclidean Geometry was for 2000 years the most elaborated mathematical theory. With Kant it was raised to an *Anschauungsform a priori* which gave it an allegedly untouchable status of eternal truth. The discovery of non-Euclidean Geometries, however, not only discredited Kant's conception, but called also the whole concept of truth in Mathematics into question. The idea of an absolute notion of truth made it hard to see how alternatives to the Euclidean Geometry could reasonably be possible. In this way, Tóth writes about Taurinus[1]:

> Nevertheless he rejected it [his non-Euclidean Geometry]. His rejection was essentially based on the belief in the dominant opinion that there could "be only *one* scientific method, *one* scientific system of Geometry". From this *conception of unicity* it follows immediately, that the Euclidean system bears a relation of excluding disjunction to the opposed system; they are alternatives. If one is accepted (as true), then the other must be rejected (as impossible); he wrote: "would the third system (*sc.* hyperbolic geometry) be the true one, so there would be no Euclidean Geometry at all."

And we still find this view in Frege who wrote[2]:

> If the Euclidean Geometry is true, then the non-Euclidean Geometry is false, and if the non-Euclidean Geometry is true, then the Euclidean Geometry is false.

Of course, one could consider the notion of truth here just as one referring to the physical world; this was the way, Gauß used it when he spoke about true geometry.[3] In fact, it took the mathematical community a long time to release Geometry from its empirical status as a theory of our physical space.[4] But the result of this process was captured by Poincaré in his famous dictum[5]:

> One geometry cannot be more true than another; it can only be more convenient.

---

[1]Toth (1980); our translation of the German original, as given in Mehrtens (1990, p. 51):

> Dennoch hat er [seine nicht-euklidische Geometrie] abgelehnt. Seine Ablehnung beruht im wesentlichen auf seinem Glauben an die herrschende Auffassung, daß es <nur *eine* wissenschaftliche Methode, *ein* wissenschaftliches System der Geometrie geben> kann. Aus dieser *Konzeption der Unizität* folgt unmittelbar, daß sich das euklidische mit dem ihm entgegengesetzten System in der Relation einer ausschließenden Disjunktion, in einer Alternative befindet; ist das eine (als wahr) akzeptiert, so muß das andere (als unmöglich) zurückgewiesen werden: <wäre das dritte System (*sc.* das der hyperbolischen Geometrie) das wahre, so gäbe es überhaupt keine Euklidische Geometrie> — schrieb er.

[2]Frege (1983, p. 183); in German:

> Wenn die euklidische Geometrie wahr ist, so ist die nichteuklidische Geometrie falsch, und wenn die nichteuklidische wahr ist, so ist die euklidische Geometrie falsch.

[3]See Volkert (2010, p. 118): " 'true' is to understand here in the sense of the classical theory of adequacy" (" 'wahr' ist hier zu verstehen im Sinne der klassischen Adäquatheitstheorie").

[4]See, for instance, Volkert (2010, §2).

[5]Poincaré (1905, p. 59), French original in Poincaré (1902, p. 66f):

And we put it as: One can be a better tool than another. And this, clearly, depends on the purpose. By analogy with our initial comparison, where a hammer is not "truer" than a screwdriver, but just a better tool to deal with nails, the Euclidean Parallel Postulate might be the more adequate to deal with certain spaces (physical or not) than its alternative (or vice versa).

Thus, is truth still an issue in Mathematics? Let us give here two citations which seem to deny this. The first one is from a mathematical survey paper on the foundations of geometry, published in 1968[6]:

> Geometry as passed on by Euclid was considered for two millennia as textbook example of a logically structured science. Its axioms were considered as evident, everything else was deduced from them by logical rules. This axiomatic viewpoint became today prevalent in all of Mathematics; just that the axioms are no longer evident truths, but arbitrary stipulations under aspects of convenience. (Lenz 1968, p. 64).[7]

The second citation is from a philosopher in a book which bears the title *Das Wahrheitsproblem und die Idee der Semantik* (The truth problem and the idea of semantics):

> If one speaks of a semantic notion of truth, then this can indeed be considered as an incomplete, abbreviating formulation. In semantics, it is not supposed that there exist a "notion of truth", but one assumes only that the predicate "true" as meaningful in relation to a concrete system *S*. (Stegmüller 1957, p. 216).[8]

What Stegmüller called *S* is, of course, just a semantic structure in the Tarskian sense like, for instance, that of the natural numbers. There is no controversy about a

(Footnote 5 continued)

> Une géométrie ne peut pas être plus vraie qu'une autre; elle peut seulement être *plus commode*.

[6]It would not be hard to find many other examples in the mathematical literature; we took this one, as its (uncritical) appearence in the first paragraph of a survey paper shows that it is assumed to be common knowledge.

[7]Our translation; the German original reads:

> Die durch Euklid überlieferte Geometrie galt zwei Jahrtausende lang als Musterbeispiel einer logisch aufgebauten Wissenschaft. Ihre Axiome wurden als selbstverständlich angesehen, alles andere wurde aus ihnen durch logisch Schlüsse abgeleitet. Dieser axiomatische Standpunkt hat sich heute in der ganzen Mathematik durchgesetzt; nur sind Axiome keine selbstvertändlichen Wahrheiten mehr, sondern willkürliche Festsetzungen nach Zweckmäßigkeitsgesichtspunkten.

[8]Our translation; the German original reads:

> Wenn von dem semantischen Wahrheitsbegriff gesprochen wird, so kann dies tatsächlich als eine unvollständige, abkürzende Formulierung angesehen werden. Es wird in der Semantik nicht vorausgesetzt, daß es einen "Wahrheitsbegriff" gibt, vielmehr wird bloß das Prädikat "wahr" in bezug auf ein bestimmtes System *S* als sinnvoll angenommen.

notion of truth relative to such structures[9]; what is at issue, is either the question of the very existence of such structures, or whether some structures are distinguished over others, often associated with the terms "standard model" and "non-standard models".

As far as the existence of such structures is concerned, one may follow Hilbert's motto[10]:

> Consistency implies Existence.

This claim is not unproblematic, and for a thorough discussion of it by Bernays, we refer to (Bernays, 1950). At least for first-order languages it can be taken for granted; but, in this case, one will be confronted with incompleteness phenomena which imply that the majority of our formal theories allow for non-standard models. This fact leads to the temptation to reserve the notion of mathematical truth for "truth in the standard model". The problem is now how to distinguish standard models from the non-standard ones.

Apparently, we have no problem distinguishing the standard (first-order) structure for the natural numbers from its competitors; thus, there is a generally accepted notion of truth for Arithmetic. For Geometry, however, a general notion of truth was abandoned, at least for the Parallel Postulate, in view of the success of non-Euclidean Geometries.

The crucial question today is the status of truth in set theory. Thus, in contemporary mathematics the primary controversy regarding truth concerns the status of the Continuum Hypothesis (CH). Knowing that it is independent of Zermelo–Fraenkel set theory with the axiom of choice (ZFC), we cannot yet see whether CH or one of its alternatives[11] is actually intended to hold in our standard set-theoretic universe.[12] At

---

[9]Tarski's "truth definition", given in Tarski (1936), was a breakthrough for the mathematical notion of truth; in particular, it paved the way for a purely mathematical (set-theoretical) treatment of truth in formalized languages. Tarskian semantics may be subject to (philosophical) criticism: hardline formalists as well as intuitionists, for instance, will reject it as such. But, as Heyting pointed out in a short note with the distinctive title *On Truth in Mathematics* (Heyting 1958), in both cases, formalism and intuitionism, the very notion of truth makes no particular sense any longer. He actually continues by giving a sketch of the Tarskian notion of truth, albeit with the caveat "to me personally the assumption of an abstract reality of any sort seems meaningless".

[10]To our knowledge, Hilbert never formulated the motto is this concise form, but it follows immediately from his argument in the letter to Frege of 29 December 1899, (Frege 1976, Letter XV/4, p. 66). This idea was later even backed by Poincaré (1914, p. 151f):

> In mathematics the word exist can only have one meaning; it signifies exemption from contradiction.

French original (Poincaré 1908, p. 162): "en mathématiques le mot exister ne peut avoir qu'un sens, il signifie exempt de contradiction".

[11]We use here the plural, as the negation of CH can be specified by concrete alternative cardinalities for the real numbers.

[12]Of course, any such intention has to be crossed-checked with the mathematical consequences one obtains when adding one or the other form as new axiom. Interestingly, nearly all properties, first of all, in connection with large cardinals, which were considered to decide CH turned out to be

a first glance one could think that the situation should be analog to the parallel axiom. Gödel's proof of the consistency of CH with ZFC and Cohen's proof of the consistency of its negation with ZFC show that we could consider different set-theoretic universes, one with CH and others with its alternatives—in the very same way that we can consider Euclidean Geometry and non-Euclidean Geometries. However, in sharp contrast to the situation in Geometry, neither Gödel's model of ZFC + CH, the *constructible universe* L, nor models of ZFC + ¬CH based on Cohen's result could be considered as the natural or intended set-theoretic universe. This leaves space to suppose that in the intended universe CH turns out to be determined as true or false.[13] The debate concerning "new axioms" in set theory, and with it the question whether one speaks of one set-theoretic universe or admits a "multiverse", is fierce.[14] We may refer, for the latter, to Shelah (2003, p. 211), who does "not agree with the pure Platonic view that the interesting problems in set theory [as CH] can be decided, that we just have to discover the additional axiom. My mental picture is that we have many possible set theories, all conforming to ZFC." For the former, Woodin just announced for his Paul Bernays Lectures, to be held in September 2016 at the ETH in Zurich, a resolution of the situation based on an "Ultimate L".[15]

Whatever the ultimate outcome of the discussion might be, the status of mathematical truth as a relative one does not seem to be affected; only the possible distinction of one particular set-theoretic universe is at issue—in L, CH will always be true; as much as it will be false in other forced universes.

(Footnote 12 continued)
independent of it; and those few which relate to CH, such as the *proper forcing axiom* or *Martin's maximum* (both implying $2^{\aleph_0} = \aleph_2$), do not have the status of "evident truths" or intuitively clear axioms (but see also Woodin's announcement for his Paul Bernays Lectures, Footnote 15).

[13]This is one motivation behind Kreisel's promotion of *informal rigour*. He writes (Kreisel 1967, p. 138f): "Informal rigour wants [...] not to leave undecided questions which can be decided by full use of evident properties of these intuitive notions." And Kreisel has a formal argument: CH might be decidable as a *second-order consequence* from additional intuitive axioms. As second-order logic is not (recursively) axiomatizable, we may simply overlook it (Kreisel 1967, p. 152): "most people in the field are so accustomed to working with the restricted [first-order] language that they may simply not succeed in taking other properties seriously". Thus, the independence of CH is, at this stage, related to the first-order nature of ZFC; this is a quite different situation compared to the Parallel Postulate in Geometry which "is not even a second order consequence of this axiom [i.e., the second-order axiom of continuity]." (Kreisel 1967, p. 151). Still, up to today, nobody has come up with an intuitive second-order property deciding CH (but, again, see the reference to Woodin's Paul Bernays Lectures in Footnote 15).

[14]See, for instance, Feferman et al. (2000) for "new axioms" and Antos et al. (2015) for "multiverse".

[15]https://www.gess.ethz.ch/en/news-and-events/paul-bernays-lectures/bernays-2016.html. One may note his ostentatious use of "true" and "false" in the abstracts of his lectures.

3 Mathematical Structures as Tools

In the following we will briefly review some mathematical structures with respect to their function as tools.

3.1 *Natural Numbers*

> A Druidic myth relates how Lucanor, coming upon the other gods as they sat at the banquet table, found them drinking mead in grand style, to the effect that several were drunk, while others remained inexplicably sober; could some be slyly swilling down more than their share? The disparity led to bickering, and it seemed that a serious quarrel was brewing. Lucanor bade the group to serenity, stating that the controversy no doubt could be settled without recourse either to blows or to bitterness. Then and there Lucanor formulated the concept of numbers and enumeration, which heretofore had not existed. The gods henceforth could tally with precision the number of horns each had consumed and, by this novel method, ensure general equity and further, explain why some were drunk and others not. "The answer, once the new method is mastered, becomes simple!" explained Lucanor. "It is that the drunken gods have taken a greater number of horns than the sober gods, and the mystery is resolved." For this, the invention of mathematics, Lucanor was given much honor.
>
> Jack Vance, *Madouc*[16]

The natural numbers are the most basic structure in mathematics. Their original function is also clear: to count. And in this function, natural numbers are nothing like a tool. Given any collection of objects (in a non-mathematical sense: finite and consisting of physical objects) we may count its elements by attributing (consecutive) numbers to every single object, not assigning the same number to two different objects.

With the help of arithmetical and logical expressive power, we are able to develop a rather sophisticated internal structure of the natural numbers, containing an endless number of number-theoretic functions, starting with addition and multiplication, as well as relations, including the natural order relations which allow comparisons.

For such functions and relations we can express propositions which should turn out true or false. Let us look at two examples:

$$2 + 11 \neq 1, \tag{1}$$

$$\mathsf{Con}_{\mathsf{PA}} \tag{2}$$

where $\mathsf{Con}_{\mathsf{PA}}$ should be a standard arithmetical consistency statement for Peano Arithmetic, given as $\neg\exists x.\mathsf{Bew}(x, \ulcorner 0 = 1 \urcorner)$ with Bew the usual proof predicate from the proof of Gödel's (first) Incompleteness Theorem.

As far as (1) is concerned, nobody will have the slightest doubt that it is a true arithmetical statement. We will see below, however, in which way its truth should indeed be seen as one relative to the structure of the natural numbers.

---

[16]The quotation is taken from Franzen (2003, p. v).

For (2) Gödel's Second Incompleteness Theorem tells us, that it is not derivable in Peano Arithmetic; still, it is definitely a true arithmetical formula.[17] Thus, it holds in the standard structure of the natural numbers. Gödel's Theorem, however, implies that there have to exist structures which satisfy the axioms of Peano Arithmetic, but this formula is false. We have also to reflect briefly on these non-standard models of PA.

Cyclic Groups

If you look at the face of your watch, do the numerals you are seeing represent natural numbers? They appear so, but, insofar counting of hours is concerned, one can clearly say that 2 h after 11 o'clock is 1 o'clock. So, (1) turns out to be false, at least for the numerals of our clock-faces.

Mathematically, it is clear what happens: numbers on a clock-face are supposed to be elements of $\mathbb{Z}/12\mathbb{Z}$, the cyclic group of 12 elements, rather than to be elements of the natural numbers. In $\mathbb{Z}/12\mathbb{Z}$, of course, (1) is false. But as (1) is stated, there is nothing which forces us to read the numbers as natural numbers, and not as elements of $\mathbb{Z}/12\mathbb{Z}$. Thus, its truth is, indeed, relativized to a structure. The structure in question is almost always clear from the context; but this simple example shows that it would be misleading to consider (1) as an absolute truth.

Of course, one could try to argue that, in (1), the numbers are supposed to be natural numbers, and a reading of (1) in $\mathbb{Z}/12\mathbb{Z}$ would be an abuse of notation. But such a presupposition would not turn (1) in an absolute truth; to the contrary, the very presupposition—that the numbers are supposed to be *natural numbers*—would just fix the relativization of (the truth of) (1) to the structure of the natural numbers. As a matter of fact, the notational "overloading" of numbers—as natural numbers or elements of a cyclic group—requires a disambiguation which just comes down to the relativization to the different structures.[18]

---

[17]Of course, we presuppose here the consistency of Peano Arithmetic. Doubts about the consistency of PA cannot be taken mathematically seriously as long as nobody presents an explicit proof of an inconsistency. In particular, doubting the consistency of Peano Arithmetic puts any kind of Arithmetic in doubt, and clearly denies the very existence of the standard structure of the natural numbers (i.e., the first-order structure including addition and multiplication). Mathematically, it is pointless to even consider such doubts; and no philosophical debate about it provided so far any conceptual insight.

[18]We like to note in passing that, for Logic, Bernays observed that the difference of classical and intuitionistic logic can attributed to an overloading of the logical connectives, especially the negation (Bernays 1979, p. 4):

> As one knows, the use of the "tertium-non-datur" in relation to infinite sets, in particular in Arithmetic, was disputed by L.E.J. BROUWER, namely in the form or an opposition of the traditional logical principle of the excluded middle. Against this opposition is to say that it is just based on a reinterpretation of the negation. BROUWER avoids the usual negation non-A, and takes instead "A is absurd". It is then obvious that the general alternative "Every sentence A is true or absurd" is not justified.

German original: "Wie man weiß, ist die Verwendung des 'tertium non datur' in bezug auf unendliche Gesamtheiten, insbesondere schon in der Arithmetik, von L.E.J. BROUWER angefochten worden, und zwar in der Form einer Opposition gegen das traditionelle logische Prinzip vom

The clock example shows also that $\mathbb{Z}/12\mathbb{Z}$ is a quite useful tool to deal with cyclic arithmetic (of period 12). Thus, there are applications where we, indeed, would like to have $2 + 11 = 1$ to be true.

While we obtain cyclic groups quite naturally (and for any positive order) out of the natural numbers, finite fields, i.e., finite structures which are equipped with addition *and multiplication*, are not so easily to define. It is, in fact, a non-trivial mathematical theorem that there only finite fields of prime power order. For somebody who rejects any kind of objectivity or any notion of truth within mathematical structures, such a limitation theorem should have an objective character, at least in its negative part: there is simply no way to define a finite field of, let us say, order 6. Thus, when we say that mathematical structures are tools, this theorem gives us a formal constraint on what kind of tools are actually possible.

Insane models of PA

As said, it follows from Gödel's Incompleteness Theorems that PA is consistent together with its own formalized inconsistency statement $\neg\mathsf{Con}_{\mathsf{PA}}$; *a forteriori* there has to be a first-order structure serving as a model for this theory. Following a terminology used by Kikuchi and Kurahashi (2016), one can call such a model *insane*. However, there is a way to understand what is going on in such an insane model—Takeuti reports how he learned it from Gödel (Yasugi and Passell 2003, p. 3):

> [Gödel's] way of teaching nonstandard models was an interesting one. It went as follows. Let T be a theory with a nonstandard model. By virtue of his Incompleteness Theorem, the consistency proof of T cannot be carried out within T. Consequently, T and the proposition "T is inconsistent" is consistent. There is, therefore, a natural number N which is the Gödel number of the proof leading to a contradiction from T. Such a number is obviously an infinite natural number.

In an insane model of PA it is indeed the case that the inconsistency statement of PA is *true*. But we do not run into a contradiction: the formula $\neg\mathsf{Con}_{\mathsf{PA}}$ has received a new meaning which simply does not deserve to be called an "inconsistency statement" any longer.[19]

According to our line of argument, insane models of PA will be tools in mathematics in the same way as the standard structure of the natural numbers. And yes, they are a tool, but, admittedly utterly useless tools: at least to our knowledge there is not a single mathematical problem which would profit from being treated by an insane model of PA.

(Footnote 18 continued)
ausgeschlossenen Dritten. Gegenüber dieser Opposition ist zu bemerken, daß sie ja auf einer Umdeutung der Negation beruht. BROUWER vermeidet die übliche Negation nicht-A, und nimmt stattdessen 'A ist absurd'. Es ist dann klar, daß eine allgemeine Alternative 'Jede Aussage A ist wahr oder ist absurd' nicht berechtigt ist.".

[19]To be a little bit more explicit: $\neg\mathsf{Con}_{\mathsf{PA}}$ is equivalent to $\exists x.\mathsf{Bew}(x, \ulcorner 0 = 1 \urcorner)$. While it is our intention that the existential quantifier should range over (standard) natural numbers, we have no formal means (in first-order logic) to prevent an interpretation where it could be instantiated by nonstandard natural numbers. Just that happens in an insane model. But out of such an interpretation, of course, we cannot build an actual (finite) proof of the inconsistency of PA, even in the insane model.

So, again, the truth of (2) holds only in the (intended) standard structure of the natural numbers; the fact that there are non-standard models of PA in which (2) is false just requires that we have to indicate which structure is under consideration.

3.2 Geometry

As discussed above, the discovery of non-Euclidean Geometry led, eventually, to the conclusion that one cannot speak about truth simpliciter in the case of the Parallel Postulate. Experience shows that the different Geometries have all their *raison d'être*; and they function as different tools for different purposes.

Even long before the discovery of non-Euclidean Geometry, there existed another example which shows how Geometries functions as a tool: as the name suggests, *Spherical Geometry* is just the adequate tool for geometric reasoning on a sphere. And it is a useful tool even in cases where no physical sphere is present: its very conception was triggered by the aim to study the stellar sphere—which, as we know now, is anything but a sphere.

3.3 Set Theory

Cantor designed set theory as a tool to give a proper fundation to the real numbers and the concept of function.[20] For this purpose, it is somehow another type of tool as the Natural Numbers and Geometry, as it is not used for applications outside of Mathematics, but rather as a tool within Mathematics. But as such, it turned out that it can be considered as a kind of *universal tool*, allowing for an encoding of essentially every other mathematical structure.[21]

In its function as a universal mathematical base, and with the appearance of the set-theoretical paradoxes, the notion of truth for set theory gains additional importance. But, as discussed above, we may consider different set-theoretic universes. One example, already mentioned, is Gödel's constructible hierarchy L, which restricts the power set operation to definable subsets. As we would like to have in our standard set-theoretic universe more subsets in a power set, L is generally not consider to be the "true set theory". Still, it has turned out to be an extraordinary

---

[20]From the *Mengenlehrebericht* of Schoenflies (1913):

> The development of *set theory* has its source in the endeavor to provide a clear analysis for two fundamental mathematical notions, namely the notions of *argument* and of *function*.

German original: "Die Entwickelung der *Mengenlehre* hat ihre Quelle in dem Bestreben, für zwei grundlegende mathematische Begriffe eine klärende Analyse zu schaffen, nämlich für die Begriffe des *Arguments* und der *Funktion*.".

[21]In Moschovakis (2006) one can find a detailed account how such an encoding works.

useful *tool* in Mathematics (first of all, in Gödel's proof of the consistency of CH with ZFC); a tool we would not like to deprive ourselves of, in particular not on the basis of the argument that it is not our intended "true set theory".

4 Final Remarks

We have argued that mathematics does not rely on any absolute notion of truth. In contrast, by studying different structures truth should be relativized to such structures while the structures themselves are used as tools in applications (outside and within Mathematics). Thus, the question of the truth of an axiom (like the Parallel Postulate or CH) could either only be asked with respect to a chosen structure or would have to be abandoned in favour of a decision concerning the usefulness of it (and the structure in which it is fulfilled) and/or its alternatives.

The view presented here is very much in line with Bernays's analysis of the existence of ideal objects in terms of "*bezogene Existenz*" (Bernays 1950).[22] In the very same way, as Bernays let existence *refer* to a structure, we would like to let truth refer to a structure.

The existence of different structures leads naturally to "different truths" in these ones. But this does not imply, by no means, that Mathematics would lose its objectivity; as existence and truth, it also refers to structures, and all we have to concede is a certain form of pluralism in Mathematics.

A question which remains concerns the very existence of a structure and its internal configuration. We will discuss this question in more detail elsewhere, but remark here only that the structures considered in our examples are *concrete* structures. In contrast to such structures Bourbaki, for instance, considers *abstract structures* which are supposed to encompass abstract properties of several, quite different, concrete structures.[23] Their ontological status differs substantially from that of the structures we have considered here, but for both sorts of structures we can say with Bourbaki (1950, p. 231) that:

> mathematics appears thus as a storehouse of abstract forms—the mathematical structures.

And Bourbaki (1950, p. 227):

> The "structures" are tools for the mathematician.

[22]Beth (1959, p. 642) translates "bezogene Existenz" as "conditional existence" which we consider slightly misleading. It would correspond to "bedingte Existenz" and suggests just a *condition* for the existence; "bezogen" is stronger in presupposing a *reference*. Thus, "referring existence" would be a more literal translation.

[23]Cf. Corry (2004) for a critical evaluation of Bourbaki's notion of structure.

References

Antos, C., S.-D. Friedman, R. Honzik, and C. Ternullo (2015). Multiverse conceptions in set theory. *Synthese 192*(8), 2463–2488.

Bernays, P. (1950). Mathematische Existenz und Widerspruchsfreiheit. In *Etudes de Philosophie des Sciences*, pp. 11–25. Neuchâtel: Éditions du Griffon. Reprinted in (Bernays, 1976, pp. 92–106).

Bernays, P. (1976). Abhandlungen zur Philosophie der Mathematik. Wissenschaftliche Buchgesellschaft.

Bernays, P. (1979). Bemerkungen zu LORENZEN's Stellungnahme in der Philosophie der Mathematik. In K. Lorenz (Ed.), *Konstruktionen versus Positionen*, Volume 1, pp. 3–16. Berlin.

Beth, E. W. (1959). *The Foundations of Mathematics*. Studies in Logic and the Foundations of Mathematics. North-Holland.

Bourbaki, N. (1950). Architecture of mathematics. *The American Mathematical Monthly 57*(4), 221–232.

Corry, L. (2004). *Modern Algebra and the Rise of Mathematical Structures* (Second revised ed.). Birkhäuser.

Feferman, S., H. M. Friedman, P. Maddy, and J. R. Steel (2000). Does mathematics need new axioms? *Bulletin of Symbolic Logic 6*(4), 401–446.

Franzen, T. (2003). *Inexhaustibility: A Non-Exhaustive Treatment*, Volume 16 of *Lecture Notes in Logic*. ASL and AK Peters.

Frege, G. (1976). *Wissenschaftlicher Briefwechsel*. Felix Meiner. Edited by G. Gabriel et al.

Frege, G. (1983). Über Euklidische Geometrie. In Gottlob Frege. *Nachgelassene Schriften* (2. ed.)., pp. 182–184. Meiner. The editors date the item back to 1899–1906?

Heyting, A. (1958). On truth in mathematics. In *Verslag van de plechtige viering van het honderdvijftigjarig bestaan der Koninklijke Nederlandse Akademie van Wetenschappen met de teksten der bij die gelegenheid gehouden redevoeringen en voordrachten*, pp. 277–279. Amsterdam.

Kikuchi, M. and T. Kurahashi (2016). Illusory models of peano arithmetic. *The Journal of Symbolic Logic 81*(3), 1163–1175.

Kreisel, G. (1967). Informal rigour and completeness proofs. In I. Lakatos (Ed.), *Problems in the Philosophy of Mathematics*, Volume 47 of *Studies in Logic and the Foundations of Mathematics*, pp. 138–186. Elsevier.

Lenz, H. (1968). Grundlagen der Geometrie. In D. Laugwitz (Ed.), *Überblicke Mathematik 1, 1968*, Volume 161/161a of *BI Hochschultaschenbücher*, pp. 63–86. Bibliographisches Institut.

Mehrtens, H. (1990). *Moderne—Sprache—Mathematik*. Suhrkamp.

Moschovakis, Y. (2006). *Notes on Set Theory* (2nd ed.). Undergraduate Texts in Mathematics. Springer.

Poincaré, H. (1902). *La science et l'hypothèse*. Paris, Flammarion.

Poincaré, H. (1905). *Science and Hypothesis*. Walter Scott. English translation of Poincaré (1908).

Poincaré, H. (1908). *Science et méthode*. Paris, Flammarion.

Poincaré, H. (1914). *Science and Method*. Thomas Nelson and Sons. English translation of Poincaré (1902).

Schoenflies, A. (1913). *Entwickelung der Mengenlehre und ihrer Anwendungen*. B.G. Teubner. Rework of the Report published in volume VIII of the *Jahresberichte der Deutschen Mathematiker-Vereinigung*.

Shelah, S. (2003). Logical dreams. *Bulletin (New Series) of the American Mathematical Society 40*, 203–228.

Stegmüller, W. (1957). *Das Wahrheitsproblem und die Idee der Semantik*. Springer, Wien.

Tarski, A. (1936). Der Wahrheitsbegriff in den formalisierten Sprachen. *Studia Philosophica 1*, 261–405.

Toth, I. (1980). Wann und von wem wurde die nichteuklidische Geometrie begründet? Bemerkungen zu Hans Reichardts *Gauß und die nichteuklidische Geometrie*. *Archives internationales d'histoire des sciences 30*, 192–205.

Volkert, K. (2010). Poincarés Konventionalismus und der Empirismus in der Geometrie. In P. E. Bour, M. Rebuschi, and L. Rollet (Eds.), *Construction. Festschrift for Gerhard Heinzmann on the occasion of his 60th birthday*, pp. 113–127. London: College Publications.
Yasugi, M. and N. Passell (Eds.) (2003). *Memoirs of a Proof Theorist*. World Scientific. English translation of a collection of essays written by Gaisi Takeuti.

Index of Names

E. Agazzi (ed.), *Varieties of Scientific Realism*,
DOI 10.1007/978-3-319-51608-0

MIX
Papier aus verantwortungsvollen Quellen
Paper from responsible sources
FSC® C105338

FSC
www.fsc.org

Printed by Libri Plureos GmbH
in Hamburg, Germany